1979

TRIGONOMETRY

A Study
of Certain
Real
Functions

TRIGONOMETRY
A Study
of Certain
Real
Functions

DONALD R. HORNER, Ph.D.
EASTERN WASHINGTON STATE COLLEGE

Holt, Rinehart and Winston
New York   Chicago   San Francisco   Atlanta
Dallas   Montreal   Toronto   London

To my children Sherry and Kyle—
they make it worthwhile

Preface

Unlike many "modern" trigonometry textbooks, the philosophy and goals of the material in this book are contemporary and express the changing needs of the calculus curriculum. Unlike many so-called "functional" approaches, the central theme here is that of the (real-valued) function (of a real variable). Unlike those standard textbooks that are in actuality modern and have in reality a functional central theme, *Trigonometry—A Study of Certain Real Functions* gears a good portion of its efforts toward helping the student understand in detail the concept of the function and some of its implications.

A course designed and taught around the first five chapters of this text is of necessity a precalculus-type course aimed at deepening the maturity and sophistication of the student while thoroughly grounding him in the basic concepts of the function. All the basic properties (that is, being 1:1, onto, bounded, periodic, and so forth) are discussed, as are inverses and compositions of functions. Over 28 figures and charts and 15 worked examples are used to help fix the idea of a function (Chapter 3).

The functional approach to the trigonometric and inverse trigonometric functions is carried farther than in any textbook the author has encountered. The trigonometric functions are defined in terms of compositions and quotients of functions, but only after the student is adequately prepared in the theory of a function.

In fact, the primary motivating factor for the order of the contents of the book, the examples chosen, and the problems given is the desire to ensure that the student is always ready to take the next step. The problems preparing a student for one section often begin two to three chapters earlier, and always two or three sections earlier. An intensive study was made in order that the problems be complete from the standpoint of application, motivation, extension, and preinstruction. The review exercises, although not always a review of the chapter's material, do cause the student to review. The advanced exercises are extensions of the text, making available to the above-average student topics not normally discussed until after the calculus sequence.

Even the index is formed so as to lead the student through a study of a few key notions.

Some unconventional notation has been used. The symbols $\sin^{-1}$, $\cos^{-1}$, and so forth, are used to represent the inverse trigonometric functions, while arcsin $\mathbf{k}$, arccos $\mathbf{k}$, and so forth, are used to denote sets introduced as solution sets for conditional trigonometric equations. The value $\sin^{-1} k$, for example, is shown to be the unique element of arcsin $\mathbf{k} \cap [-\pi/2, \pi/2]$. The author believes that conditional equations form a natural introduction to the topic, while the concept of an inverse trigonometric function is a natural result of the detailed study of functions given in Chapter 3. The terminology Arcsin, Arccos, and so forth, is not used. The author has used all the notation of the text since 1964 and feels that the notation settled upon causes the least confusion and the greatest understanding. Various forms of the manuscript have been class-tested since February of 1965.

The original version of the text was less than 60 percent as voluminous as the present version. Virtually all the added material has been in the form of worked examples, diagrams, applications, and the highly extensive lists of exercises. There are 189 figures; many, many worked examples and exercises, greater in number and more diverse in character than can be required.

Applications are made to angles, triangles, geometry, complex numbers, polar coordinates, vector spaces, geometric vectors, and other areas. Consequently, the basic precalculus material (Chapters III, IV, and V) can easily be taught in a two-semester hour or three-quarter hour course, while ample extensions and applications are present to fill courses running to five quarter-hours in length. Because of the nature of the approach and the material involved, no one using this text need apologize for the length of his trigonometry course.

It is recommended that the order of events be taken as found in the book. Chapters 1 through 5 form the basic theory and should be given in all courses. It is anticipated that some will use Chapters 1 through 3, and Section 4.1 followed by Chapter 7. It is recommended that anyone following this pattern then cover the remaining portion of Chapter 4 together with Chapter 5. Chapter 6 may follow either Chapter 5 or 7, while Chapter 8 must follow Chapter 7.

The author wishes to acknowledge the helpful comments of Stanley Ball of the University of Texas at El Paso and Ray Strauss of Foothills Junior College. The author's wife Donna achieved her always excellent production of the typed manuscript from pieced-together and generally unintelligible articles of communication.

D. H.

Cheney, Washington
January 1968

Contents

TRIGONOMETRY
A Study
of Certain
Real
Functions

Chapter 1
Goals
and
Background

Nowadays in a course in trigonometry, one must look for objectives involving far more than the fundamental concepts and mechanics of the subject. In many colleges and universities, trigonometry is considered remedial, and relative to this subject, the predominant trend appears to be one of deemphasis. However, the not so fortunate student who for various and sundry purposes must absorb trigonometry in his curriculum has a right to expect subject material that will not only show him the basic notions but strive to be valuable by commencing his search (consciously or not) for the following:

1. mathematical maturity,
2. sophistication in the use of mathematical symbolism,
3. insight into the concept of functions, and
4. an appreciation for an analytic approach to trigonometry and geometry.

These goals are not all to be obtained during the course of study of this text. Items 1 and 2 are cumulative in nature and are achieved only with the passing of time and the expending of great quantities of energy. Those of 3 and 4 can be achieved to a satisfactory level in a comparatively short span of time.

1.2 SOME SET THEORY

Symbolism in mathematics is both a system of shorthand and a descriptive language. To illustrate a useful symbolism, we shall introduce items from set theory. Examine the following symbolism (capital letters will be used to represent sets and lower-case letters will be used to signify elements of sets).

$$x \in A$$

The above symbol is read in any of the following ways:

1. x is an element of A,
2. x is a member of A,
3. x belongs to A, or
4. x is in A.

The reader can probably exhibit numerous other phrases depicting this same element-set relationship.

The symbol $x \notin A$ is read "x is not an element of A." The meaning of the slash as introduced here is somewhat standard in mathematics.

It will be well to observe that the terms *element* and *set* have not been defined. Neither will they be defined, for each is an example of the use of *undefined terms*. The mathematician does this to prevent "circular" definitions.

The set-builder symbolism is particularly important in mathematics today. An example of this is given below.

$$A = \{x : x \text{ has property } p\}$$

The symbolic statement is read, "A is the set of all elements x such that x has property p." A more vivid example follows.

Example 1.2.1

$$A = \{x : x \text{ is a dog or } x \text{ is a cat}\}$$

The set A is the set of all objects x, where x is either a dog or a cat. In particular, A is the set of all dogs and cats.

A similar arrangement may be used to list the elements of a set.

$$B = \{p, \$, z, \pi, >\}$$

In this setting B is the set consisting precisely of the objects p, \$, z, π, and $>$.

Consider at this time the following relationship (between two sets) as symbolized by either:

$$A \subset B \quad \text{or} \quad B \supset A$$

We read the sentence: A *is a* **subset** *of* B. This statement means that every element of A is also an element of B. In this situation we might say descriptively that A is *contained* in B or that B *contains* A.

Example 1.2.2 Let the sets A, B, and C be

$$A = \{1, 2, 3, 4, 5\}$$
$$B = \{0, 1, 3, 5, 9\}$$
$$C = \{1, 3, 4\}$$

Which of the following are true and which are false?

1. $C \subset A$.
2. $A \subset C$.
3. $A \subset B$.
4. $B \subset C$.
5. $C \subset B$.

Examine first the solution of 1. Is $C \subset A$? To answer this question, we must compare the elements of C with those of A. For C to be a subset of A, it must be true that each element of C is also an element of A. Is this true? Yes, for each of the elements 1, 3, and 4 of C is found also in A. Thus, 1 is true.

The remaining four statements are all false. Statement 2 is false, since $2 \in A$ but $2 \notin C$. This condition violates the meaning of $A \subset C$.

Now, $2 \in A$ but $2 \notin B$, implying that $A \subset B$. Also, since $9 \in B$ but not in A, $B \not\subset A$. Finally, $4 \in C$ but $4 \notin B$, which is to say that $C \not\subset B$. Necessarily, solutions 2 to 5 are false as predicted.

In viewing the inclusion (subset) relationship, you should also question in your mind the meaning of equality between sets. Intuitively, two sets are to be *equal* when they have precisely the same elements. One very usable definition of equality is: two sets are **equal** whenever each is a subset of the other. We shall use this definition throughout the text. Symbolically,

$$A = B \quad \text{means} \quad A \subset B \quad \text{and} \quad B \subset A$$

Example 1.2.3 Are any of the sets A, B, and C from Example 1.2.2 found to be equal? The answer is no. This is because no two of the sets have exactly the same elements.

Viewing the question from the standpoint of the subset relation, $A \neq B$, since $A \not\subset B$ (nor is $B \subset A$). Furthermore, $A \neq C$, since $A \not\subset C$. Finally, $B \neq C$, since neither is a subset of the other.

The next few paragraphs will be devoted to sets that are easily constructed as combinations of other sets.

$$A \cup B = \{x : x \in A \quad \text{or} \quad x \in B\}$$
$$A \cap B = \{x : x \in A \quad \text{and} \quad x \in B\}$$
$$A - B = \{x : x \in A \quad \text{but} \quad x \notin B\}$$

The symbol $A \cup B$ represents the **union** of A and B. The union is proclaimed to be the set of all elements belonging to *at least one* of A and B. The phrase "at least one" is a result of the word *or* (rather, its interpretation).

The expression $A \cap B$ symbolizes the **intersection** (common part) of A and B. The intersection of A and B is the set of all elements belonging to *both* sets.

Finally, $A - B$ is the **difference** of A and B or the complement of B with respect to A. This difference is the set of all elements of A that do *not* belong to B.

Example 1.2.4 Continuing with the exploration of the sets A, B, and C as used in Examples 1.2.2 and 1.2.3, we determine:

$$A \cup B = \{0, 1, 2, 3, 4, 5, 9\}$$
$$A \cap B = \{1, 3, 5\}$$
$$A - B = \{2, 4\}$$
$$B - A = \{0, 9\}$$
$$A \cup C = A$$
$$A \cap C = C$$
$$A - C = \{2, 5\}$$

The union of A and B was computed by listing within one set of braces the elements of A and the elements of B. The intersection was described by listing the elements found in common. The difference of A and B ($A - B$) was formed by deleting from A those elements that belong to B. The other sets are formed analogously.

Example 1.2.5 In Example 1.2.4, how do we describe the set $C - A$? If we delete from C all the elements found also in A, we necessarily must delete all the elements (remember that $C \subset A$). Thus, $C - A$ does not have a single element.

It now appears useful to define: $\varnothing$ represents any set not having elements and $\varnothing$ is said to be **empty, null,** or **void.**

Using this concept, we write (in Example 1.2.5): $C - A = \varnothing$.

There are simple diagrams that can be used to aid intuition (when discussing sets). These diagrams are called **Venn-Euler** or simply Venn diagrams. Figures 1.2.1 to 1.2.5 show the possible relationships existing between two sets P and Q. The sets are taken to be the respective circles, together with their interiors.

Figure 1.2.1
$P \cap Q = \varnothing$

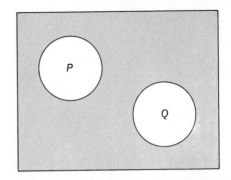

Figure 1.2.2
$P \cap Q \neq \varnothing$

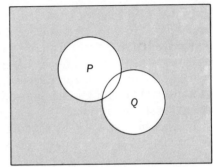

Figure 1.2.3
$P \supset Q$

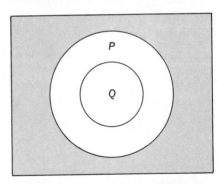

Figure 1.2.4
$Q \supset P$

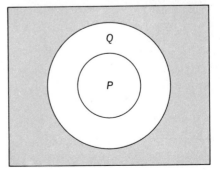

Figure 1.2.5
$P = Q$

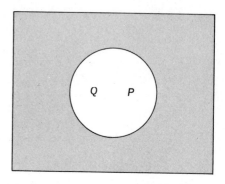

In Fig. 1.2.1 the two sets do not have any element in common (the intersection is empty). Two such sets are said to be **disjoint**.

Horizontal and vertical shading can be used to illustrate some of the aforementioned combinations of sets (union, intersection, and so forth). Figures 1.2.6 to 1.2.10 reproduce the conditions of Figs. 1.2.1 to 1.2.5, respectively, but with the shading added. The set P is shaded horizontally while Q is shaded vertically.

In what manner does the shading display certain combinations of sets? Any element covered by a particular shading belongs to the set whose shad-

Figure 1.2.6
$P \cap Q = \varnothing$

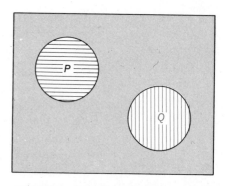

Figure 1.2.7
$P \cap Q \neq \varnothing$

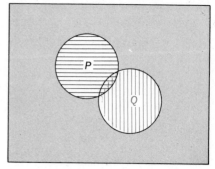

Figure 1.2.8
$P \supset Q$

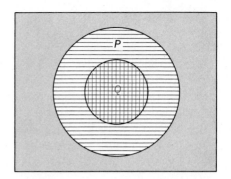

Figure 1.2.9
$Q \supset P$

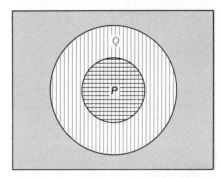

Figure 1.2.10
$P = Q$

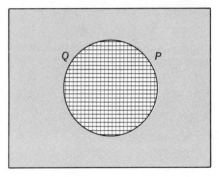

ing covers it. Remembering, for instance, that $P \cup Q$ is the set of elements belonging to at least one of P and Q, we realize that the union is given (in shading) by the total area shaded (with at least one type in shading).

How is $P \cap Q$ described? This is the set shaded with both horizontal and vertical shading. In other words, the intersection is represented by the crosshatched area.

Other sets are described in an analogous manner.

In order that you may better understand the abstract use of set theoretic ideas, we now present two theorems from set theory.

Theorem 1.2.1 If $A \subset B$ and $B \subset C$, then $A \subset C$. (This is called the *transitive* property of "$\subset$".)

PROOF   What is to be shown? On the hypothesis that $A \subset B$ and $B \subset C$, we are to show that $A \subset C$. How can this be accomplished?

The definition of $A \subset C$ declares we need only show that each $x \in A$ belongs also to C. The following argument, making use of the hypothesis, is designed to do just that.

Since $A \subset B$, each x in A is in B. Since $B \subset C$, each $x \in B$ satisfies $x \in C$. Thus, each $x \in A$ is also found in C, whence, we conclude that, whenever $x \in A$, $x \in C$ (that is, $A \subset C$).

Theorem 1.2.2 If $A \subset B$, then $A \cup B = B$.

PROOF   To show that $A \cup B = B$, it is sufficient to show that (1) $A \cup B \supset B$ and (2) $B \supset A \cup B$. The proof of (1) is left as an exercise for the reader. We now proceed to the proof of (2).

Let x be in $A \cup B$. It follows that x is in A or x is in B. However, since $A \subset B$, $x \in A$ yields $x \in B$. Necessarily, $x \in B$ whenever $x \in A \cup B$. This illustrates the fact that $B \supset A \cup B$.

The two inclusion statements (1) and (2) give (by definition) the desired equality.

The reader is asked to prove some uncomplicated statements in the exercises that follow. Advanced problems are given in the advanced section at the end of the chapter.

■ **EXERCISES 1.2**

1. Let $A = \{1, 2, 3, 4, 5\}$, $B = \{1, 3, 5\}$, $C = \{2, 4\}$, $D = \{0, 1, 3\}$, $E = \{0, 2, 4\}$, and $F = \{0, 1, 2, 3, 4, 5\}$. Answer the following by true or false.

a. $A \subset A$.	**m.** $C \subset A$.	**y.** $E \subset A$.
b. $A \subset B$.	**n.** $C \subset B$.	**z.** $E \subset B$.
c. $A \subset C$.	**o.** $C \subset C$.	**a'.** $E \subset C$.
d. $A \subset D$.	**p.** $C \subset D$.	**b'.** $E \subset D$.
e. $A \subset E$.	**q.** $C \subset E$.	**c'.** $E \subset E$.
f. $A \subset F$.	**r.** $C \subset F$.	**d'.** $E \subset F$.
g. $B \subset A$.	**s.** $D \subset A$.	**e'.** $F \subset A$.
h. $B \subset B$.	**t.** $D \subset B$.	**f'.** $F \subset B$.
i. $B \subset C$.	**u.** $D \subset C$.	**g'.** $F \subset C$.
j. $B \subset D$.	**v.** $D \subset D$.	**h'.** $F \subset D$.
k. $B \subset E$.	**w.** $D \subset E$.	**i'.** $F \subset E$.
l. $B \subset F$.	**x.** $D \subset F$.	**j'.** $F \subset F$.

2. After having answered the questions in Exercise 1, you have probably formed an opinion about the truth of $P \subset P$. What conclusion have you drawn? Justify your conclusion.

3. Using the sets from Exercise 1, find each of the following sets.

a. $A \cup A$.	**j.** $C \cup B$.	**s.** $D \cup D$.
b. $A \cup B$.	**k.** $B \cup D$.	**t.** $D \cup E$.
c. $B \cup A$.	**l.** $B \cup E$.	**u.** $D \cup F$.
d. $A \cup C$.	**m.** $B \cup F$.	**v.** $E \cup D$.
e. $A \cup D$.	**n.** $C \cup C$.	**w.** $E \cup E$.
f. $A \cup E$.	**o.** $C \cup D$.	**x.** $E \cup F$.
g. $A \cup F$.	**p.** $D \cup C$.	**y.** $F \cup E$.
h. $B \cup B$.	**q.** $C \cup E$.	**z.** $F \cup F$.
i. $B \cup C$.	**r.** $C \cup F$.	

4. From your experience with Exercise 3, what do you find $P \cup P$ to be? Justify your conclusion. What is the relation between $P \cup Q$ and $Q \cup P$? Justify this answer also.

5. Find all the sets formed by replacing "$\cup$" by "$\cap$" in Exercise 3.

6. Answer all the questions from Exercise 4 after replacing "$\cup$" by "$\cap$."

7. Find all the sets formed by replacing "$\cup$" by "$-$" in Exercise 3.

8. Answer the questions from Exercise 4 by replacing "$\cup$" by "$-$."

9. In Exercise 1 the set F contains all elements under discussion. In this sense, F is the **universal set** of the discussion. For each subset P of F, $F - P = P'$ is called the **complement** of P (relative to F). The expression "relative to F" is dropped when the set F is the universal set of the discussion. The universal set is, for all practical purposes, the set of *all* elements (remember that this is relative to the individual discussion). Thus, we can say that P' *is the set of all elements not in P*. From Exercise 1, find the following sets.

a. A'.	**d.** E'.
b. B'.	**e.** F'.
c. D'.	**f.** $\varnothing'$.

10. The following are in relation to Fig. 1.2.11. Describe each set in terms of the shadings. Note that parentheses are used to describe an order in performing the set combinations. For example, to compute $(P \cup Q) - S$, compute $P \cup Q$ first, and then compute the difference of $P \cup Q$ and S.

a. A.	**h.** $A \cap B$.	**o.** $C - A$.
b. B.	**i.** $A \cap C$.	**p.** $B - C$.
c. C.	**j.** $B \cap C$.	**q.** $C - B$.
d. $A \cup B$.	**k.** $A \cap B \cap C$.	**r.** $(A \cup B) - C$.
e. $A \cup C$.	**l.** $A - B$.	**s.** $(A \cup C) - B$.
f. $B \cup C$.	**m.** $B - A$.	**t.** $(B \cup C) - A$.
g. $A \cup B \cup C$.	**n.** $A - C$.	

Figure 1.2.11

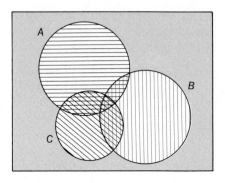

Let the interior of the entire rectangle of Fig. 1.2.11 represent the universal set (denoted here by X). Describe (in terms of the shadings) the following sets.

u. A'.	**x.** $(A \cup B)'$.	**a'.** $(A \cap B)'$.
v. B'.	**y.** $(B \cup C)'$.	**b'.** $(A \cap C)'$.
w. C'.	**z.** $(A \cup C)'$.	**c'.** $(B \cap C)'$.

11. Show that $A \subset A \cup B$ and $A \supset A \cap B$.

1.3 SOME PROPERTIES OF REAL NUMBERS

The system of real numbers forms a basic set of elements to be used throughout the remainder of our investigation. Various characteristics of these numbers are important to us and we wish to examine some of these together with several mechanical operations involving the real numbers. We shall denote the set of real numbers by R.

The set R together with the concepts of equality ($=$), order ($<$), addition ($+$), and multiplication ($\cdot$) form an algebraic system called an **ordered field.** This system satisfies (1) to (9) below.

1. If $a, b \in R$, $a + b$, $a \cdot b = ab \in R$.
2. If $a, b \in R$, $a + b = b + a$ and $ab = ba$.
3. If $a, b, c \in R$, $(a + b) + c = a + (b + c)$ and $(ab)c = a(bc)$.
4. If $a, b, c \in R$, $a(b + c) = ab + ac$.
5. There is a $0 \in R$ with $a + 0 = a$ for all $a \in R$.
6. If $a \in R$, there is an element $-a \in R$ with $a + (-a) = 0$.
7. There is a $1 \in R$ with $a \cdot 1 = a$ for all $a \in R$.
8. If $a \in R$, $a \neq 0$, then there is an $a^{-1} \in R$ with $aa^{-1} = 1$.
9. If $a, b \in R$, exactly one of the following holds:

   (a) $a = b$,
   (b) $a < b$, or
   (c) $b < a$.

In terms of (9) we say that each $a < 0$ is **negative,** while each $b > 0$ is **positive.** The following properties of inequalities (statements involving "$<$") hold.

10. $a < b$ means $b - a$ is positive.
11. $a < b$ implies $a + c < b + c$.
12. $a < b$ implies $ac < bc$ when c is positive.
13. $a < b$ implies $ad > bd$ when d is negative.
14. $a < b$ and $b < c$ implies that $a < c$.

Various subsets of R are of interest and should be pointed out. The set $N = \{1, 2, 3, \cdots\}$ is called the set of **natural** numbers or **positive integers.** The set $Z = \{\cdots, -3, -2, -1, 0, 1, 2, 3, \cdots\}$ is the set of integers, while the set $Q = \{m/n : m, n \in Z, n \neq 0\}$ of all quotients of integers is the set of **rational** numbers. The set $I = R - Q$ is the set of **irrational** numbers. We see that $N \subset Z \subset Q \subset R$, $I \subset R$, $I \cap Q = \varnothing$, and $I \cup Q = R$.

The rational numbers are those that can be expressed (in decimal form) as either *terminating* decimals (.25, for example) or *infinite, periodic* decimals (.625151$\cdots$, where the concatenation 51 is indicated as being repeated indefinitely). The irrational numbers, then, are those with infinite nonperiodic decimal expansions. See Exercise 1.3.6.

The concept of the **exponent** is of interest to us. We define x^1 to be x and for each integer $n > 1$, $x^n = x \cdot x^{n-1}$. We say this definition is an *inductive* one (that is, its meaning must be determined by a chain of events). The following illustrates the definition.

Example 1.3.1

$$x^1 = x$$
$$x^2 = x \cdot x^1 = x \cdot x$$
$$x^3 = x \cdot x^2 = x \cdot x \cdot x$$
$$x^4 = x \cdot x^3 = x \cdot x \cdot x \cdot x$$

If $a^n = x$, a is said to be an **nth root** of x while x is said to be an **nth power** of a (n is a positive integer).

Example 1.3.2   If $x = 16$, $4^2 = 16 = (-4)^2$, whence 4 and -4 are square (second) roots of x. Since $2^4 = (-2)^4 = 16$, 2 and -2 are fourth roots of x. The fact that $2^3 = 8$ implies that 2 is a cube (third) root of 8. Similarly, -2 is a cube root of -8.

We refer you to Exercises 1.3.9 and properties 21 to 26 listed therein.

The absolute value of a real number is useful and has particular geometric significance. The **absolute value** of a real number x, symbolized $|x|$, is defined by:

$$|x| = \begin{cases} x \text{ if } x \geqslant 0 \text{ (that is, } x > 0 \text{ or } x = 0) \\ -x \text{ if } x < 0 \end{cases}$$

The absolute value satisfies the following properties.

15. $|x| \geqslant 0$.
16. $|x| = |-x| \geqslant x$.
17. $|x + y| \leqslant |x| + |y|$.
18. $|x \cdot y| = |x| \cdot |y|$.
19. $|x| < a$ if and only if $-a < x < a$.
20. $|x| > a$ if and only if $x < -a$ or $x < a$.

It is to be observed that we have not discussed subtraction or division. We shall simply consider these operations as **inverses** of addition and subtraction, respectively (that is, $x - y = x + (-y)$ and $x \div y = x \cdot y^{-1} = x(1/y)$). Since zero has no defined multiplicative inverse (recall property 8), we cannot allow division by zero.

■ **EXERCISES 1.3**

1. Without referring to the text, answer the following true or false.

a. $N \subset Z$.	**h.** $Q \subset I$.	**o.** $R \cap I = \varnothing$.
b. $N \subset Q$.	**i.** $Q \subset R$.	**p.** $Q \cup I = R$.
c. $N \subset I$.	**j.** $I \subset Q$.	**q.** $\{x \in R : x^2 = -1\} = \varnothing$.
d. $N \subset R$.	**k.** $I \subset R$.	**r.** $0 \in N$.
e. $Z \subset Q$.	**l.** $N \cap I = \varnothing$.	**s.** $0 \in Z$.
f. $Z \subset I$.	**m.** $Z \cap I = \varnothing$.	**t.** $0 \in Q$.
g. $Z \subset R$.	**n.** $Q \cap I = \varnothing$.	**u.** $0 \in R$.

2. If x is a real number, $(-x)$ is the *additive inverse* or *negative* of x (see property 6). Find the negative of each of the following.

a. 2. **d.** -3.
b. 31. **e.** -21.
c. 0.

3. Find the absolute value of each number in Exercise 2.
4. Show that:

a. $7 < 10$.
b. $8 < 19$.
c. $-20 < -18$.

5. In (a) to (c), find the set of all values of x satisfying the inequality. Use properties 10 to 14.

a. $x + 2 > 5$.
b. $3x - 1 > x + 5$.
c. $3 - 2x < 15 + 4x$.

6. Let $x = 0.1212121212\cdots$. A fractional expression for x can be found by the following technique. Multiplying x by 100, we have $100x = 12.12121212\cdots$. Both x and $100x$ have the same expressions to the right of the decimal point. Notice, then, that the act of subtracting equals from equals yields:

$$
\begin{aligned}
100x &= 12.12121212\cdots \\
-x &= -0.12121212\cdots \\
\hline
99x &= 12
\end{aligned}
$$

This is to say, $x = \frac{12}{99} = \frac{4}{33}$. Using some variation of the illustrated technique, find the fractional equivalent of each of the following.

a. 0.125. **d.** $0.141414\cdots$. **g.** $0.99999\cdots$.
b. 1.0625. **e.** $0.230230230\cdots$. **h.** $0.012121212\cdots$.
c. 1.666. **f.** $3.11111\cdots$. **i.** $0.24999\cdots$.

7. The symbol $a^{1/2} = \sqrt{a}$ is defined to be the *nonnegative square root of* x whenever one exists. (Remember that b is a square root of a means that $b^2 = a$.) Show that:

a. If $x > 0$, $\sqrt{x^2} = x$.
b. If $x = 0$, $\sqrt{x^2} = 0$.
c. If $x < 0$, $\sqrt{x^2} = -x$.

This shows that $\sqrt{x^2} = |x|$.

8. Let x^+ be the larger of x and 0 with x^- the smaller of the two. Prove the following. (Hint: The three cases as given in Exercise 7 will have to be used.)

a. $x = x^+ + x^-$.
b. $|x| = x^+ - x^-$.

9. We define $y^{1/n}$ to be the *principal nth root of* y (that is, the positive nth root of y if one exists and any nth root otherwise). Remember, y may not have an nth root (-1 has no real square root). Furthermore, if $m \in N$, define $y^{m/n}$ to be $(y^{1/n})^m$. The following illustrate some of the laws of the combination of exponents. (Each expression is assumed to represent a real number.)

21. $x^a x^b = x^{a+b}$.
22. $x^a/x^b = x^{a-b}$.
23. $(x^a)^b = x^{ab}$.
24. $x^{-a} = 1/x^a$ and $x^0 = 1$ if $x \neq 0$. (This is a definition.)
25. $x^a y^a = (xy)^a$.
26. $x^a/y^a = (x/y)^a$.

Using (21) to (26), find the value of each of the following.

a. 3^4.

b. $0^{1/2}$.

c. $(-\frac{1}{2})^3$.

d. $(4a)^2$.

e. $(\frac{1}{6})^4$.

f. $9^{1/2}$.

g. $81^{1/4}$.

h. $16^{1/4}$.

i. $8^{1/3}$.

j. 8^{-2}.

k. 3^{-5}.

l. $(\frac{1}{4})^{1/2}$.

m. $(\frac{1}{81})^{1/4}$.

n. $(1/-27)^{-1/3}$.

o. $(-3125)^{1/5}$.

p. $2 \div 3^4$.

q. $2 - 3^4$.

r. $(2-3)^4$.

s. $(x^n \cdot x^m)^2$.

t. $2^n + 2^n$.

u. $3^n + 3^n$.

v. $(\frac{2}{3})^2 (\frac{3}{2})^{-2}$.

w. $(x^3/y^3)^{1/3}$.

x. $\left(\dfrac{10^{-3}a^4}{640}\right)^{1/2}$.

y. $\dfrac{x^{-2} - y^{-2}}{x^{-1} - y^{-1}}$.

10. Simplify the following expressions or perform the indicated operation, whichever is appropriate.

a. $\frac{1}{3} + \frac{2}{5}$.

b. $-\frac{5}{9} + \frac{1}{8}$.

c. $\frac{1}{3} + \frac{3}{4}$.

d. $\frac{3}{15} + \frac{3}{25}$.

e. $\frac{15}{7} - \frac{1}{3}$.

f. $-\frac{1}{5} - \frac{1}{3}$.

g. $\frac{1}{19} - \frac{1}{10}$.

h. $\frac{2}{3} - (-1)/6$.

i. $(\frac{2}{3}) \cdot (\frac{20}{1})$.

j. $(-\frac{3}{4}) \cdot (\frac{7}{2})$.

k. $(\frac{3}{5}) \div (\frac{1}{2})$.

l. $(-\frac{1}{3}) \div (\frac{5}{3})$.

m. $(\frac{1}{2} - 3)/(5 - \frac{2}{3})$.

n. $(\frac{1}{3} + \frac{1}{6})/(\frac{1}{4} + \frac{2}{3})$.

o. $\dfrac{\frac{1}{3} + \frac{1}{4}}{\frac{1}{3} + [1/(1 + \frac{1}{4})]}$.

p. $\dfrac{\frac{1}{3} + \frac{1}{5} - \frac{1}{7}}{\frac{1}{21} + \frac{1}{3}}$.

q. $\dfrac{\frac{1}{3} + \dfrac{1}{6 + \frac{1}{7}}}{\frac{1}{5} - \dfrac{1}{7 + [1/(6 + \frac{1}{2})]}}$.

11. The set of numbers given by $\{x \in R : a \leqslant x \leqslant b\}$ is called the **closed interval** ab and is symbolized by $[a, b]$. The **open interval** ab is the set given by

$$(a, b) = \{x \in R : a < x < b\}$$

The following are **half-open, half-closed intervals:**

$$[a, b) = \{x \in R : a \leqslant x < b\}$$
$$(a, b] = \{x \in R : a < x \leqslant b\}$$

In addition there are the four types of **infinite intervals:**

$$[a, \infty) = \{x \in R : a \leqslant x\}$$
$$(a, \infty) = \{x \in R : a < x\}$$
$$(-\infty, a) = \{x \in R : a > x\}$$
$$(-\infty, a] = \{x \in R : a \geqslant x\}$$

Find each of the following sets.

a. $[a, b] \cup [b, c]$.
b. $[a, b] \cap (a, b)$.
c. $[a, b] - (a, b)$.
d. $[a, b) \cap (c, d]$ where $a < c < b < d$.
e. $(a, b] \cap [c, d)$ where $a < c < b < d$.
f. $(a, \infty) \cup (-\infty, a)$.
g. $(a, \infty) \cap (b, \infty)$ where $a < b$. (Note: In each of the intervals, it is tacitly assumed that $a < b$.)

1.4 REVIEW EXERCISES

1. Write a complete sentence describing each indicated set.

a. $A = \{x : x$ is a trigonometry student$\}$.
b. $B = \{x : x$ is an algebra student$\}$.
c. $C = A \cup B$.
d. $D = A \cap B$.
e. $E = A - B$.
f. $F = B - A$.
g. $G = (A - B) \cup (B - A)$.

2. Draw Venn diagrams for the problems in Exercise 1, shading A vertically and B horizontally. Describe each of the sets in Exercise 1 in terms of the shadings.

3. Let $X = \{2n : n$ is a positive integer$\}$, $Y = \{4n : n$ is a positive integer$\}$, and $A = \{3n : n$ is a positive integer$\}$. Answer the following true or false.

a. $12 \in Y$.
b. $12 \in A$.
c. $12 \in X$.
d. $Y \subset A$.
e. $Y \subset X$.
f. $A \subset X$.
g. $A \subset Y$.
h. $X \subset A$.
i. $X \subset Y$.
j. $X \cap Y = Y$.
k. $X \cap A = \emptyset$.
l. $Y \cap A \neq \emptyset$.
m. $X \cup Y = X$.
n. $6 \in X$.

4. Answer the following true or false.

a. $N \subset Z$.
b. $Z \subset Q$.

   c. $Q \subset R$.
   d. If $x \in N$, x has no negative in N.
   e. The set Z is closed relative to subtraction (that is, subtraction is always possible).
   f. Division is always possible in Z.
   g. Q is closed relative to division.
   h. $Q \cap I = \emptyset$.
   i. $Q \cup I = R$.
   j. If $x \in I$, we can find elements in Q arbitrarily "close" to x (that is, elements in Q that differ very little from x).
   k. If $a, b \in R$, $a \leqslant b$, and $b \leqslant a$, then $a = b$.

5. From the collection $\{N, Z, Q, I, R\}$, give a complete list of these sets containing each of the following numbers.

a. 3.	**e.** ¾.	**i.** $(\frac{1}{2})\sqrt{5}$.
b. 0.	**f.** 2.	**j.** .05131313⋯.
c. -2.	**g.** 3.	
d. ⅓.	**h.** -4.	

6. Give the negative of each number in Exercise 5.
7. Give the absolute value of each number in Exercise 5. Each number in Exercise 6.
8. If $x < y$, both x and y are real, show that the **average** $(x + y)/2$ is such that $x < (x + y)/2 < y$.
9. The result in Exercise 8 shows in particular that "between" any two rational numbers there is another rational number. The average is one such rational number. (Note: $x < y$ and z between x and y means $x < z < y$.) This property of the rationals is sometimes referred to as "being dense." Using this information, answer the following questions.

   a. Is there a largest rational number x so that $x < 0$? Why or why not?
   b. Is there a smallest rational number y so that $y > 0$? Why or why not?

10. Find a fractional expression for each of the following repeating decimals.

   a. 0.01999999⋯.
   b. 0.011011011011011⋯.
   c. 1.55555⋯.
   d. 0.01233333⋯.

1.5 ADVANCED EXERCISES

1. Prove that the following set relations are true.

a. $A \subset A \cup B$.
b. $A \supset A \cap B$.
c. $(A - B) \cap B = \emptyset$.
d. $(A - B) \subset A$.
e. $(A - B) \cap (B - A) = \emptyset$.
f. $\emptyset \subset A$ for any set A.
g. $A - (B \cup C) = (A - B) \cap (A - C)$.
h. $A - (B \cap C) = (A - B) \cup (A - C)$.
i. $A \cap (B \cup C) = (A \cap B) \cup (A \cap C)$.
j. $A \cup (B \cap C) = (A \cup B) \cap (A \cup C)$.
k. If $A \cap B = \emptyset, A - B = A$.
l. If $(A - B) \cup B = A, B \subset A$.
m. If $B \subset A, (A - B) \cup B = A$.
n. If $A \cap B = \emptyset, (A \cup B) - B = A$.
o. If $(A \cup B) - B = A, A \cap B = \emptyset$.

2. Given that the product of two positive real numbers or two negative real numbers is positive, while the product of a positive with a negative is negative, and using property 10 as a definition, prove:

a. $a < b$ implies $a + c < b + c$ and $a - c < b - c$.
b. $a < b$ and $c > 0$ imply $ac < bc$ and $a/c < b/c$.
c. $a < b$ and $c < 0$ imply $ac > bc$ and $a/c > b/c$.
d. $a < b$ and $c < d$ imply $a + c < b + d$.

3. Show each of the following.

a. $|a| \leqslant b$ if and only if $-b \leqslant a \leqslant b$.
b. $-|a| \leqslant a \leqslant |a|$.
c. $|a + b| \leqslant |a| + |b|$ [Hint: Use (a) and (b)].
d. $||a| - |b|| \leqslant |a - b|$.   [Hint: Use (a), (b), and (c)].
e. $|ab| = |a| \cdot |b|$.
f. $|a/b| = |a|/|b|$.

4. Show that:

a. "$<$" is not **reflexive** (that is, $a < a$ is not true),
b. "$<$" is not **symmetric** (that is, $a < b$ does not imply $b < a$), and
c. "$<$" is **transitive** (that is, prove, using property 10, that property 14 holds).

5. If $A \subset R$, and if there is a number m so that $x \leqslant m$ for all $x \in A$, m is said to be an *upper bound* of A. Any number p so that $x \geqslant p$ for all $x \in A$ is said to be a *lower bound* of A. To say that A is *bounded above (below)* means that A has an upper (lower) bound. To say that A is *bounded* means that A is bounded above and below. A **least upper bound** (l.u.b.) of A is any upper bound q of A so that $q \leqslant m$ for any upper bound m of A. A **greatest lower bound** r of A is any lower bound where $r \geqslant p$ for all lower bounds p. *The real numbers have the property that every subset that is bounded above (below) has a l.u.b. (g.l.b.).* An ordered field with this property is called a **complete ordered field.**

Prove the following.

a. If $A \subset R$, A bounded, there is a closed interval $[a, b]$ so that $A \subset [a, b]$.

b. If $A \subset R$, A bounded, there is a closed interval $[a, b]$ so that $A \subset [a, b]$ and if $[c, d]$ contains A also, then $[a, b] \subset [c, d]$. In some sense, $[a, b]$ is the *smallest* closed interval containing A.

c. If m is an upper bound of A and $s \geqslant m$, then s is an upper bound of A.

d. If p is a lower bound of A and $q \leqslant p$, then q is a lower bound of A.

e. If p is a l.u.b. of A, p is the only l.u.b. of A.

f. Solve (e) with g.l.b. replacing l.u.b.

6. Give l.u.b. and g.l.b. where applicable (see Exercise 5). If a set is unbounded (not bounded) above or below, specify.

a. $A = \{1/n : n \in N\}$.

b. N.

c. $B = \{1/n : n \in Z, n \neq 0\}$.

d. $C = \{(n + 1)/n : n \in Z, n \neq 0\}$.

e. $\{n : n \text{ is a negative integer}\}$.

f. $\{x : x = (-1)^n \text{ for } n \in N\}$.

7. Does the l.u.b. (Exercise 5) of a set necessarily belong to the set? Does the g.l.b.?

8. Let
$$\bigcup_{i=1}^{n} (A_i) = A_1 \cup A_2 \cup \cdots \cup A_n$$

Show that, if each of the $A_i = A$, the union is A. (Use mathematical induction.)

Let
$$\bigcap_{i=1}^{n} (A_i) = A_1 \cap A_2 \cap \cdots \cap A_n$$

Under the same conditions as above, show that the intersection is A also.

Chapter 2
Geometry

2.1 SOME SPACES

The reader more than likely thinks of trigonometry and geometry as being inescapably related–and rightfully so. It will be well, then, to acquire a "space" in which to work and exhibit trigonometric and other properties.

The text will be primarily concerned with two of these so-called spaces. The first space is simply the set R of real numbers and will be thought of as a geometric entity via the following discussion.

The real line, a model of R for example, is viewed intuitively as a straight line (horizontal seems customary) with zero arbitrarily designated. We must keep in mind that the concept of a straight line is (at this point) a strictly intuitive one. The positive real numbers are located to the right of zero according to the ordering induced by "$<$." Similarly, the negative real numbers are found to the left of zero by the same ordering. In the discussions to follow, the spacing of the numbers on the line will be uniform in the sense that the markings on a ruler are laid uniformly. Figure 2.1.1 illustrates the model being discussed.

Figure 2.1.1
The real line

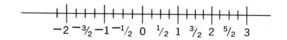

$$-2 \ -\tfrac{3}{2} \ -1 \ -\tfrac{1}{2} \ \ 0 \ \ \tfrac{1}{2} \ \ 1 \ \ \tfrac{3}{2} \ \ 2 \ \ \tfrac{5}{2} \ \ 3$$

The term space has been used coincident with R, since the numbers are taken to be points of the space. In other words, the points of the line are identified with the real numbers, and vice versa. This identification is such that each point has a uniquely associated real number, and, conversely, each real number is uniquely associated with a point of the line.

We can think intuitively of the separation of points by a distance. For example, consider the points 3 and 6. How far "apart" are these numbers (or points)? The urgent question asked here is: What does a distance mean? We can define a distance in various ways. A seemingly logical choice here is the difference found by subtracting the smaller from the larger. This does correspond to length as usually conceived when a ruler is employed. Then, given two arbitrary points x and y, what is the desired distance between x and y? It is either $x - y$ or $y - x$, whichever is nonnegative. From the discussions in Chapter 1, we see that such a distance may be expressed as $|x - y|$, $|y - x|$, $((x - y)^2)^{1/2}$, or $((y - x)^2)^{1/2}$. This **distance** will be symbolized by $d(x, y)$. Thus, we have defined: $d(x, y) = ((x - y)^2)^{1/2} = |x - y|$ for $x, y \in R$.

The second space to be employed is that of the Euclidean **plane,** sometimes denoted by E_2, R^2, or $R \times R$, the **Cartesian product** of R with itself. The plane is defined as the set of all ordered pairs of real numbers, that is to say, $R^2 = \{(x, y) : x \in R \text{ and } y \in R\}$. The word "order" is important in the sense that $(1, 3)$ and $(3, 1)$ are not the same point. We define $(a, b) = (c, d)$ if and only if $a = c$ and $b = d$.

The a and b in (a, b) are called the **coordinates** of (a, b), with a the *first* coordinate and b the *second* coordinate. Because the plane is to be the most widely used space (in a geometric sense), we call our subject *plane trigonometry*.

The plane is pictured by a horizontal copy of the real line and a vertical copy of the same. The point of intersection of the two is called the **origin** (the point $(0, 0)$). The horizontal line is *axis number 1* (in our applications, axis No. 1 will be called the x axis), while the vertical line is *axis number 2* (or y axis, as we shall often call it). Points are then located via their first and second coordinates. Figure 2.1.2 illustrates the plane and the location of a particular point therein.

When we consider a pair of points in the plane, we conceivably think in terms of their separation (the distance between them), just as we did in the case of the real line.

The *distance* between $P_1 = (x, y)$ and $P_2 = (a, b)$ is defined and symbolized by $d(P_1, P_2) = ((x - a)^2 + (y - b)^2)^{1/2}$.

Figure 2.2.2

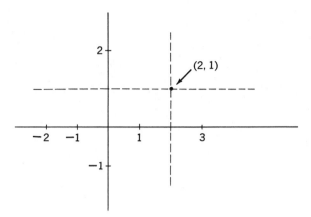

The definition shows that, to calculate $d(P_1, P_2)$, we:

1. determine the differences between the respective coordinates,
2. square each difference as determined,
3. sum these squares, and, finally,
4. calculate the principal square root of this sum.

The distance between points of the line R was motivated intuitively. Can there be such a motivation for this case?   The answer is in the affirmative.   Figure 2.1.3 shows that the distance concept might be thought of as being a statement of the *Pythagorean Theorem*.

Assuming that the triangle in Fig. 2.1.4 is a right triangle, we deduce (from the Pythagorean Theorem) that $c^2 = a^2 + b^2$.   Since $a = 5$ and

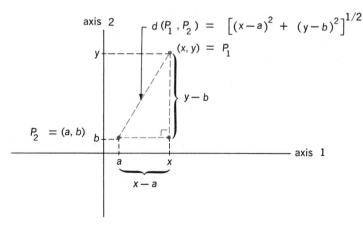

Figure 2.1.3
Distance and the Pythagorean Theorem

Figure 2.1.4
$c^2 = a^2 + b^2$

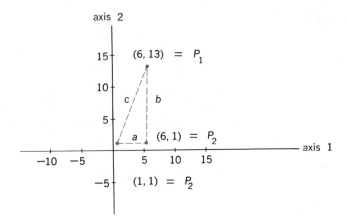

$b = 12$, it follows that $c^2 = 169$ or $c = 13$. Now, by the distance formula, $d(P_1, P_2) = ((6 - 1)^2 + (13 - 1)^2)^{1/2} = (169)^{1/2} = 13$. Thus, in R^2, the Pythagorean Theorem and the distance concept are compatible (the distance between two points is apparently what one would like the length of the line segment connecting them to be). This condition is not necessary and no formal justification has really been given the correctness of this result.

■ **EXERCISES 2.1**

1. Find the distance between each of the pairs of points in R.

   a. 0 and 5.   **b.** 0 and -5.   **c.** -5 and 5.

2. Find the distance between the following pairs of points.

   a. $(0, 0)$ and $(3, 4)$.   **e.** $(5, -4)$ and $(-1, 2)$.
   b. $(0, 0)$ and $(-2, 3)$.   **f.** $(-1, -1)$ and $(-2, -2)$.
   c. $(-1, 2)$ and $(0, 3)$.   **g.** $(1, 4)$ and $(4, 1)$.
   d. $(3, 5)$ and $(1, 2)$.   **h.** $(1, 0)$ and $(5, 0)$.

3. Each set of three points can be considered as the set of vertices of a triangle. Tell whether each of the following triples of points form a right triangle (according to the converse of the Pythagorean Theorem), an isosceles triangle (two sides of equal length), or an equilateral triangle (all sides of equal length).

   a. $(-1, 2)$, $(7, 2)$, and $(7, 17)$.   **d.** $(-1, 0)$, $(1, 0)$, and $(0, 3)$.
   b. $(0, 12)$, $(1, 2)$, and $(5, 10)$.   **e.** $(0, 0)$, and $(1, 0)$, and $(0, 1)$.
   c. $(5, 12)$, $(5, 8)$, and $(9, 12)$.   **f.** $(0, 0)$, $(1, 0)$, and $(½, \sqrt{3}/2)$.

4. What property characterizes all points of the plane that lie on axis 1? Axis 2?

5. How can you categorize the points of R^2 lying "below" axis 1? "Above" that same axis? To the right of axis 2? To the left of axis 2? To the right of axis 2 and above axis 1? To the left of axis 2 and above axis 1? To the left of axis 2 and below axis 1? To the right of axis 2 and below axis 1?

6. Let A and B be sets. The Cartesian product $A \times B$ of A and B is defined to be the following set: $A \times B = \{(a, b) : a \in A$ and $b \in B\}$ with $(a, b) = (c, d)$ if and only if $a = c$ and $b = d$. Determine the Cartesian products below (write out the list of elements in each case), where $A = \{0, 1, 2\}$; $B = \{3, 4, 5\}$; and $C = \{1, 3\}$.

a. $A \times A$.
b. $A \times B$.
c. $B \times A$.
d. $A \times C$.

e. $C \times A$.
f. $B \times C$.
g. $(A \times B) \times C$.
h. $A \times (B \times C)$.

2.2 STRAIGHT LINES

Consider now the plane (the terms plane and R^2 will be used interchangeably). A **straight line** in the plane is defined to be the set of all points (x, y) that satisfy some linear equation $ax + by + c = 0$ (a, b, and c being real numbers at least one of which is different from zero). The concept of a straight line is now backed by a precise statement of definition:

If ℓ is a straight line, there are constants a, b, and c (not all zero) such that $\ell = \{(x, y) \in R^2 : ax + by + c = 0\}$.

A *vertical* line is a straight line whose determining equation is of the form $x = c$. A *horizontal* line is one whose determining equation is of the form $y = c$. Figure 2.2.1 depicts a case of each of the two types.

These two special cases having been examined, let us turn our attention to the general case $ax + by + c = 0$, where $a \neq 0$ and $b \neq 0$. The equation may be rewritten in the form

$$y = \left(\frac{-a}{b}\right)x - \frac{c}{b}$$

The quantity $-a/b$ is called the **slope** of the line. Furthermore, setting $x = 0$, we see that $y = -c/b$ results. Hence $(0, -c/b)$ is one point of the line; namely, it is the **intercept** on the y axis (axis 2). Thus, $-c/b$ is called the y *intercept* of the line. For this reason, the form $y = mx + p$ is known as the *slope-intercept* form of the equation of a straight line. The term slope is highly suggestive.

Let (x_1, y_1) and (x_2, y_2) be two distinct points of the line in question.

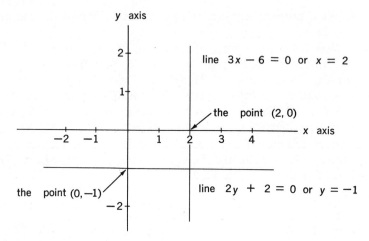

Figure 2.2.1
Two lines of a special nature

Since the line is neither vertical nor horizontal, $x_1 \neq x_2$ and $y_1 \neq y_2$. Figure 2.2.2 portrays such a situation.

Define Δy to be $y_2 - y_1$ and Δx to be $x_2 - x_1$. Notice that, in some sense, Δy represents the *directed y distance from* (x_1, y_1) to (x_2, y_2) while Δx represents the corresponding *directed x distance*. Examine the ratio of these two quantities.

$$\frac{\Delta y}{\Delta x} = \frac{y_2 - y_1}{x_2 - x_1} = \frac{\text{rise}}{\text{run}}$$

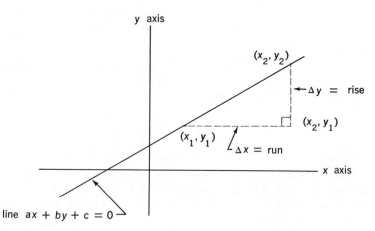

Figure 2.2.2
Interpretation of slope

Since (x_1, y_1) and (x_2, y_2) are on the line, they must satisfy the equation of the line:

$$y_1 = \frac{-ax_1}{b} - \frac{c}{b}$$

$$y_2 = \frac{-ax_2}{b} - \frac{c}{b}$$

Subtracting y_1 from y_2, we note that

$$y_2 - y_1 = \left(\frac{-a}{b}\right)(x_2 - x_1)$$

or

$$\frac{y_2 - y_1}{x_2 - x_1} = \frac{-a}{b}$$

Substituting into the above ratio $\Delta y/\Delta x$, we conclude that

$$\frac{\Delta y}{\Delta x} = \frac{-a}{b}$$

Consequently, the slope as defined earlier has this geometric significance: the slope is a measure of the inclination of the line relative to the x axis (axis 1). Further, the slope is defined independently of any given point on the line. Necessarily $\Delta y/\Delta x$ is the same, regardless of the two points chosen. This fact then allows us to (1) create the equation of *a* (the) line passing through two known points or (2) find the equation of a line having a given slope and passing through a particular point.

We have inadvertently (perhaps) inferred the existence of precisely one line through a given pair of points. See Exercise 2.2.6.

We might be concerned about whether or not the slope can be zero. In our construction above, zero slope was not the case ($y_1 \neq y_2$). However, in the form $y = mx + b$, $m = 0$ makes sense, implying that the line is horizontal. Conversely, any horizontal line is of the form $y = b$, which is to say $m = 0$. Thus, we can *associate horizontal lines with a slope of zero.*

We cannot define a slope for vertical lines, and hence we say that *a vertical line has no associated slope.* If we say that we want the equation of the line of slope m passing through the point P, we automatically imply that the line is not vertical.

Example 2.2.1   As an example of one of the situations above, find the equation of the line passing through (x_0, y_0) and having slope m.

Let (x, y) be any point (of the desired line) different from the fixed point (x_0, y_0). Then $\Delta y = y - y_0$ and $\Delta x = x - x_0$, whereby $m = \Delta y/\Delta x = (y - y_0)/(x - x_0)$. Since $x - x_0 \neq 0$, the equation may be altered to appear in the form

$$y - y_0 = m(x - x_0)$$

The equation developed in Example 2.2.1 is known as the *point-slope* form of the equation of a line. It is well to observe that (x_0, y_0) satisfies the equation even though (x_0, y_0) cannot be substituted into the expression $m = (y - y_0)/(x - x_0)$.

Example 2.2.2 Find an equation depicting a line through $(6, 4)$ with slope 2.

Letting (x, y) denote a general point on the line, we find that

$$\frac{y - 4}{x - 6} = 2$$

or

$$y - 4 = 2(x - 6)$$

whence

$$y = 2x - 8$$

The next to last form of the equation may be found directly from the result of Example 2.2.1 by setting $(x_0, y_0) = (6, 4)$ and $m = 2$.

Example 2.2.3 Should we be given two distinct points (x_0, y_0) and (x_1, y_1) (instead of a single point and a slope), we ought not despair since

$$m = \frac{y_0 - y_1}{x_0 - x_1}$$

It has been tacitly assumed that $x_0 \neq x_1$. If this were not the case, the equation would be immediate, as a vertical line results. Hence, in the nonvertical case, the point-slope form becomes the *two point* form

$$y - y_0 = \frac{(y_0 - y_1)(x - x_0)}{x_0 - x_1}$$

Example 2.2.4 Find an equation of the line through the points $(1, 5)$ and $(2, 8)$.

The slope m is determined from the two points by

$$\frac{8 - 5}{2 - 1} = 3$$

Thus, if (x, y) is an arbitrary point on the line,

$$\frac{y - 5}{x - 1} = 3$$

or

$$y - 5 = 3(x - 1)$$

whence

$$y = 3x + 2$$

We might also note that the equation can be derived by using the slope and the point $(2, 8)$ rather than the point $(1, 5)$.

$$\frac{y - 8}{x - 2} = 3$$
$$y - 8 = 3(x - 2)$$
$$y = 3x + 2$$

■ EXERCISES 2.2

1. Draw a picture of the model of R^2 given in the text and then locate each of the following points.

a.	$(0, 0)$.	**e.**	$(0, -3)$.	**i.**	$(-3, 3)$.
b.	$(3, 0)$.	**f.**	$(3, 3)$.	**j.**	$(1, 2)$.
c.	$(0, 3)$.	**g.**	$(-3, 3)$.	**k.**	$(2, 1)$.
d.	$(-3, 0)$.	**h.**	$(3, -3)$.	**l.**	$(3, 5)$.

2. Write the equation of each indicated straight line.

a. Passing through $(-3, 4)$ with slope $-\frac{1}{2}$.
b. Passing through $(1, 0)$ with slope 6.
c. Passing through $(1, 1)$ and $(2, 1)$.
d. Slope 3 and y intercept -2.
e. Passing through $(1, 2)$ with slope 4.
f. Passing through $(\frac{1}{2}, \frac{1}{2})$ with slope 1.
g. Passing through $(-1, -4)$ with slope 2.
h. Passing through $(\frac{1}{3}, -1)$ with slope -1.
i. Passing through $(1, 1)$ and $(3, 5)$.
j. Passing through $(-1, 1)$ and $(0, 0)$.
k. Passing through $(1, -2)$ and $(2, -2)$.
l. Slope -1 and y intercept 1.
m. y intercept 1 and x intercept 1.

3. The **midpoint** between (a, b) and (c, d) is defined to be the point (x, y) where $x = (a + c)/2$ and $y = (b + d)/2$. Show that (a) the midpoint is on the line determined by (a, b) and (c, d), and (b) the midpoint is equidistant from (a, b) and (c, d).

4. Given the points $(5, 6)$ and $(11, -2)$, find the equation of the line passing through the midpoint between the two (see Exercise 3) and having slope the negative reciprocal of the slope of the line containing the original two points. Show that any point (x, y) on this new line is equidistant from the two original points $(5, 6)$ and $(11, -2)$. Can you think of another name for this line?

5. How do you characterize **parallel** lines (lines that do not intersect) using the concept of slope? What, then, can you say about the slopes of lines that do intersect?

6. Show that two points determine precisely one straight line. [Hint: Let one line be given by $ax + by + c = 0$ and the other by $dx + ey + f = 0$. Show that either there is one point (x, y) satisfying both equations, no point satisfying both, or that each point satisfying one equation satisfies both. Algebraically, we say that the equations are, respectively, *simultaneous, inconsistent,* or *dependent.* Answer the question by attempting to solve the two equations simultaneously.]

2.3 LOCUS PROBLEMS

Another typical problem we encounter is that of providing an analytic description for some geometrically defined locus. Consider the "usual" definition of a circle. A **circle** C of center $P_0 = (h, k)$ and radius $r > 0$ is the set of all points $P = (x, y)$ where $d(P_0, P) = r$. Symbolically,

$$C = \{P = (x, y) : d(P_0, P) = r\}$$

The problem is that of finding an equation (involving x and y) which describes the set of points on this circle. Such an equation is

$$d(P_0, P) = r$$
Alternatively, $\quad ((x - h)^2 + (y - k)^2)^{1/2} = r$
or $\quad (x - h)^2 + (y - k)^2 = r^2$

It is not difficult to show conversely that each point satisfying the above equation is on the circle C. The beauty of this last form of the equation is that, at a single glance, we know the center and radius of the circle. These two items completely characterize or determine the circle. The latter form is called the *standard form* of the equation of a circle in the plane.

Example 2.3.1 Analyze the following equation:

$$(x - 3)^2 + (y - 2)^2 = 36$$

This must evidently be a circle of center $(3, 2)$ and radius 6.

Example 2.3.2 One circle that will be used extensively is given by

$$x^2 + y^2 = 1$$

and is known as the **unit circle.** The name is derived from the fact that the circle has unit radius $(r = 1)$ and center $(0, 0)$.

■ EXERCISES 2.3

1. Write the standard form of the equation of each of the following circles.

   a. Center $(1, 1)$ and radius 10.
   b. Center $(-1, -1)$ and radius 4.
   c. Center $(1, -1)$ and radius 3.
   d. Center $(0, 0)$ and radius 5.
   e. Center $(4, -2)$ and radius 2.
   f. Center $(1, 1)$ passing through $(1, 4)$.
   g. Center $(2, 3)$ and passing through $(6, 6)$.

2. Write an equation (as was done for the circle) depicting the set of points (x, y) that are equidistant from the two points $(1, 3)$ and $(-1, 2)$.

3. Follow the instructions for Exercise 2 where the points (x, y) have the property that the distance from (x, y) to $(1, 1)$ added to the distance from (x, y) to $(-1, 1)$ is 4.

2.4 REVIEW EXERCISES

1. Consider the "space" of real numbers. What is the distance between the following pairs of points (real numbers)?

   a. 0 and 5.   e. 5 and -5.
   b. 0 and -5.   f. 3 and 1.
   c. 5 and 7.   g. 2 and -4.
   d. -5 and -7.   h. 6 and -8.

2. Are each of the following properties held by the distance concept for real numbers?

   a. $d(x, y) \geqslant 0$. (Examine the definition of absolute value.)
   b. $d(x, y) = 0$ if and only if $x = y$.
   c. $d(x, y) = d(y, x)$.
   d. $d(x, y) \leqslant d(x, z) + d(z, y)$.

[Note: A distance with the four properties (a) to (d) is called a **metric**; a space together with such a metric is called a **metric space.**]

3. Find the distance between the following pairs of points in the plane.

   a. $(1, 0)$, $(\sqrt{3}/2, \frac{1}{2})$.   f. $(1, 0)$, $(0, 1)$.
   b. $(\sqrt{3}/2, \frac{1}{2})$, $(\frac{1}{2}, \sqrt{3}/2)$.   g. $(0, 1)$, $(-1, 0)$.
   c. $(\frac{1}{2}, \sqrt{3}/2)$, $(0, 1)$.   h. $(-1, 0)$, $(0, -1)$.
   d. $(1, 0)$, $(1/\sqrt{2}, 1/\sqrt{2})$.   i. $(0, -1)$, $(1, 0)$.
   e. $(1/\sqrt{2}, 1/\sqrt{2})$, $(0, 1)$.

4. What is the relationship between the distances found in 3(a), (b), and (c)? 3(d) and (e)? 3(f), (g), (h), and (i)?

5. In Exercise 3, each pair of points together with the origin $(0, 0)$ determines a triangle. Which of the triangles are right triangles? Isosceles triangles? Equilateral triangles?

6. Show that each of the points in Exercise 3 is on the unit circle $x^2 + y^2 = 1$.

7. What conclusion do you draw (using intuition if you wish) concerning that part of the unit circle between the pairs of points in Exercises 3(a), (b), and (c)? (Hint: Use the results of Exercise 4.) Answer the same question for the pairs of points in Exercises 3(d) and (e) for the pairs in (f), (g), (h), and (i).

8. Find the equations of the following straight lines.

   a. Passing through $(6, 1)$ and $(6, 2)$.
   b. Passing through $(6, 1)$ and $(5, 1)$.
   c. Passing through $(2, 3)$ and $(3, 5)$.
   d. Passing through $(1, 7)$ and $(3, 1)$.
   e. Passing through $(3, -2)$ with slope $-\frac{2}{3}$.
   f. Passing through $(-1, 2)$ with slope $\frac{4}{3}$.
   g. Slope $\frac{1}{2}$ and y intercept 2.
   h. Slope $\frac{2}{3}$ and x intercept -4.

9. Find the equations of the following circles.

   a. Center $(0, 0)$ and radius 5.
   b. Center $(4, 5)$ and radius 3.
   c. Center $(4, -5)$ and radius 3.
   d. Center $(-4, 5)$ and radius 3.
   e. Center $(-4, -5)$ and radius 3.

10. Find in each case the point(s) of intersection of the given line and circle.

   a. Line $y = 3x - 2$, circle $x^2 + y^2 = 4$.
   b. Line $y + 2 = x$, circle $x^2 + (y - 1)^2 = 25$.
   c. Line through $(5, 10)$ with slope 1 and circle of center $(5, 10)$ and radius 1.

11. Let (a, b) and (c, d) be two points. The point (x, y) is *between* (a, b) and (c, d) if there is a $\lambda \in R$ with $0 < \lambda < 1$ and $(x, y) = (a + \lambda(c - a), b + \lambda(d - b)) = (\lambda c + (1 - \lambda)a, \lambda d + (1 - \lambda)b)$. Show that each point between (a, b) and (c, d) lies on the line containing the two points. Furthermore, if (x, y) is between (a, b) and (c, d), then $d((a, b), (x, y)) + d((x, y), (c, d)) = d((a, b), (c, d))$.

12. Refer to Exercise 11 for definitions. The *ray emanating from* (a, b)

and passing through (c, d) is the set

$$\{(x, y) : (x, y) = (a, b), \ (x, y) \text{ is between } (a, b) \text{ and } (c, d), \text{ or}$$
$$(c, d) \text{ is between } (a, b) \text{ and } (x, y)\}$$

Show that the ray is a subset of the line through (a, b) and (c, d).

13. Use Exercise 12 to show that, if a ray emanates from the origin and passes through (c, d), and if (x, y) is on the ray, x and c are both positive or both negative. Also, d and y are both positive or both negative (that is, x and c have the same sign, while d and y have the same sign).

14. The **line segment** between (a, b) and (c, d) is the set

$$\{(x, y) : (x, y) = (a, b), \ (x, y) = (c, d), \text{ or}$$
$$(x, y) \text{ is between } (a, b) \text{ and } (c, d)\}$$

The line segment between (a, b) and (c, d) is "very nearly" the intersection of two rays. Describe this line segment in terms of the intersection of two rays. (There are two ways of doing it.)

15. Refer to Exercise 14. Describe the line segment between each pair of points in terms of two rays, describing the rays completely.

a. $(0, 1)$ and $(0, 5)$.
b. $(1, 1)$ and $(2, 2)$.
c. $(-5, 4)$ and $(-5, 12)$.
d. $(3, 5)$ and $(9, 11)$.
e. $(2, 4)$ and $(1, 6)$.
f. $(-2, -3)$ and $(-4, 1)$.

16. The removal of a straight line ℓ from the plane results in two disjoint collections of points. We define: P_1 and P_2 (neither on ℓ) lie on the *same side* of ℓ if the line segment between P_1 and P_2 does not contain a point of ℓ. Show that, if P_1 and P_2 lie on the same side of ℓ, and if P_2 and P_3 lie on the same side of ℓ, then P_1 and P_3 lie on the same side of ℓ. (This shows that the relation "lies on the same side of ℓ as" is an equivalence relation on $R^2 - \ell$.) If P_1 and P_2 (neither on ℓ) do not lie on the same side of ℓ, they lie on *opposite sides* of ℓ. Show that ℓ has exactly two sides and that these sides may be determined by (1) selecting P_1 and P_2 (neither on ℓ) on opposite sides and (2) showing that

$$A = \{P_3 : P_3 \text{ and } P_1 \text{ lie on the same side of } \ell\}$$
$$B = \{P_3 : P_3 \text{ and } P_2 \text{ lie on the same side of } \ell\}$$

satisfy $A \cap B = \varnothing$ and $A \cup B = R^2 - \ell$.

17. Show that, if ℓ is given by the equation $y = mx + b$,

then
and

$$A = \{(x, y) : y < mx + b\}$$
$$B = \{(x, y) : y > mx + b\}$$

form the two sides of ℓ.

2.5 ADVANCED EXERCISES

1. The following is a symbolism widely used in mathematics.

$$\sum_{i=1}^{n} (a_i) = a_1 + a_2 + a_3 + \cdots + a_n$$

This is read, "The sum of the a_i as i goes from 1 to n." In other words, the *index i* is allowed to take on all integral values from the lower value 1 to the upper value n. The *sum* of all expressions obtained in this process is the value indicated by the Σ notation on the left. As an example,

$$\sum_{i=1}^{5} \left(\frac{1}{i}\right) = \frac{1}{1} + \frac{1}{2} + \frac{1}{3} + \frac{1}{4} + \frac{1}{5} = \frac{137}{60}$$

Evaluate each of the following.

a. $\sum_{i=0}^{10} (i)$.

b. $\sum_{i=1}^{7} (i^2)$.

c. $\sum_{i=1}^{4} (i^3)$.

d. $\sum_{i=1}^{5} [(i^3 - 1)/i^2]$.

e. $\sum_{j=1}^{3} (j^2 - j)$.

f. $\sum_{k=1}^{5} [(k) - (k + 1)]$.

2. Let ℓ and ℓ' be lines of slopes m and m', respectively (it is implied that m and m' are real numbers).

a. Show that, if $mm' = -1$, ℓ and ℓ' intersect. (Recheck Exercise 2.2.6.)

b. Let the situation in (a) be the case, and let $P_1 = (a, b)$ be the point of intersection. Let $P_2 = (x, y)$ be on ℓ but different from P_1 with $P_3 = (p, q)$ on ℓ' different from P_1. Show that the triangle formed by these three points is a right triangle in the sense of the converse of the Pythagorean Theorem.

c. To examine a converse of (b), let $P_1 = (a, b)$ be the intersection of ℓ and ℓ' with slopes m and m', respectively. Suppose that, whenever $P_2 = (x, y) \in \ell$ and $P_3 = (p, q) \in \ell'$ with neither the same as P_1, it follows that the triangle formed is a right triangle in the sense of the Pythagorean Theorem. Prove, then, that $mm' = -1$. [Note: The lines in (b) and (c) are usually said to be **perpendicular.**]

d. Find an equation of the line perpendicular to $y = 3x + 2$ at the point $(0, 2)$.

e. Find an equation of the perpendicular to the line passing through $(1, 5)$ and $(2, 7)$ such that the perpendicular passes through the origin.

f. A **tangent** to a circle at a point (a, b) is the line through (a, b) perpendicular to the radius line drawn through the center of the circle and (a, b). Find the line tangent to the circle of center $(1, 3)$ and radius 17 at the point $(0, 7)$.

g. Find the equation of a tangent to the circle $(x - 2)^2 + (y + 3)^2 = $

25 so that the tangent is horizontal (has slope zero). There are two such tangents.

h. Find the equation of two tangents to the circle $(x - h)^2 + (y - k)^2 = r^2$, where the tangents do not have a defined slope.

3. a. Given the line $x = -5$ and the point $(5, 0) = P_0$, find the equation representing the points $(x, y) = P$ where $d(P, P_0)$ is equal to the distance from (x, y) to $(-5, y)$. The distance from (x, y) to $(-5, y)$ is seen to be the distance along the line perpendicular to $x = -5$ and passing through (x, y). By "distance along the perpendicular" is meant the distance from the given point to the intersection of the given line with the constructed perpendicular. [Note: The curve generated in this exercise is called *a parabola*. The curve in Exercise 2.3.2 is the *perpendicular bisector* of the line segment from $(1, 3)$ to $(-1, 2)$.]

The curve of Exercise 2.3.3 is an *ellipse*. The curve in the following exercise is a *hyperbola*.

b. Consider the two points $P_1 = (3, 0)$ and $P_2 = (-3, 0)$. Write an equation in x and y depicting the set of points

$$\{P = (x, y) : |d(P, P_1) - d(P, P_2)| = 2\}$$

4. Consider the point $(1, 3)$. How can we describe by equation the set of all circles of radius 5 that pass through $(1, 3)$? [Hint: Each circle has equation $(x - h)^2 + (y - k)^2 = 25$. What are the conditions describing h and k?]

5. Let P_1 be (x_1, x_2) and P_2 be (z_1, z_2). Find the midpoint between P_1 and P_2 and call it P_3. Find the midpoints P_4 and P_5 between P_1 and P_3 and P_3 and P_2, respectively. Write the equation of ℓ_1 passing through P_1 and P_2. Then write the equations of the following lines with slope m, $m/2$, and $m/4$, respectively: the line ℓ_2 passing through P_3, ℓ_3 passing through P_4, and ℓ_4 passing through P_5. Under what conditions will ℓ_2, ℓ_3, and ℓ_4 be parallel? Can two of these lines be parallel without all three satisfying the parallel condition?

6. The spaces R^1 and R^2 have been introduced in the text material. These are two special cases of a more general concept, namely the *Euclidean n space*, R^n. The generalization is completely analogous to what has been done. A point in R^1, for example, was a single real number, while a point in R^2 was an ordered pair of real numbers. What, then, is a point in R^3? It must be an ordered triple of real numbers, one such being $(3, -1, 0)$. The number 3 is the first coordinate value, -1 the second, and 0 the third. The concept of order in the triple is the same as that in the ordered pair of the plane. Then, R^3 is the set of all ordered triples of real numbers.

For the completely general case, let n be any positive integer. The following symbol is known as an *n-tuple* of real numbers:

$$(x_1, x_2, \cdots x_n)$$

The designation *n*-tuple comes from the fact that there are *n* of the sub-scripted *x*'s appearing. The term "real numbers" in the name is a result of each subscripted *x*'s being a real number. For clarification:

1. $(1, -2, 3)$ is an ordered triple of real numbers.
2. $(-\frac{1}{2}, 14)$ is an ordered pair.
3. $(-1, 6, 15, -3, \frac{1}{2}, 56, -90, 12)$ is an ordered 8-tuple.

Each position in the symbol (position being given by the subscript or by counting from left to right) is called a *coordinate* of the *n*-tuple. The quadruple (x_1, x_2, x_3, x_4) has first coordinate value x_1, second coordinate value x_2, and so forth. The values of the numbers in the coordinate posi-tions are also called *projections* (onto the respective axis). In the quad-ruple just above, x_1 is the projection onto the first axis, x_2 the projection onto the second axis, and so on. The significance (geometric and intui-tive) of the axis will be seen shortly.

Two *n*-tuples are said to be *equal* if and only if they are equal coordi-nate by coordinate. This means that $(1, 2, 3)$ and $(3, 2, 1)$ are not the same (equal) pairs, since they differ in the first and third coordinates. In this sense, the order of the numbers (order of appearance) in the *n*-tuple is important. For this reason, such a definition makes the *n*-tuples into *ordered n*-tuples of real numbers.

Now, R^n denotes $\{(x_1, x_2, \cdots, x_n) : (x_1, x_2, \cdots, x_n)$ is an ordered *n*-tuple of real numbers}. The ordered *n*-tuples will be, as one might sup-pose, the points in the space. The *n* in R^n denotes the dimension of the space. The mathematician thinks nothing of working in a space of dimen-sion 3000000000000000. He may have no suggestive model, but the con-sequences of such a deficiency are not of major proportions.

A distance concept can be extended in a manner completely analogous to that for R^1 and R^2. Using the notation of Exercise 2.5.1, we may define *distance* relative to R^n. Let

$$P_1 = (x_1, x_2, x_3, \cdots, x_n) \quad \text{and} \quad P_2 = (y_1, y_2, \cdots, y_n)$$

Then
$$d(P_1, P_2) = \left[\sum_{i=1}^{n} (x_i - y_i)^2 \right]^{1/2}$$

This distance is calculated in the same manner as done in R^1 and R^2. We simply

1. compute the difference between the points coordinate by coordinate,
2. square each difference,
3. sum the squares, and, finally,
4. compute the principal square root of this sum.

For illustration, find the distance between P_1 and P_2 where $P_1 = (1, -1, 3, 0, 1)$ and $P_2 = (2, 0, 0, 1, 3)$. (Note: P_1 and P_2 are in R^5.)

$$d(P_1, P_2) = [(1 - 2)^2 + (-1 - 0)^2 + (3 - 0)^2 + (0 - 1)^2 + (1 - 3)^2]^{1/2}$$

$$= [1 + 1 + 9 + 1 + 4]^{1/2} = \sqrt{16} = 4$$

Several fundamentals should be observed.

1. $d(P_1, P_2) \geqslant 0$, since the princpal square root is indicated.
2. $d(P_1, P_2) = 0$ if and only if $P_1 = P_2$.
3. $d(P_1, P_2) = d(P_2, P_1)$.
4. $d(P_1, P_2) \leqslant d(P_1, P_3) + d(P_3, P_2)$, P_3 any point in R^n.

The proof of the validity of (1) is self-evident from the definition of principal square root.

a. Prove (2).
b. Prove (3).

A proof of (4) is quite complicated in nature and will not be asked of the student. A proof can be found in many texts covering more advanced analysis. Property (4) is referred to quite often as the *Triangle Inequality*. The geometric reason for this is clear.

Find the distances between the following pairs of points. (Make sure both points in each pair are from the same dimension space.)

c. $(1, 0, 3)$, $(2, -3, -1)$.
d. $(\frac{1}{2}, \frac{1}{3}, \frac{1}{4}, -2)$, $(-\frac{1}{2}, -\frac{2}{3}, -\frac{3}{4}, -1)$.
e. $(5, 0, 6, -2, 3, 7)$, $(1, 1, 1, 1, 1, 1)$.
f. $(3, 0, 5, 6)$, $(4, 0, 2, 7, 1)$.
g. For each point in (c) to (f), give the value of the projection onto the first axis. The projection onto the third axis. The projection onto the fifth axis where appropriate.

Figure 2.5.1 shows a two-dimensional picture of a model of R^3. There

Figure 2.5.1
R^3

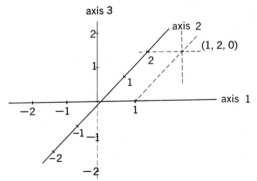

are three axes (the same as the number of the dimension). A point $(1, 2, -1)$ is located. The procedure is analogous to that for locating points in a model of the plane R^2.

Sketch a picture of a model of R^3 and locate the following points.

h. $(0, 0, 0)$.   **k.** $(0, 0, 1)$.
i. $(1, 0, 0)$.   **l.** $(-1, -1, -1)$.
j. $(0, 1, 0)$.   **m.** $(4, 2, 3)$.

7. In the plane define a new *distance*. Let $P_1 = (x, y)$ and $P_2 = (a, b)$. Define the distance between P_1 and P_2 by

$$t(P_1, P_2) = |x - a| + |y - b|$$

This is sometimes called the "taxicab metric." As an example, the distance, in this context, between $(1, 5)$ and $(2, 6)$ is given by $|1 - 2| + |5 - 6| = 1 + 1 = 2$.

Find the distances (in the taxicab metric) between the following pairs of points.

a. $(1, 6)$, $(2, 8)$.
b. $(0, 0)$, $(3, 5)$.
c. $(3, 5)$, $(2, -3)$.
d. $(2, 1)$, $(-3, -6)$.
e. Show that the taxicab metric has all the properties of a metric as given in Exercise 2.4.2.

Chapter 3
Functions

3.1 INTRODUCTION TO FUNCTIONS

As mentioned at the very onset of this text, the formulation of a concept of the function is a primary desire. Although there are several ways to introduce this idea, we shall use two: one because of its intuitive ramifications, and the other because of its value in the sense of graphing and abstraction.

Let each of X and Y be a set. A **function** f from X into Y, written $f:X \longrightarrow Y$, is a rule or correspondence which relates to each x in X a unique y in Y. Now, y will be denoted by $f(x)$. That is, $y = f(x)$.

There are several facts to observe in connection with a function $f:X \longrightarrow Y$.

1. If $x \in X$, there is a $y \in Y$ with $y = f(x)$. Moreover, there is only one such y.
2. There may be some $y \in Y$ with $y \neq f(x)$ for any x in X. Intuitively, Y may not be entirely "used up" by the correspondence.
3. The correspondence may join more than one x to the same y. That

is to say, it may be that $x_1 \neq x_2$ (that is, x_1 and x_2 are different) but $f(x_1) = f(x_2)$.

The following examples will illustrate.

Example 3.1.1   Let $X = \{0, 1, 2, 3, 4\}$ and $Y = \{a, b, c, d, e\}$. Define f by the following correspondence:

$$
\begin{array}{ll}
0 \xrightarrow{f} a & \text{that is, } f(0) = a \\
1 \longrightarrow b & \text{that is, } f(1) = b \\
2 \longrightarrow c & \text{that is, } f(2) = c \\
3 \longrightarrow a & \text{that is, } f(3) = a \\
4 \longrightarrow b & \text{that is, } f(4) = b
\end{array}
$$

In Fig. 3.1.1 we have drawn a sketch representing the manner in which f "carries" the elements of X over to elements of Y. Observe that there is one and only one arrow emanating from each element of X. Each arrow terminates at an element of Y. The correspondence f may be thought of as being the set of arrows. Is f a function from X into Y? The answer is yes, because:

1. Every $x \in X$ is mated to some y in Y. That is, for every x there is an $f(x)$.
2. For no x in X do there exist two choices for $f(x)$.

Thus, we conclude that to every $x \in X$, f mates one element y from Y. The definition of a function is satisfied.

Example 3.1.2   Let X be as in Example 3.1.1 with $Z = \{a, b, c\}$. Let f be the same rule given in the example. Again, f is a function ($f: X \longrightarrow Z$). Note, however, that although Y was not entirely used up (d and e are not used) in the correspondence, Z is used. (See Fig. 3.1.2.)

Whenever the function is defined so that the set *into which* it maps is used up, we say that the function is an **onto function.** Such a function is $f: X \longrightarrow Z$.

Figure 3.1.1
$f: X \longrightarrow Y$

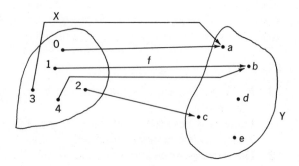

Figure 3.1.2
$f:X \longrightarrow Z$

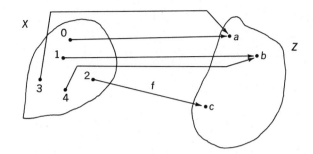

Example 3.1.3 Let X and Y be as in Example 3.1.1. Define a correspondence g by the following:

$$g(0) = a, \, g(1) = b, \, g(2) = c, \, g(3) = d, \text{ and } g(4) = e$$

You should verify that g is a function. Figure 3.1.3 displays the correspondence. The function g has a very special property. The correspondence mates to each point of Y *precisely one* element from X. Figure 3.1.3 shows that no more than one arrow terminates at any given element of Y. Figure 3.1.2 shows this not to be the case for f. As an example (from Fig. 3.1.2), two arrows terminate at b.

Whenever no element of Y is used more than once, the function is said to be **1 : 1**. Clearly, g is $1:1$. In this situation, g happens to be "onto" also. However, a $1:1$ function need *not* be an onto function.

Example 3.1.4 Let $P = \{a, b, c, d, e, f, g\}$. Let g be the same correspondence as given immediately above. Now $g:X \longrightarrow P$ is a function. (See Fig. 3.1.4.) No element of P is the image of more than one element of X under g. Thus, $g:X \longrightarrow P$ is $1:1$. Since P is not exhausted by g, g is not an onto function.

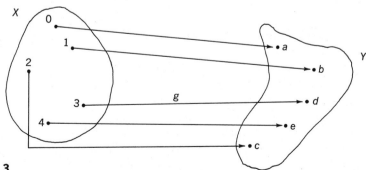

Figure 3.1.3
$g:X \longrightarrow Y$

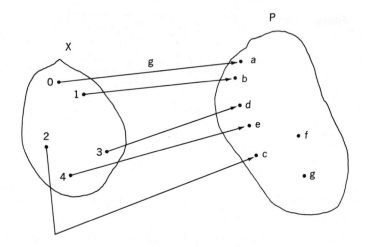

Figure 3.1.4
$g:X \longrightarrow P$

One interesting note is that, whenever a function $h:X \longrightarrow Y$ is $1:1$ and an onto function, it intuitively suggests that X and Y have (in some sense) the same number of elements. As a matter of record, this is generally the technique used to define such a condition. See Exercise 3.8.7.

■ EXERCISES 3.1

1. Let $X = \{0, 1, 2, 3, 4, 5\}, A = \{0, 1, 2\}, Y = \{a, b, c, d, e\}, B = \{a, c, e\}$.
Let f, g, and h be the following correspondences.

$0 \xrightarrow{f} a$	$0 \xrightarrow{g} a$	$0 \xrightarrow{h} a$
$1 \longrightarrow c$	$1 \longrightarrow a$	$1 \longrightarrow c$
$2 \longrightarrow e$	$2 \longrightarrow a$	$2 \longrightarrow e$
$3 \longrightarrow b$	$3 \longrightarrow a$	
$4 \longrightarrow d$	$4 \longrightarrow a$	
$5 \longrightarrow a$	$5 \longrightarrow a$	

a. Give reasons to justify the statement that f is a function with $f:X \longrightarrow Y$.
b. Is f an onto function? $1:1$?
c. Why is $g:X \longrightarrow Y$ a function? Note that g is an example of a **constant function** (every element of X mated with the same element of Y).
d. Is $h:A \longrightarrow Y$ a $1:1$ function? An onto function?
e. Is $h:A \longrightarrow B$ a $1:1$ function? An onto function?

2. Follow the instructions and answer the questions of Exercise 1, where $X = \{0, -1, 1, 2, -2, -3, 3, 4, -4\}, A = \{0, 1, 2, 3, 4\}, Y = \{a, b, c, d, e, i, j, k, m, n\}, B = \{a, c, d, i, k, n\}$, and f, g, and h are the following correspondences.

$$0 \xrightarrow{f} a$$
$$1 \longrightarrow b$$
$$-1 \longrightarrow c$$
$$2 \longrightarrow d$$
$$-2 \longrightarrow e$$
$$3 \longrightarrow i$$
$$-3 \longrightarrow j$$
$$4 \longrightarrow k$$
$$-4 \longrightarrow n$$

$$0 \xrightarrow{g} a$$
$$1 \longrightarrow b$$
$$-1 \longrightarrow b$$
$$2 \longrightarrow d$$
$$-2 \longrightarrow d$$
$$3 \longrightarrow i$$
$$-3 \longrightarrow i$$
$$4 \longrightarrow k$$
$$-4 \longrightarrow k$$

$$0 \xrightarrow{h} a$$
$$1 \longrightarrow b$$
$$2 \longrightarrow c$$
$$3 \longrightarrow i$$
$$4 \longrightarrow k$$

Answer the following questions also.

a. Is any of the three correspondences a constant function?

b. If $k:P \longrightarrow Q$ is a function such that $k(a) = k(-a)$ for each $a \in P$, k is an **even** function. Which among f, g, and h is an even function?

3. Using the correspondences of Exercise 1, find the value of the following.

a. $f(0)$.	**e.** $g(1)$.
b. $f(2)$.	**f.** $g(3)$.
c. $f(4)$.	**g.** $h(1)$.
d. $f(5)$.	**h.** $h(2)$.

4. Using the correspondences of Exercise 2, find the value of each of the following.

a. $f(0)$.	**e.** $g(1)$.	**i.** $h(0)$.
b. $f(-1)$.	**f.** $g(2)$.	**j.** $h(2)$.
c. $f(3)$.	**g.** $g(-3)$.	**k.** $h(3)$.
d. $f(-4)$.	**h.** $g(4)$.	**l.** $h(4)$.

3.2 INVERSE FUNCTIONS

Upon examining Fig. 3.1.3, we should note that, if the arrows were reversed, a function from Y to X would result (Fig. 3.2.1). The fact that reversing the arrows leads to a new function holds only for functions that are both 1:1 and onto functions. Figure 3.2.2 shows the situation occurring when the arrows of Fig. 3.1.1 are reversed.

Two observable things happen. First, there are elements in Y from which no arrow emanates; second, two arrows originate from both a and b. Each situation is indicative of the fact that the described correspondence is not a function.

Figure 3.2.3 illustrates the reversal of the arrows of Fig. 3.1.2. At least

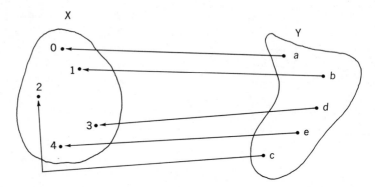

Figure 3.2.1
Reversing the arrows

one arrow radiates from each element of Z, but, as before, two arrows start at each of a and b. Thus, the new correspondence does not yield a function according to our definition.

If a function $h: X \longrightarrow Y$ is 1:1 and an onto function, the correspondence h^{-1} given by $h^{-1}(y) = x$ where $h(x) = y$ is a function called the **inverse function** for h. This is a mathematical description of the act of reversing the arrows. The idea of an inverse function will be used frequently throughout the text. Some specific examples of inverse functions follow.

Example 3.2.1 Let $X = \{0, 1, 2, 3\}$ and $Y = \{r, s, t, u\}$ with f defined by

$$
\begin{array}{ccc}
0 & \xrightarrow{\;f\;} & s \\
1 & \longrightarrow & u \\
2 & \longrightarrow & t \\
3 & \longrightarrow & r
\end{array}
$$

Figure 3.2.2
No function

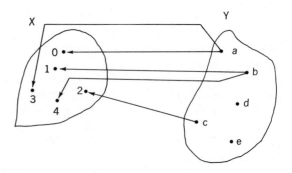

Figure 3.2.3
No function

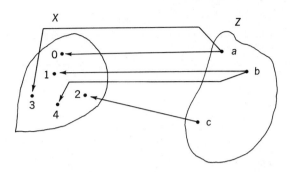

Now, f is $1:1$ and "onto" as a function from X to Y. The inverse function f^{-1} is given by

$$s \xrightarrow{\ f^{-1}\ } 0$$
$$u \longrightarrow 1$$
$$t \longrightarrow 2$$
$$r \longrightarrow 3$$

Example 3.2.2   Using X as in Example 3.2.1, and letting $Z = \{r, s, t\}$, define h by the following:

$$0 \xrightarrow{\ h\ } r$$
$$1 \longrightarrow s$$
$$2 \longrightarrow t$$
$$3 \longrightarrow r$$

Does h have an inverse?   No, h is not $1:1$ (it is an onto function, though).   In trying to reverse the diagram for h, we have

$$r \xrightarrow{\ h^{-1}\ } 0$$
$$s \longrightarrow 1$$
$$t \longrightarrow 2$$
$$r \longrightarrow 3$$

Thus, h^{-1} takes r to both 0 and 3, whence it is not a function.

Example 3.2.3   Let X be as in Example 3.2.1, and let $W = \{r, s, t, u, v\}$. Also, let f be the correspondence of that example.   That is,

$$0 \xrightarrow{\ f\ } s$$
$$1 \longrightarrow u$$
$$2 \longrightarrow t$$
$$3 \longrightarrow r$$
$$v$$

Observe that $f: X \longrightarrow W$ is not "onto" (although it is 1:1). In this case, then, f does not have an inverse. Reversing the diagram above, we see that

$$s \xrightarrow{f^{-1}} 0$$
$$u \longrightarrow 1$$
$$t \longrightarrow 2$$
$$r \longrightarrow 3$$
$$v$$

Since f^{-1} does not mate v with an element of X, f^{-1} is not a function (W must be "used up").

■ **EXERCISES 3.2**

1. Tell which of the following functions have inverses and which do not. Describe each inverse function; if a function does not have an inverse, tell why.

a. $a \xrightarrow{f} 1$
$b \longrightarrow 2$
$c \longrightarrow 3$
$d \longrightarrow 4$
$e \longrightarrow 5$

b. $a \xrightarrow{g} 1$
$b \longrightarrow 2$
$c \qquad 3$
$d \longrightarrow 4$
$e \longrightarrow 5$

c. $a \xrightarrow{h} 1$
$b \longrightarrow 2$
$c \qquad 3$
$d \qquad 4$
$e \qquad 5$

d. $a \xrightarrow{t} 1$
$b \qquad 2$
$c \qquad 3$
$d \longrightarrow 4$
$e \longrightarrow 5$

e. $a \xrightarrow{u} 1$
$b \qquad 2$
$c \qquad 3$
$d \qquad 4$
$e \qquad 5$
$\qquad 6$

f. $a \xrightarrow{k} 1$
$b \qquad 2$
$c \qquad 3$
$d \qquad 4$
$e \qquad 5$

2. Answer the following problems relative to (1) Exercise 3.1.1 and (2) Exercise 3.1.2.

    a. Does f have an inverse? Why or why not?
    b. Does g have an inverse? Why or why not?
    c. Does $h: A \longrightarrow B$ have an inverse? If so, what is the correspondence?
    d. Does $h: A \longrightarrow Y$ have an inverse? If so, what is the correspondence?

Figure 3.3.1
Domains and codomains

Example	Function	Domain	Codomain
3.1.1	f	X	Y
3.1.2	f	X	Z
3.1.3	g	X	Y
3.2.1	f	X	Y
3.2.2	h	X	Z
3.2.3	f	X	W

3.3 MORE ON FUNCTIONS

Some further terminology and symbolism will prove useful and descriptive. Given a function $f:X \longrightarrow Y$, X is called the **domain** of f, while Y is a **codomain.** Examples of domains and codomains are given in the chart in Fig. 3.3.1.

That subset of the codomain Y "used up" under the correspondence is denoted as the **range** of f and is written $f[X]$. Symbolically, for $f:X \longrightarrow Y$,

$$f[X] = \{y \in Y : y = f(x) \text{ for some } x \in X\}$$

The ranges for the functions in the six previous examples are listed in Fig. 3.3.2.

We are aware that a function is called "onto" if and only if its range is the prescribed codomain, that is, $k:X \longrightarrow Y$ is "onto" if and only if $k[X] = Y$. We could have taken this for the definition of an onto function.

We can generalize the concept of the range. Given that $f:X \longrightarrow Y$ is a function with $A \subset X$, while $B \subset Y$, define the following descriptive sets:

image of $A = f[A] = \{y \in Y : y = f(a) \text{ for some } a \in A\}$
inverse image of $B = f^{-1}[B] = \{x \in X : f(x) \in B\}$

Figure 3.3.2
Ranges

Example	Function	Range
3.1.1	f	$\{a, b, c\}$
3.1.2	f	$\{a, b, c\} = Z$
3.1.3	g	$\{a, b, c, d, e,\} = Y$
3.2.1	f	$\{r, s, t, u\} = Y$
3.2.2	h	$\{r, s, t\}$
3.2.3	f	$\{s, t, u, r\}$

Figure 3.3.3
$f[A]$

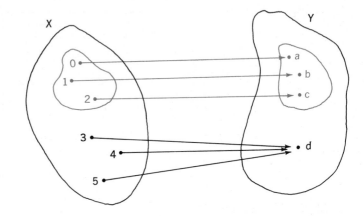

Example 3.3.1   Let $X = \{0, 1, 2, 3, 4, 5\}$, $A = \{0, 1, 2\}$, $Y = \{a, b, c, d\}$, and $B = \{a, c, d\}$.   Consider the correspondence f given below.

$$0 \xrightarrow{f} a$$
$$1 \longrightarrow b$$
$$2 \longrightarrow c$$
$$3 \longrightarrow d$$
$$4 \longrightarrow d$$
$$5 \longrightarrow d$$

Figure 3.3.3 shows the situation described by the correspondence f. Now, $f[A]$ is that part of Y used up as f "takes" the elements of A over to Y. The figure shows this set to be $\{a, b, c\}$, that is, $f[A] = \{a, b, c\}$.

Now examine the set B. What elements of X are carried into B via the correspondence f?   Figure 3.3.4 shows this set to be $\{0, 2, 3, 4, 5\}$. According to the definition, $f^{-1}[B] = \{0, 2, 3, 4, 5\}$.

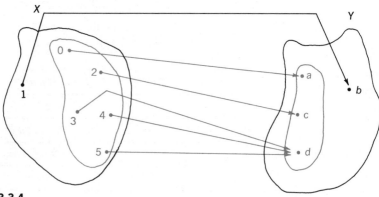

Figure 3.3.4
$f^{-1}[B]$

In this usage, the f^{-1} is *not* to be construed as denoting an inverse function. Rather, $f^{-1}[B]$ should be considered a symbol representing a subset of the domain of the function f. In fact, f^{-1} may not even exist (as a function).

Two functions f and g are said to be **equal** if they have the same domain and represent the same correspondence (that is, $f(x) = g(x)$ for each x in the common domain).

Example 3.3.2 $X = \{0, 1, 2, 3\}$, $Y = \{a, b, c\}$, and $W = \{a, b\}$.   Define $f, g, h,$ and k as below.

$f:X \longrightarrow Y$	$g:X \longrightarrow W$	$h:X \longrightarrow Y$	$k:Y \longrightarrow W$
$0 \xrightarrow{f} a$	$0 \xrightarrow{g} a$	$0 \xrightarrow{h} a$	$a \xrightarrow{k} a$
$1 \longrightarrow b$	$1 \longrightarrow b$	$1 \longrightarrow b$	$b \longrightarrow b$
$2 \longrightarrow b$	$2 \longrightarrow b$	$2 \longrightarrow c$	$c \longrightarrow a$
$3 \longrightarrow a$	$3 \longrightarrow a$	$3 \longrightarrow a$	

Now, f, g, and h have the same domain. Is $f = g$? $f = h$? $g = h$? It is true that $f = g$, since the correspondences are the same. However, $f(2) \neq h(2)$, whence $f \neq h$. Similarly, $g \neq h$.

None of the three (f, g, and h) is equal to k, since the domain of k is Y and not X.

Suppose that $f:X \longrightarrow Y$ and $g:W \longrightarrow Z$ are functions with $f[X] \subset W$. Then we can define a function $h:X \longrightarrow Z$ by using f and g. The function so obtained is known as the **composition** of g with f and is written $h = g \circ f$. The composition function $g \circ f$ is defined by $(g \circ f)(x) = g(f(x))$.

Figure 3.3.5 illustrates the construction of the composition. Selecting any x from X, we automatically acquire a y in $Y(y = f(x))$. Having obtained y, we are directed (by g) to a unique element z ($z = g(y)$) in the set Z. Thus, the choosing of any x in X results in the mating of x to a single z in Z. This, then, shows that $g \circ f$ is a function.

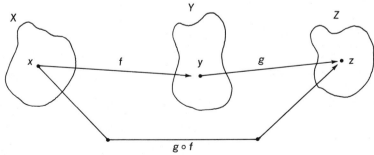

Figure 3.3.5
Composition of functions

That this is true *is due entirely to the fact that the range of f is contained in the domain of g.* This is important.

The following example shows the construction of a composition of two functions.

Example 3.3.3 Let $X = \{0, 1, 2, 3, 4, 5\}$, $Y = \{6, 7, 8, 9\}$, and $Z = \{10, 11, 12, 13\}$. Define $f:X \longrightarrow Y$ and $g:Y \longrightarrow Z$ by the following:

$$
\begin{array}{ll}
f(0) = 6 & f(5) = 7 \\
f(1) = 7 & g(6) = 10 \\
f(2) = 8 & g(7) = 11 \\
f(3) = 9 & g(8) = 12 \\
f(4) = 6 & g(9) = 10
\end{array}
$$

Hence, $g \circ f$ is the correspondence below.

$$
\begin{array}{l}
g \circ f(0) = g(f(0)) = g(6) = 10 \\
g \circ f(1) = g(f(1)) = g(7) = 11 \\
g \circ f(2) = g(f(2)) = g(8) = 12 \\
g \circ f(3) = g(f(3)) = g(9) = 10 \\
g \circ f(4) = g(f(4)) = g(6) = 10 \\
g \circ f(5) = g(f(5)) = g(7) = 11
\end{array}
$$

(See Fig. 3.3.6.)

Again, the item of importance ensuring that the composition is a function is that the range of f is a subset of the domain of g.

Example 3.3.4 To illustrate the last remark above, let $X = \{1, 2, 3, 4, 5\}$, $Y = \{a, b, c\}$, and $Z = \{0, -1, -2\}$ with f and g the following functions.

$$
\begin{array}{ll}
1 \overset{f}{\longrightarrow} a & 1 \longrightarrow 0 \\
2 \longrightarrow b & 2 \longrightarrow -1 \\
3 \longrightarrow c & 3 \longrightarrow -2 \\
4 \longrightarrow a & 4 \longrightarrow 0 \\
5 \longrightarrow b & 5 \longrightarrow 0
\end{array}
$$

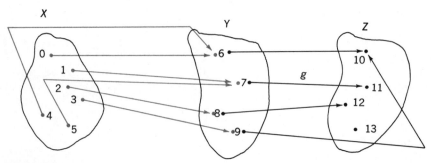

Figure 3.3.6

What is $g \circ f$? The answer is that $g \circ f$ does not make sense. Why?   Try to calculate $g \circ f(1)$, for example.   If $g \circ f(1)$ can be found, $g \circ f(1) = g(f(1)) = g(a)$.   However, *a is not in the domain of g* and $g(a)$ is therefore not defined.   This points out the necessity that f's range be a subset of g's domain.

■ EXERCISES 3.3

1. Let $X = \{0, 1, 2, 3, 4\}$, $Y = \{a, b, c, d\}$, and $Z = \{½, ⅓, ¼, ⅕\}$, with f and g as follows.

$$
\begin{array}{ll}
0 \xrightarrow{\;f\;} a & \qquad a \xrightarrow{\;g\;} ½ \\
1 \longrightarrow b & \qquad b \longrightarrow ⅓ \\
2 \longrightarrow c & \qquad c \longrightarrow ¼ \\
3 \longrightarrow a & \qquad d \longrightarrow ⅕ \\
4 \longrightarrow d &
\end{array}
$$

   a. What is the correspondence $g \circ f$?
   b. Does $f \circ g$ make sense?   Why or why not?

2. From Exercise 1, find each of the following.

a. $f(0)$.	**g.** $g(c)$.
b. $f(1)$.	**h.** $g(d)$.
c. $f(3)$.	**i.** $g \circ f(0)$.
d. $f(4)$.	**j.** $g \circ f(1)$.
e. $g(a)$.	**k.** $g \circ f(3)$.
f. $g(b)$.	**l.** $g \circ f(4)$.

3. Using X, Y, and Z as in Exercise 1, define f, g, h, and k as follows.

$$
\begin{array}{llll}
f\colon X \longrightarrow Y & g\colon X \longrightarrow Y & h\colon Y \longrightarrow Z & k\colon Z \longrightarrow Y \\
0 \longrightarrow a & 0 \longrightarrow c & a \longrightarrow ½ & ½ \longrightarrow a \\
1 \longrightarrow b & 1 \longrightarrow d & b \longrightarrow ⅓ & ⅓ \longrightarrow c \\
2 \longrightarrow c & 2 \longrightarrow a & c \longrightarrow ¼ & ¼ \longrightarrow d \\
3 \longrightarrow d & 3 \longrightarrow b & d \longrightarrow ⅓ & ⅕ \longrightarrow a \\
4 \longrightarrow a & 4 \longrightarrow b & &
\end{array}
$$

Which of the following make sense?

a. $f \circ g$.	**i.** $h \circ k$.
b. $f \circ h$.	**j.** $k \circ f$.
c. $f \circ k$.	**k.** $k \circ g$.
d. $g \circ f$.	**l.** $k \circ h$.
e. $g \circ h$.	**m.** $f \circ f$.
f. $g \circ k$.	**n.** $g \circ g$.
g. $h \circ f$.	**o.** $h \circ h$.
h. $h \circ g$.	**p.** $k \circ k$.

Show (via diagrams) the compositions that make sense. What must be true for $f \circ f$ to make sense?

3.4   REAL FUNCTIONS

The class of functions with which most of the remaining material will be concerned is comprised of functions

$$f: X \longrightarrow Y$$

where both X and Y are subsets of R, the set of real numbers. Such a function is called a *real-valued function of a real variable* (real function).

In this vein, an equation may portray a functional relationship.

Example 3.4.1   For instance, let $k \in R$ and $f(x) = k$ for each real number x. Since, for each real number $x, f(x)$ is uniquely determined, a suitable domain for f (and the largest such) is R itself. Thus, we may consider f in the sense $f: R \longrightarrow R$. Our function f is a constant function, every element of R mated to k.

Example 3.4.2   Consider now the equation $g(x) = 1/x$. Is there a domain rendering g a function? Can such a domain be R? The answer to the latter question is in the negative, since $0 \in R$ and $g(0)$, being interpreted as $1/0$, is not defined.

Thus, g cannot mate a real number to 0, whence 0 cannot be in any domain of g. Every nonzero real number has a unique reciprocal, however, meaning that $R - \{0\}$ is the (largest) valid choice for the domain of g. We may consider g a function from $R - \{0\}$ into R (or $R - \{0\}$).

Now, examine the linear relationship given by $f(x) = mx + b; m, b \in R$. We may wish to write $f(x) = y$ and examine $y = mx + b$. Is a function implied by this equation?

Because of the properties of addition and multiplication, for each real number $x, mx + b$ is a unique real number; that is, every real number qualifies as a domain element for f and we see that $f: R \longrightarrow R$ makes sense. We say that f is a **linear function.**

We might say that R is a *natural* domain for f. The natural domain will be taken to be the "largest" set that is an acceptable domain (that is, under which the relationship represents a function).

We feel that we "know" the correspondence f, since given a real number $x, y = f(x) = mx + b$ is easily determined. It can be shown that, if $m \neq 0, f$ is both 1:1 and an onto function. Can we determine, then, what f^{-1} ought to be?

We can make such a determination by discovering what x looks like in terms of y. Before, given a domain element x, we expressed in terms of x

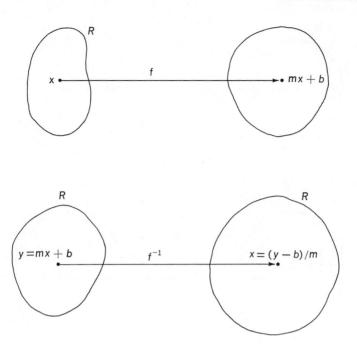

Figure 3.4.1
Forming an inverse

the corresponding range element.   Now, we are essentially given a range element and must express in terms of this range element the corresponding domain element (the domain element to which it corresponds).   See Fig. 3.4.1.

In our case $y = mx + b$ can be converted to $x = y/m - b/m = (y - b)/m$.   Thus, $f^{-1}(y) = (y - b)/m$.

What is $f \circ f^{-1}(y)$ and $f^{-1} \circ f(x)$?   If we give thought to the relationship between f and f^{-1}, we should see immediately that the respective answers are y and x.   To check this, we compute:

$$f \circ f^{-1}(y) = f\left(\frac{y - b}{m}\right) = m\left(\frac{y - b}{m}\right) + b = y$$

$$f^{-1} \circ f(x) = f^{-1}(mx + b) = \left(\frac{(mx + b) - b}{m}\right) = x$$

In Example 3.4.2, $g(x) = 1/x$; solving for x, we see that $x = 1/g(x)$. In the notation $y = g(x)$, $x = 1/y$.   Thus $g^{-1}(y) = 1/y$, whence g is its own inverse.   (See Fig. 3.4.2.)   Again, $g \circ g^{-1}(y) = y$ and $g^{-1} \circ g(x) = x$.

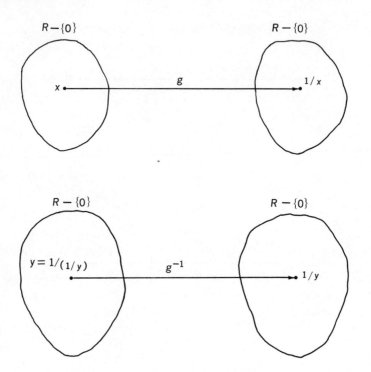

Figure 3.4.2
Forming another inverse

■ EXERCISES 3.4

1. Consider the equation $f(x) = 2x + 1$. Is $f: R \longrightarrow R$ a function (that is, for each real number x, is it true that $f(x) = 2x + 1$ is a unique real number)? Find the values of each of the following.

a. $f(0)$.
b. $f(\frac{1}{2})$.
c. $f(-\frac{1}{2})$.
d. $f(2)$.

e. $f(1)$.
f. $f(-1)$.
g. $f(6)$.
h. $f(24)$.

2. Does f in 1 have an inverse? If so, state the inverse in terms of an equation.

3. If $f(x) = x^2$ describes a real function whose natural domain is R and whose range is $f[R]$, what is $f[R]$?

a. Is f a 1:1 function?

b. Is f a 1:1 function if the domain is the set of all nonnegative real numbers?

c. If either of the above two are answered in the affirmative, give an equation showing an inverse functional relationship.

4. Using the function of Exercise 3, find the following.

a. $f(0)$.   **d.** $f(2)$.
b. $f(1)$.   **e.** $f(-2)$.
c. $f(-1)$.   **f.** $f(3)$.

5. Each of the following is to be considered as implying a functional relationship, the function being in each case a real function. Give the largest possible domain in each instance and then give the image of the function under that domain.

a. $g(x) = x^{1/2}$.   **b.** $\eta(x) = x$.   **c.** $\lambda(x) = 3x - 1$.

6. In (c) of Exercise 5, λ is 1:1 and "onto." Determine an equation describing λ^{-1}. In other words, $\lambda^{-1}(x) = ?$.

7. From Exercise 5, find the following values.

a. $g(4)$.   **g.** $\eta(0)$.   **m.** $\lambda(\frac{1}{3})$.
b. $g(-2)$.   **h.** $\eta(1)$.   **n.** $\lambda(0)$.
c. $g(0)$.   **i.** $\eta(-1)$.   **o.** $\lambda(4)$.
d. $g(\frac{9}{4})$. **j.** $\eta(2)$.   **p.** $\lambda(2)$.
e. $g(7)$.   **k.** $\eta(-2)$.   **q.** $\lambda(-1)$.
f. $g(2)$.   **l.** $\eta(-4)$.   **r.** $\lambda(-3)$.

8. From your answer to Exercise 6, find the following.

a. $\lambda^{-1}(0)$.   **c.** $\lambda^{-1}(1)$.   **e.** $\lambda^{-1}(-7)$.
b. $\lambda^{-1}(3)$.   **d.** $\lambda^{-1}(5)$.   **f.** $\lambda^{-1}(2)$.

9. Find the natural domain for each of the following.

a. $f(x) = 1/(x - 1)$.   **c.** $g(x) = x/x$.
b. $h(x) = x/(x + 1)$.   **d.** $k(x) = 1/(x^2 + 5x + 6)$.

10. From Exercise 9, find each of the following.

a. $f(2)$.   **g.** $g(-1)$.
b. $f(1)$.   **h.** $g(51)$.
c. $h(2)$.   **i.** $k(0)$.
d. $h(1)$.   **j.** $k(-1)$.
e. $h(-1)$.   **k.** $k(3)$.
f. $h(2)$.   **l.** $k(-2)$.

11. Let f and g be given as in each case. Write an equation showing the functional relationship describing $g \circ f$.

a. $f(x) = x + 1$ and $g(x) = 1/x$.
b. $f(x) = x^2$ and $g(x) = x^2$.
c. $f(x) = x^2$ and $g(x) = \sqrt{x}$.
d. $f(x) = 2x + 1$ and $g(x) = x - 1$.

12. Find the following values of $g \circ f$ for each case in Exercise 11.

a. $g \circ f(0)$.
b. $g \circ f(1)$.
c. $g \circ f(-2)$.
d. $g \circ f(-3)$.
e. $g \circ f(4)$.
f. $g \circ f(\frac{1}{2})$.

The remaining exercises form a section on elementary functions encountered in the study of calculus.

13. Let n be a positive integer. Each equation of the form $f(x) = x^n$ denotes a function whose domain is R.

a. Which values of n render f 1:1?
b. Which values of n render f "onto"?
c. For which n is it true that $f(x) = f(-x)$? Such functions are said to be **even.**
d. For which n is it true that $f(x) = -f(-x)$? Such functions are said to be **odd.**

14. If n is a nonnegative integer with $a_0, a_1, a_2, \cdots, a_n$ real numbers, $a_n \neq 0$, $f(x) = a_0 + a_1x + a_2x^2 + \cdots + a_nx^n$ is a **polynomial** (over R) *of degree n.* Determine the degree of each of the following polynomials.

a. $f(x) = 6$.
b. $g(x) = 3x + 2$.
c. $h(x) = 4x^2 + 3x + 1$.
d. $k(x) = 4x^3 + 8x^2 - x - 2$.

15. Find the following values where $f, g, h,$ and k are as found in Exercise 14.

a. $f(2)$.
b. $g(1)$.
c. $h(5)$.
d. $k(3)$.
e. $f(-2)$.
f. $g(-1)$.
g. $h(-5)$.
h. $k(-3)$.

16. If f is a polynomial, the real numbers x where $f(x) = 0$ are called **zeros** of the polynomial. Find the zeros for each polynomial in Exercise 14.

17. Which polynomials are even functions? Odd functions?

18. If f is the quotient of two polynomials, f is said to be a **rational func-**

tion. The domain for f is $R - A$, where A is the set of zeros of the denominator. The elements of A are called **poles** of f. The **zeros** of f are the domain elements that are also zeros of the numerator. Find the domains, zeros, and poles for the rational functions below.

  a. $f(x) = x/(x + 1)$.
  b. $g(x) = x/x^2$.
  c. $h(x) = (x + 1)/(x^2 + 4x + 4)$.
  d. $k(x) = x^2/x$.
  e. $\mu(x) = x$.

Is $k = \mu$?

19. Evaluate each of the following using the functions from Exercise 18.

  a. $f(1)$.   **e.** $g(-2)$.
  b. $f(0)$.   **f.** $h(4)$.
  c. $f(-1)$.   **g.** $h(-2)$.
  d. $g(1)$.   **h.** $k(7)$.

20. Under what conditions can a rational function be an even function? An odd function?

21. Let $b \in R$, $b > 0$, $b \neq 1$. Then, if $f(x) = b^x$, f is said to be an **exponential function** with base b.

  a. Show that $f(x)f(y) = f(x + y)$.
  b. Show that $f(x)/f(y) = f(x - y)$.
  c. Show that $(f(x))^n = f(nx)$.

22. Let $f(x) = 2^x$ and $g(x) = 10^x$. Evaluate each of the following.

  a. $f(0)$.   **j.** $g(-3)$.
  b. $g(0)$.   **k.** $f(-3)$.
  c. $f(1)$.   **l.** $g(-3)$.
  d. $g(1)$.   **m.** $f(\frac{1}{2})$.
  e. $g(-1)$.   **n.** $g(\frac{1}{2})$.
  f. $g(-1)$.   **o.** $f(\frac{3}{2})$.
  g. $f(2)$.   **p.** $g(\frac{3}{2})$.
  h. $g(-2)$.   **q.** $g(-\frac{1}{2})$.
  i. $f(3)$.   **r.** $g(-\frac{1}{2})$.

23. In Exercise 21, f is 1:1 and an onto function if we consider f in the sense $f:R \longrightarrow \{a \in R:a > 0\}$. Thus, f^{-1} is defined by $f^{-1}:\{a \in R:a > 0\} \longrightarrow R$ and $f^{-1}(y) = x$ where $b^x = y$. In other words, $f^{-1}(y)$ is the exponent for b giving y. We are accustomed to the notation $f^{-1}(y) = \log_b y$, $\log_b y$ being the **logarithm** of y with respect to b. Evaluate each of the following (f and g as in Exercise 22).

a. $f^{-1}(1)$.	**k.** $\log_2 1$.
b. $g^{-1}(1)$.	**l.** $\log_{10} 1$.
c. $f^{-1}(2)$.	**m.** $\log_2 2$.
d. $g^{-1}(10)$.	**n.** $\log_{10} 10$.
e. $f^{-1}(\sqrt{2})$.	**o.** $\log_2 \sqrt{2}$.
f. $g^{-1}(10)$.	**p.** $\log_{10} 10$.
g. $f^{-1}(4)$.	**q.** $\log_2 4$.
h. $g^{-1}(100)$.	**r.** $\log_2 8$.
i. $f^{-1}(8)$.	**s.** $\log_{10} 100$.
j. $g^{-1}(1000)$.	**t.** $\log_{10} 1000$.

24. Show that for $f(x) = b^x$ (use the ideas of Exercise 21):

a. $f^{-1}(xy) = f^{-1}(x) + f^{-1}(y)$.
b. $f^{-1}(x/y) = f^{-1}(x) - f^{-1}(y)$.
c. $f^{-1}(x^n) = nf^{-1}(x)$.
d. $\log_b (xy) = \log_b x + \log_b y$.
e. $\log_b (x/y) = \log_b x - \log_b y$.
f. $\log_b x^n + n \log_b x$.

25. Using Exercises 21, 23, and 24, evaluate the following.

a. $b^{\log_b 2}$. **c.** $b^{\log_b a}$.
b. $\log_b b^2$. **d.** $\log_b b^a$.

3.5 FUNCTIONS, RELATIONS, AND ORDERED PAIRS

In the previous section, the concept of a function was described in terms of words like "relate," "rule," "correspond," "mate," and so forth. The basic idea was that each x in the domain was paired with some unique element $f(x)$ in the codomain. Figure 3.5.1 illustrates various ways of picturing one such function.

The final diagram of Fig. 3.5.1 shows the *function as a collection of ordered pairs* $(x, f(x))$. The first element in each pair is the domain element, while the second element is its image under the correspondence. The function f is seen to be described in detail by listing all such ordered pairs. Examples 3.1.1 to 3.1.3 of Section 3.1 can be restated (the functions can be redesignated) in the following way.

Examples 3.1.1 and 3.1.2:

$$f = \{(0, a), (1, b), (2, c), (3, c), (4, b)\}$$

Example 3.1.3:

$$g = \{(0, a), (1, b), (2, c), (3, d), (4, e)\}$$

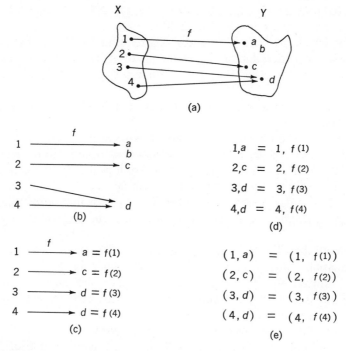

Figure 3.5.1
Means of expressing functional relationships

In this notation, the set of all first elements from the ordered pairs is the domain, while the set of all second elements forms the range of the function. The more general codomain as given before does not (generally) appear explicitly in the new setting. An exception to this statement occurs when the function being discussed is an onto function.

Thus, we can define a **function** as a collection of ordered pairs such that each element that is a first element in some ordered pair is a first element in exactly one ordered pair. That is, if (x, y) and (x, z) are ordered pairs in some function f, then y must be the same as (equal to) z. Alternatively, no two different ordered pairs have the same first element. (Two ordered pairs are the same if and only if they have the same first element and the same second element.)

Before continuing this discussion, let us interrupt ourselves for a moment to discuss a more general concept.

If A is a set, a **relation** on A is merely a subset of the Cartesian product $A \times A = \{(a, b) : a \in A, b \in A\}$.

Example 3.5.1 Let $A = \{0, 1, 2, 3\}$. Let $B = \{(0,0), (1,1), (2,2), (3,3)\}$,

$C = \{(0,1),(1,2),(2,3),(3,0)\}, D = \{(1,2),(2,1)\}, E = \{(1,2),(1,3),$
$(2,1),(3,1)\}$, and $F = \{(4,0),(2,3)\}$.

Now, which of B, C, D, E, and F is a relation on A? All except F are collections of ordered pairs of elements of A (that is, each except F is a subset of $A \times A$). $F \not\subset A \times A$ since $4 \notin A$. Hence, all but F are relations on A.

The most obvious relation on A is $A \times A$ itself, while $\varnothing$ is another obvious choice.

A relation V on a set A is **reflexive** if whenever $x \in A$, (x, x) is in V. The set $\{(x, x) : x \in A\}$ is called the **diagonal** of A and will be denoted by δ_A. In this vein, a relation on A is reflexive if and only if it contains the diagonal. Among the relations in Example 3.5.1, which contain the diagonal? A quick observation shows $B = \delta_A$. Furthermore, none of the relations C, D, and E contains δ_A. Hence, B is the only reflexive relation listed.

We note that $A \times A$ itself is a reflexive relation on A.

A relation V on A is said to be **symmetric** if whenever $(a, b) \in V$, $(b, a) \in V$ also. Intuitively, if V is symmetric, we can pick any element of V, reverse the order of the first and second elements, and the resulting pair is still an element of V.

Which of the relations in Example 3.5.1 is symmetric? The diagonal B is clearly symmetric. The relations D and E are symmetric also. That C is not symmetric is seen by $(0, 1) \in C$, but $(1, 0) \notin C$. The relation $A \times A$ also has the property of symmetry.

A relation V on A is **transitive** if (a, b) and (b, c) in V implies $(a, c) \in V$. The primary observation to make is that the first element of one pair is the second element of the other (b is the element indicated).

Are any of the relations from Example 3.5.1 seen to be transitive? The following are not: C, D, and E. Why?

$$(0, 1) \in C, (1, 2) \in C, \text{ but } (0, 2) \notin C$$
$$(1, 2) \in D, (2, 1) \in D, \text{ but } (1, 1) \notin D$$
$$(1, 3) \in E, (3, 1) \in E, \text{ but } (1, 1) \notin E$$

The diagonal is transitive, since (a, b) and (b, c) in B implies that $(a, c) \in B$. (In the diagonal, $a = b = c$ would necessarily be the case.) The Cartesian product $A \times A$ is also transitive.

A relation V is said to be an **equivalence** relation on A if V is reflexive, symmetric, and transitive.

In Example 3.5.1, B and $A \times A$ are equivalence relations. The diagonal is the smallest equivalence relation while $A \times A$ is the largest.

Is a function $f : A \longrightarrow A$ a relation? The answer is affirmative, since f is (can be thought of as) a collection of ordered pairs of elements of A (that is, $f \subset A \times A$).

Is a relation necessarily a function? No! The relation E (from Example 3.5.1), for instance, is not a function. The diagram below indicates why this is the case.

$$1 \xrightarrow{\;E\;} 2$$
$$1 \longrightarrow 3$$
$$2 \longrightarrow 1$$
$$3 \longrightarrow 1$$
$$0$$

E *carries* 1 to both 2 and 3. Disregarding this, we see that E is still not a function with domain A, since 0 is not used in the mating. Hence, not all relations are functions.

One useful concept arising from the ordered pair approach to functions is that, whenever f is a real function, f can be thought of as being a collection of ordered pairs of real numbers, hence a subset of R^2, the plane. This gives us a means of picturing the function. This picture is generally called a *graph*.

We might interpret the graph and the function as being the same set of points of R^2, one being a pictorial version. It is not physically possible to draw a graph of every function, though. Students should not acquire the habit of relying exclusively on graphs to yield information about functions. Algebraic techniques of analysis are more nearly accurate and valid.

Example 3.5.2 Graph the following real function.

$$f(x) = \begin{cases} 0 \text{ for } x \leqslant 0 \\ x \text{ for } 0 < x < 1 \\ 1 \text{ for } 1 \leqslant x \end{cases}$$

What is f? In the ordered pair concept, f is the following collection of points in R^2.

$$f = \{(x, 0) : x \leqslant 0\} \cup \{(x, x) : 0 < x < 1\} \cup \{(x, 1) : 1 \leqslant x\}$$

That this is true is seen by a three fold analysis. If $x \leqslant 0$, f mates x to 0. Thus the collection of ordered pairs resulting is the set of ordered pairs of the form $(x, 0)$ where $x \leqslant 0$ (f is constant on the nonpositive numbers).

On the other hand, if $0 < x < 1$, f mates x with itself, a condition which can be symbolized by $x \longrightarrow x$. This yields the second set of points of the form (x, x).

Finally, if $x \geqslant 1$, x is mated with 1, giving the third set of points (f is constant on the real numbers greater than 1). See Fig. 3.5.2.

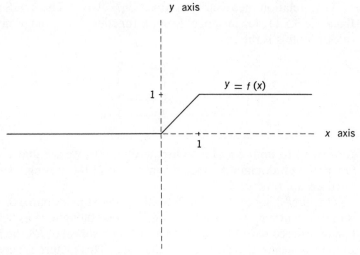

Figure 3.5.2
Graph of the function in Example 5.3.2

■ **EXERCISES 3.5**

1. List all possible relations on A where $A = \{0, 1\}$.
2. Do Exercise 1 where $A = \{a, b, c\}$.
3. List all the reflexive relations from:

  a. Exercise 1.
  b. Exercise 2.

4. List all the symmetric relations from:

  a. Exercise 1.
  b. Exercise 2.

5. List all the transitive relations from:

  a. Exercise 1.
  b. Exercise 2.

6. List all the equivalence relations from:

  a. Exercise 1.
  b. Exercise 2.

7. Tell which of the collections (a) to (c) represent a function. For each function, state the domain and the range and tell whether or not the function is 1:1. For each 1:1 function give the collection representing the inverse function.

  a. $f = \{(a, 1), (b, 2), (c, 3), (d, 1), (e, 3), (f, 0), (e, 2)\}$.
  b. $g = \{(1, 1), (0, 1), (3, 1), (4, 5), (2, 3)\}$.
  c. $h = \{(1, 1), (2, 3), (3, 3), (4, 5), (5, 6)\}$.

  8. Write each of the following functions as a collection of ordered pairs.

  a. f of Example 3.1.1.
  b. g of Example 3.1.3.
  c. f of Example 3.2.1.
  d. h of Example 3.2.2.
  e. f of Example 3.2.3.
  f. f of Example 3.3.1.
  g. f, g, h, and k of Example 3.3.2.
  h. $g \circ f$ of Example 3.3.2.

  9. If any of the functions from Exercise 2 has an inverse, write the inverse in ordered pair notation.
  10. In Section 3.3 the function f given by $f(x) = mx + b$ was examined. In the ordered pair notation, f may be given as follows. For m and b in R, each fixed,

$$
\begin{aligned}
f &= \{(x, f(x)) : x \in R, f(x) = mx + b\} \\
&= \{(x, y) : x \in R, y = mx + b\} \\
&= \{(x, mx + b) : x \in R\}
\end{aligned}
$$

  Write each of the following in ordered pair symbolism.

  a. f in Exercise 3.4.1.
  b. g, η, and λ of Exercise 3.4.5.
  c. λ^{-1} of Exercise 3.4.5.
  d. f and g from each of parts (a), (b), (c), and (d) of Exercise 3.4.11.
  e. $g \circ f$ of each of parts (a), (b), (c), and (d) of Exercise 3.4.11.
  f. f^{-1} of Exercise 3.4.1.
  g. f, g, h, and k from Exercise 3.4.9.

  11. Graph each function listed below.

  a. f from Exercise 10(a).
  b. f and f^{-1} from Exercise 10(f).
  c. g, η, and λ of Exercise 10(g).
  d. λ^{-1} of Exercise 10(c).
  e. f, g, h, and k from Exercise 10(g).
  f. f and g from each part of Exercise 10(d).
  g. $g \circ f$ of each part of Exercise 10(e).
  h. f in Exercise 3.4.1.
  i. Each of the functions in Exercise 3.4.5.
  j. Each of the functions in Exercise 3.4.9.

k. Each of the functions in Exercise 3.4.11.
l. Each of the functions in Exercise 3.4.12.
m. Each of the functions in Exercise 3.4.14.
n. Each of the functions in Exercise 3.4.18.
o. f and g of Exercise 3.4.22.
p. f^{-1} and g^{-1} of Exercise 3.4.22.

12. Suppose f is a collection of ordered pairs of real numbers and $x = 6$ is the first element of some ordered pair in f. Suppose further that the vertical line through $x = 6$ intersects the graph of f twice. Is f a function? Why or why not?

13. Suppose f is a real function. How many times may a horizontal line intersect the graph of f if f is $1:1$?

3.6 SOME SPECIAL FUNCTIONS

Now that the function concept has been introduced, let us examine some new and associated ideas. Suppose a function $f: X \longrightarrow Y$ is given and $A \subset X$. The function f also defines a function from A into Y. This function is called the **restriction** of f to A, and is written $f|A$. To say explicitly that $g = f|A$ means $g: A \longrightarrow Y$ is such that $g(a) = f(a)$ for each $a \in A$. Conversely, we might say that f is an **extension** of the function g.

In the ordered pair notation for functions, $f|A$ is the set of ordered pairs from f having elements of A as their first elements.

Example 3.6.1   Let $X = \{0, 1, 2, 3, 4, 5\}$, $A = \{0, 1, 2\}$, and $Y = \{a, b, c, d\}$ with f given by:

$$f(0) = a$$
$$f(1) = b$$
$$f(2) = c$$
$$f(3) = d$$
$$f(4) = d$$
$$f(5) = d$$

(See Fig. 3.3.1.)   The restriction of f to A, $f|A$, consists of those arrows originating from the subset A of X. Thus, $f|A$ is as pictured in Fig. 3.6.1 and is given by

$$0 \longrightarrow a \qquad \text{that is, } f|A(0) = a$$
$$1 \longrightarrow b \qquad \text{that is, } f|A(1) = b$$
$$2 \longrightarrow c \qquad \text{that is, } f|A(2) = c$$

In the ordered pair notation, $f = \{(0, a), (1, b), (2, c), (3, d), (4, d), (5, d)\}$, while $f|A = \{(0, a), (1, b), (2, c)\}$.

Figure 3.6.1
$f|A$

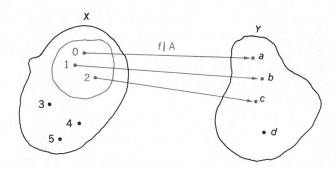

Example 3.6.2 Let $f: R \longrightarrow R$ be given by $f(x) = x$. Let N, as usual, denote the set of positive integers. What is $f|N$? Clearly, $f(n) = n$ for each $n \in N$. Then, in the ordered pair notation, $f|N$ is precisely that set of ordered pairs of natural numbers where the first and second elements are the same. Figure 3.6.2 shows a graph of the function f. The large dots represent the points belonging to $f|N$.

The function given in Example 3.6.2 is a very important one. It is known as an **identity** function. The name is self-explanatory: the identity function takes each x to itself, that is, $x \longrightarrow x$. Observe that the function is $1:1$, an onto function, and is its own inverse.

Another function of importance is the *projection function*. Recall that, whenever a point (a, b) of the plane is given, the real number a is called the first coordinate, while b is called the second coordinate. Each corre-

Figure 3.6.2
A function and a restriction

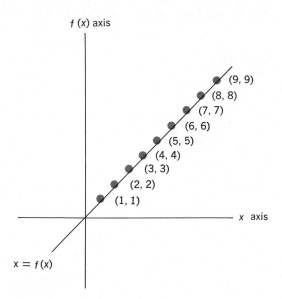

spondence $(a, b) \longrightarrow a$ and $(a, b) \longrightarrow b$ denotes a functional relationship. The function taking $(a, b) \longrightarrow a$ is denoted by p_1, while that mating $(a, b) \longrightarrow b$ is p_2. The function p_1 is the **first projection** function, and p_2 the **second projection** function.

The definitions of these two functions will be given in a more rigorous statement.

$$p_1 : R^2 \longrightarrow R \text{ is given by } p_1((a, b)) = a$$
$$p_2 : R^2 \longrightarrow R \text{ is given by } p_2((a, b)) = b$$

The projection functions simply "pick out" the appropriate coordinate values.

Example 3.6.3 Find the values of the projection functions at each of the following points: $(2, 3)$, $(-1, 2)$, $(\frac{1}{2}, 3)$, $(0, 5)$.

$$p_1((2, 3)) = 2$$
$$p_1((-1, 2)) = -1$$
$$p_1((\frac{1}{2}, 3)) = \frac{1}{2}$$
$$p_1((0, 5)) = 0$$

Now, the p_2 value for each of the points is given by

$$p_2((2, 3)) = 3$$
$$p_2((-1, 2)) = 2$$
$$p_2((\frac{1}{2}, 3)) = 3$$
$$p_2((0, 5)) = 5$$

Figure 3.6.3 shows a geometric interpretation of p_1 and p_2. The point in the plane under consideration is $(2, 5)$. It is noticed that a vertical line

Figure 3.6.3
Projections

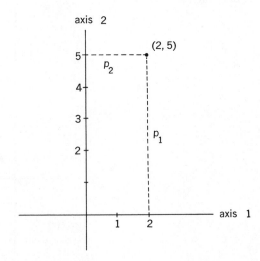

through $(2, 5)$ crosses the first axis with intercept 2, while a horizontal line through $(2, 5)$ has intercept 5 on axis 2. These are the values of $p_1((2, 5))$ and $p_2((2, 5))$, respectively.

■ EXERCISES 3.6

1. Consider Exercise 3.1.1. Is $h = f|A$?
2. Consider Exercise 3.1.3. Is $h = f|A$?
3. Describe $g|A$ in Exercises 3.1.1 and 3.1.2.
4. Let $f(x) = |x|$ and $g(x) = x$, the identity function.

   a. If P is the set of nonnegative reals, is $f|P = g|P$? Verify.
   b. Is $f|(R - P) = g|(R - P)$?

5.a. Given that $f = \{(0, 0), (1, 2), (3, 4), (5, 6), (\frac{1}{2}, 7), (\frac{1}{4}, 8)\}$ find $f|A$, where $A = \{0, 3, \frac{1}{2}\}$.
   b. Does f^{-1} exist?
   c. If so, what is f^{-1}?
   d. What is $(f|A)^{-1}$?

6. Let f be given by

$$f(x) = \begin{cases} 0 \text{ for } x < 0 \\ -1 \text{ for } x = 0 \\ 0 \text{ for } x > 0 \end{cases}$$

Let A be the closed interval $[0, 1]$.

   a. Graph the function f.
   b. Describe $f|A$.
   c. Describe $f|(R - A)$.

7. Let $g: R \longrightarrow R$ be given by

$$g(x) = \begin{cases} x \text{ for } x \in Q \\ 0 \text{ for } x \in I \end{cases}$$

   a. Describe $g|Q$.
   b. Describe $g|I$.

8. Let $\Phi(x) = |x|$. Find the largest set $A \subset R$ so that $\Phi: A \longrightarrow R$ is the identity function (that is, $\Phi(x) = x$ for all $x \in A$).

9. Find each of the following.

   a. $p_1((6, 4))$.
   b. $p_1((2, 5))$.
   c. $p_1((\frac{1}{2}, 3))$.
   d. $p_1((1, 2))$.
   e. $p_1((1, 3))$.
   f. $p_1((1, 6))$.
   g. $p_2((1, 3))$.
   h. $p_2((2, 3))$.
   i. $p_2((-1, 3))$.
   j. $p_2((0, \frac{1}{2}))$.
   k. $p_2((-1, -1))$.
   l. $p_2((2, 2))$.

10. Let $\eta:R \longrightarrow R^2$ by $\eta(x) = (x, 2x)$. For instance, $\eta(1) = (1, 2)$ and $\eta(2) = (2, 4)$. Find each of the following.

a. $p_1 \circ \eta(1)$.   **e.** $p_1 \circ \eta(x)$.

b. $p_2 \circ \eta(1)$.   **f.** $p_2 \circ \eta(x)$.

c. $p_1 \circ \eta(2)$.   **g.** $\eta \circ (p_1 \circ \eta)(x)$.

d. $p_2 \circ \eta(2)$.   **h.** $\eta \circ (p_2 \circ \eta)(x)$.

3.7 REVIEW EXERCISES

1. Let $X = \{0, 1, 2, 3, 4, 5, 6\}$ and $Y = \{a, b, c, d, e\}$.

a. Can there exist a function $f:X \longrightarrow Y$ so that f is 1:1? Why or why not?

b. Can there exist a function $g: Y \longrightarrow X$ such that g is a 1:1 function?

c. Can there exist a function $h:X \longrightarrow Y$ with h an onto function?

d. Can there exist a function $k: Y \longrightarrow X$ with k an onto function?

e. Can there exist a 1:1 and onto function $t:X \longrightarrow Y$?

f. Can there exist a 1:1 and onto function $s: Y \longrightarrow X$?

g. If the answer to any of (a) to (f) is yes, construct an example of such a function.

Let $A = \{3, 5, 6\}$ and $B = \{a, b, e\}$. Let u and v be correspondences as given by:

$$
\begin{array}{ll}
0 \xrightarrow{u} a & \quad 0 \xrightarrow{v} c \\
1 \longrightarrow c & \quad 1 \longrightarrow b \\
2 \longrightarrow e & \quad 2 \longrightarrow e \\
3 \longrightarrow b & \quad 3 \longrightarrow c \\
4 \longrightarrow c & \quad 4 \longrightarrow b \\
5 \longrightarrow a & \quad 5 \longrightarrow e \\
6 \longrightarrow e & \quad 6 \longrightarrow a
\end{array}
$$

h. Is each of u and v a function? If so, give the domain and range of such functions.

i. What is $u[X]$? $v[X]$? $u[A]$? $v[A]$? $u^{-1}[Y]$? $v^{-1}[Y]$? $u^{-1}[B]$? $v^{-1}[B]$? $u^{-1}[u[X]]$? $v^{-1}[v[X]]$? $u^{-1}[u[A]]$? $v^{-1}[v[A]]$? $u[u^{-1}[B]]$? $v[v^{-1}[B]]$?

j. Describe the correspondence $u|A$. $v|A$.

k. Is $u|A = v|A$? Why or why not?

l. Is u 1:1 and an onto function? v?

m. Is $u|A$ 1:1? $v|A$?

n. Is $u|A$ 1:1 and an onto function? $v|A$?

o. Write u and v as collections of ordered pairs.

p. Write $u|A$ and $v|A$ as collections of ordered pairs.

q. If any of u, v, $u|A$, or $v|A$ is 1:1, write a corresponding inverse for each such, using ordered pair notation.

Find the value of each of the following.

r. $u(0)$.	**u.** $u(5)$.	**x.** $v(4)$.
s. $u(1)$.	**v.** $u(6)$.	**y.** $v(2)$.
t. $u(2)$.	**w.** $v(3)$.	**z.** $v(6)$.

2. Let $f(x) = 3x$ represent a functional relationship $(f: R \longrightarrow R)$. Let $g(x) = x/3$. Now $g: R \longrightarrow R$ is a function also.

a. Show that $g \circ f$ and $f \circ g$ are both the identity function. That is, show that $g \circ f(x) = x = f \circ g(x)$ for all real x.

b. For what $x \in R$ is $f(x) = g(x)$?

Compute the following values.

c. $f(0)$.	**i.** $f(\frac{1}{3})$.	**o.** $f \circ g(-1)$.
d. $f(1)$.	**j.** $f(-\frac{1}{3})$.	**p.** $g \circ f(0)$.
e. $f(-1)$.	**k.** $g(3)$.	**q.** $g \circ f(1)$.
f. $g(0)$.	**l.** $g(-3)$.	**r.** $g \circ f(-1)$.
g. $g(1)$.	**m.** $g \circ f(0)$.	
h. $g(-1)$.	**n.** $f \circ g(1)$.	

s. From (a) one can conclude that $g = f^{-1}$ and $f = g^{-1}$ (f and g are inverses of each other). Show directly that $f^{-1}(x) = g(x)$ and $g^{-1}(x) = f(x)$.

3. Using f and g from Exercise 2, compute each of the following.

a. $f(x) + g(x)$.

b. $f(x) - g(x)$.

c. $f(x) \cdot g(x)$.

d. $f(x) \div g(x)$. (For what x does this quotient become undefined?)

e. $f(0) + g(0)$.	**m.** $f(-1) \cdot g(-1)$.
f. $f(1) + g(1)$.	**n.** $f(0) \div g(0)$.
g. $f(-1) + g(-1)$.	**o.** $f(1) \div g(1)$.
h. $f(0) - g(0)$.	**p.** $f(-1) \div g(-1)$.
i. $f(1) - g(1)$.	**q.** $6(f(0))$.
j. $f(-1) - g(-1)$.	**r.** $6(f(1))$.
k. $f(0) \cdot g(0)$.	**s.** $6(f(-1))$.
l. $f(1) \cdot g(1)$.	

4. Let f and g be two real functions. The **sum, difference, product,** and **quotient,** respectively, of f and g are symbolized and defined by:

$$(f + g)(x) = f(x) + g(x)$$
$$(f - g)(x) = f(x) - g(x)$$
$$(fg)(x) = f(x)g(x)$$
$$\left(\frac{f}{g}\right)(x) = \frac{f(x)}{g(x)}$$

In order that each of the above terms make sense, x must be in the domains of both f and g. For that reason, the domains of $f + g$, $f - g$, and fg are all given as the set $D_f \cap D_g$, where D_f and D_g are the domains of f and g, respectively. This is nearly the case for f/g, except that $g(x) = 0$ makes $f(x)/g(x)$ undefined. Thus, the domain of f/g is the set $(D_f \cap D_g) - \{x \in R : g(x) = 0\}$. We must "take out" the zeros of g. As a further example of one function generated by another, define for any real number a,

$$(af)(x) = a(f(x))$$

We call af a **scalar multiple** of f.

Compare the definitions given here with Exercise 3. Let

$$f(x) = x + 1 \text{ and } g(x) = x - 1$$

Give equations expressing:

a. $f \circ g(x)$.   **e.** $(fg)(x)$.
b. $g \circ f(x)$.   **f.** $(f/g)(x)$.
c. $(f + g)(x)$.   **g.** $f^{-1}(x)$.
d. $(f - g)(x)$.   **h.** $g^{-1}(x)$.

Give the domain of each of the following:

i. f.   **m.** $f + g$.
j. g.   **n.** $f - g$.
k. $g \circ f$.   **o.** fg.
l. $f \circ g$.   **p.** f/g.

Give the range of each of the following:

q. f.   **u.** $f + g$.   **y.** g^{-1}.
r. g.   **v.** $f - g$.   **z.** f^{-1}.
s. $g \circ f$.   **w.** fg.
t. $f \circ g$.   **x.** f/g.

5.a. Graph f and g of Exercise 2.
  b. Graph f and g of Exercise 4.
  c. Graph $g \circ f$ and $f \circ g$ from Exercise 2.
  d. Graph $f \circ g$ and $g \circ f$ from Exercise 4.
  e. Graph f^{-1} and g^{-1} from Exercise 4.
  f. Graph $f + g$, $f - g$, fg, and f/g from Exercise 4.

6. Let f be a real function. The **reciprocal** function $1/f$ is given by

$$\left(\frac{1}{f}\right)(x) = \frac{1}{f(x)}$$

The domain of $1/f$ is the domain of f, except for the numbers x where $f(x) = 0$. Compute the following where f and g are the functions of Exercise 4.

a. $(1/f)(x)$.

b. $(1/g)(x)$.

c. $(1/g)(0)$.

d. $(1/g)(0)$.

e. $(1/f)(1)$.

f. $(1/g)(1)$.

g. $(1/f)(-1)$.

h. $(1/g)(-1)$.

i. Find the domains of $1/f$ and $1/g$.

j. What are the ranges of $1/f$ and $1/g$?

7. Consider the following 16 points on the unit circle.

a. $(1, 0)$.

b. $(\sqrt{3}/2, \frac{1}{2})$.

c. $(1/\sqrt{2}, 1/\sqrt{2})$.

d. $(\frac{1}{2}, \sqrt{3}/2)$.

e. $(0, 1)$.

f. $(-\frac{1}{2}, \sqrt{3}/2)$.

g. $(-1/\sqrt{2}, 1/\sqrt{2})$.

h. $(-\sqrt{3}/2, \frac{1}{2})$.

i. $(-1, 0)$.

j. $(-\sqrt{3}/2, -\frac{1}{2})$.

k. $(-1/\sqrt{2}, -1/\sqrt{2})$.

l. $(-\frac{1}{2}, \sqrt{3}/2)$.

m. $(0, -1)$.

n. $(\frac{1}{2}, -\sqrt{3}/2)$.

o. $(1/\sqrt{2}, -1/\sqrt{2})$.

p. $(\sqrt{3}/2, -\frac{1}{2})$.

Notice that in some sense these points traverse the unit circle in a counterclockwise direction in a "uniform way."

a. Find the value of p_1 at each of the 16 points.

b. Find the value of p_2 at each of the 16 points.

c. Find the value of $1/p_1$ at each of the 16 points. (See Exercise 6.)

d. Find the value of $1/p_2$ at each of the 16 points. (See Exercise 6.)

e. Find the value of p_2/p_1 at each of the 16 points. (See Exercise 6.)

f. Find the value of p_1/p_2 at each of the 16 points. (See Exercise 6.)

Let (x, y) be any point on the unit circle.

g. What is $(p_1((x, y)))^2 + (p_2((x, y)))^2$?

h. Show that $-1 \leqslant p_1((x, y)) \leqslant 1$.

i. Show that $-1 \leqslant p_2((x, y)) \leqslant 1$.

j. From (h) show that $|(1/p_1)((x, y))| \geqslant 1$.

k. From (i) show that $|(1/p_2)((x, y))| \geqslant 1$.

l. What is $p_1[C]$ where C is the unit circle? (Hint: Graph the unit circle and mentally project each point of C onto the x axis.)

m. What is $p_2[C]$ for C the unit circle?

8. If f is a real function (thinking of f as a set of ordered pairs), what is:

a. $p_1[f]$?

b. $p_2[f]$?

9. Let A be the relation on $R^2 \times R^2$ given by:

$$A = \{(P_1, P_2) \neq (0,0) : P_1, P_2 \in R^2 \text{ and } P_1 \text{ and } P_2 \text{ are on the same ray emanating from the origin}\}$$

Show that $A \cup \{0, 0\}$ is an equivalence relation. Thus, the plane with the origin removed is broken into disjoint sets of the form $\rho - \{(0,0)\}$ where ρ is a ray emanating from the origin. Thus,

$$R^2 = \{(0,0)\} \cup \{\rho - \{0,0\} : \rho \text{ is a ray emanating from } (0,0)\}$$

The sets $\rho - \{0, 0\}$ are called **equivalence classes.**
10. Which of the following denote polynomial functions? Rational functions? Find the zeros of all the functions and the poles of the rational functions.

a. $f(x) = 3x^5 + 3x^2$.
b. $g(x) = (x^2 + 1)/(x - 1)$.
c. $h(x) = 4x^2 + x - 1$.
d. $k(x) = (x^2 + 5x + 6)/[(x + 2)(x - 1)]$.
e. $\eta(x) = (x^2 + x + 1)/(x^3 - 1)$.
f. $\mu(x) = x^2$.
g. $\lambda(x) = 5^x$.
h. $v(x) = 3x$.

Compute the degree of each polynomial above.
11. What type of function is λ in Exercise 10? v in Exercise 10? λ^{-1} for λ in Exercise 10? v^{-1} in Exercise 10?
12. Compute the functional values for each function in Exercise 10 at the following points.

a. 0.   **c.** 2.   **e.** 4.
b. 1.   **d.** 3.   **f.** 5.

3.8 ADVANCED EXERCISES

1. Show that the linear function given by $f(x) = mx + b$, $m \neq 0$ is a 1:1 and an onto real function with both range and domain R.
2. a. Verify that the following indicates a functional relationship.

$$f(x) = \begin{cases} 2x + 2 \text{ for } x < -1 \\ (1 - x^2)^{1/2} \text{ for } -1 \leqslant x \leqslant -1/\sqrt{2} \\ x + \sqrt{2} \text{ for } -1/\sqrt{2} \leqslant x \leqslant 0 \\ 2x + \sqrt{2} \text{ for } 0 \leqslant x \leqslant 10 \\ -4x + 60 + \sqrt{2} \text{ for } x \geqslant 10 \end{cases}$$

b. Graph the function f above.
c. What is the domain of f? Range of f?

3. Let $i:X \longrightarrow X$ be the identity function on X. Show that i is **bijective** (1:1 and an onto function) and that $i^{-1} = i$.

4. Given $f:X \longrightarrow Y$ and $g:Y \longrightarrow Z$, show that:

    a. $g \circ f$ is 1:1 implies f is 1:1.

    b. $g \circ f$ is an onto function shows g is an onto function.

5. $f:X \longrightarrow Y$ has an inverse implies that f is 1:1 and an onto function. Moreover, f^{-1} is 1:1 and an onto function when it exists.   Prove these statements.

6. If $f:X \longrightarrow Y$ and $g:Y \longrightarrow Z$, show each of the following.

    a. $A \subset B \subset X$ implies $f[A] \subset f[B]$.

    b. A and $B \subset X$ show $f[A \cup B] = f[A] \cup f[B]$.

    c. $A \subset B \subset Y$ implies $f^{-1}[A] \subset f^{-1}[B]$.

    d. A and $B \subset Y$ show $f^{-1}[A \cup B] = f^{-1}[A] \cup f^{-1}[B]$.

    e. With the hypothesis of (b), $f[A \cap B] \subset f[A] \cap f[B]$.

    f. With the hypothesis of (d), $f^{-1}[A \cap B] = f^{-1}[A] \cap f^{-1}[B]$.

    g. $A \subset Y$ implies $f^{-1}[Y - A] = f^{-1}[Y] - f^{-1}[A] = X - f^{-1}[A]$.

    h. $A \subset Z$ shows $(g \circ f)^{-1}[A] = f^{-1}[g^{-1}[A]]$.

7. Sets X and Y are said to be **equivalent,** written $X \sim Y$, if and only if there is a 1:1 and onto function $f:X \longrightarrow Y$. Show that:

    a. $X \sim X$. (There is an obvious 1:1 and onto function.)

    b. $X \sim Y$ implies $Y \sim X$. (Use an inverse function.)

    c. $X \sim Y$ and $Y \sim Z$ show $X \sim Z$. (Composition of functions.)

We realize that "$\sim$" is an equivalence relation on sets. Let A be the relation given by $A = \{(X, Y):X, Y \text{ are sets and } X \sim Y\}$. A is the relation "induced" by "$\sim$."

8. Let F be the relation defined on $N \times N$ by: $(a, b) \in F$ if and only if each of a and b have the same remainder upon division by 7. An example of an element if F is (7, 14), since both have a remainder of 0 upon division by 7. The element (1, 9) of $N \times N$ is not in F, since 1 has a remainder of 1 upon division by 7 and 9 has a remainder of 2.

    a. Is F an equivalence relation on N? Verify.

F can be divided into **equivalence classes** (seven such) by grouping together all equivalent pairs (that is, pairs having a common remainder for their elements). For instance,

    let $\boxed{0} = \{(x, y) \in F: \text{the remainder upon division of } x \text{ by 7 is 0}\}$
    let $\boxed{1} = \{(x, y) \in F: \text{the remainder upon division of } x \text{ by 7 is 1}\}$

Continue this process for $\boxed{2}$, $\boxed{3}$, $\boxed{4}$, $\boxed{5}$, and $\boxed{6}$.

    b. Show that $F = \boxed{0} \cup \boxed{1} \cup \boxed{2} \cup \boxed{3} \cup \boxed{4} \cup \boxed{5} \cup \boxed{6}$.

Let $\boxed{a}$ be the equivalence class containing (a, a).

c. Show that $\boxed{a - b} = \boxed{0}$, where a and b are chosen from

$$\boxed{0}, \boxed{1}, \boxed{2}, \boxed{3}, \boxed{4}, \boxed{5}, \boxed{6}, \text{ with } \boxed{a} = \boxed{b}$$

Define: $\boxed{a + b} = \boxed{a} + \boxed{b}$, where $\boxed{a + b}$ is the equivalence class to which $(a + b, a + b)$ belongs. For example: $\boxed{3} + \boxed{6} = \boxed{9} = \boxed{2}$, since $(9, 9) \in \boxed{2}$, 9 having a remainder of 2 upon division by 7. Furthermore, we can define $\boxed{a}\,\boxed{b} = \boxed{ab}$ as the equivalence class to which (ab, ab) belongs.

$$\boxed{3}\,\boxed{3} = \boxed{9} = \boxed{2}$$

d. Make complete addition and multiplication tables for this Modulo 7 system. Define:

so that
$$\boxed{a} - \boxed{b} = \boxed{c}$$
$$\boxed{b} + \boxed{c} = \boxed{a}$$

(if such a c exists).
Next define $\boxed{a} \div \boxed{b} = \boxed{d}$, where $\boxed{b}\,\boxed{d} = \boxed{a}$, provided a **unique** such $\boxed{d}$ exists:

since
$$\boxed{3} - \boxed{6} = \boxed{4}$$
$$\boxed{6} + \boxed{4} = \boxed{10} = \boxed{3}$$
Furthermore,
$$\boxed{3} \div \boxed{6} = \boxed{4}$$
since
$$\boxed{6}\,\boxed{4} = \boxed{24} = \boxed{3}$$

e. Make subtraction tables to show that subtraction is closed.

f. Division is well defined, except that $\boxed{a} \div \boxed{0}$ is not valid for precisely the same reasons. We cannot divide by zero in the set of real numbers. Make a table for division to show that division exists, except for this special case. Perform the following operations.

g. $(\boxed{3} + \boxed{5})(\boxed{4} + \boxed{5}) - \boxed{3}$.

h. $(\boxed{3} - \boxed{5})\boxed{4} + \boxed{2}$.

i. $(\boxed{3}\,\boxed{5} - \boxed{2})\boxed{6} - \boxed{4}$.

9. Go through the entire development of a system like that derived in Exercise 8 except that the remainder be that remainder upon division by 5 instead of 7.

Chapter 4
Trigonometric Functions

4.1 ARC LENGTH

Attention will now be turned to the unit circle. (Its equation has been previously given as $x^2 + y^2 = 1$.) Given a circle we (in geometry) attach a positive real number as the "distance around the circle." This number is called the **circumference** and is $2\pi r$ where r is the radius of the circle and π is an irrational number slightly larger than 3. (See Exercise 4.9.2 in the advanced exercises at the end of this chapter.)

Intuitively, if we place one end of a string on any point of the circle and proceed to wrap the string around the circle, the shortest length of string needed to return to the original point is $2\pi r$.

What is the circumference of the unit circle? It must be 2π, since $r = 1$ in this instance.

An **arc** of a circle is considered to be all of the circle contained "between" two of its points. Actually, two arcs are determined by any pair of points on the circle. Since an arc can be thought of as being a portion of the circle, we intuitively expect to determine a positive number for the length of the arc. This number is fixed by that portion of the circumference described by the arc; that is, if the arc is one fourth of a circle of radius r, the arc's length is $\frac{1}{4}(2\pi r) = \frac{1}{2}(\pi r)$.

The reader should remember that all the discussion concerning arc length is on an intuitive basis. There is no elementary method of measuring arc length. There is even no elementary method of determining that such a number should exist. (See Exercise 4.9.2 of the advanced exercises for this material.) For the purposes of this material, intuitive reckoning will be relied upon for the acquisition and existence of numbers given as lengths of particular arcs.

Example 4.1.1   What is the length of the arc (see Fig. 4.1.1) between $(1, 0)$ and $(0, 1)$, where the arc is described by traversing the circle in a counter-clockwise direction from $(1, 0)$ to $(0, 1)$?

Several observations lead to an answer. First, the circle is the unit circle so that the circumference is 2π. Secondly, $d((1, 0), (0, 1)) = d((0, 1), (-1, 0)) = d((-1, 0), (0, -1)) = d((0, -1), (1, 0))$. We recall from geometry that **chords** (line segments connecting two points on a circle) of equal length determine equal arcs. The chords in this case are the line segments drawn in the figure. Apparently the axes partition the circle into four arcs of equal length. Thus, the length of the arc in question ought to be $\frac{1}{4}(2\pi) = \pi/2$.

Example 4.1.2   Consider the problem of finding the length of the shorter arc from $(1, 0)$ to $(1/\sqrt{2}, 1/\sqrt{2})$. Recalling a previous exercise, we know that $d((1, 0), (1/\sqrt{2}, 1/\sqrt{2})) = d((1/\sqrt{2}, 1/\sqrt{2}), (0, 1))$, whereby the arc under question has a length one half that of the arc discussed in Example 4.1.1. Thus, the length of our arc under examination is $\frac{1}{2}(\pi/2) = \pi/4$. (See Fig. 4.1.2.)

Example 4.1.3   What is the length of the arc traversed counterclockwise in going from $(1, 0)$ to $(-1/\sqrt{2}, 1/\sqrt{2})$? Figure 4.1.3 depicts this arc.

Figure 4.1.1
Special arcs

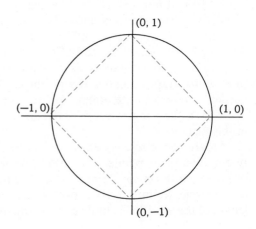

Figure 4.1.2
Determining a special arc length

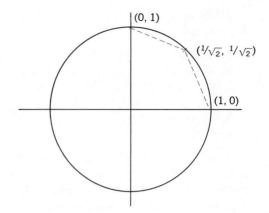

Now, the length of the shorter arc from $(0, 1)$ to $(-1, \sqrt{2}, 1/\sqrt{2})$ is equal to the length of the arc previously mentioned [the shorter arc from $(1, 0)$ to $(1/\sqrt{2}, 1/\sqrt{2})$]. This arc then must have length $\pi/4$. However, that part of the arc from $(1, 0)$ to $(0, 1)$ has length $\pi/2$, giving a sum total length of $3\pi/4$ for the desired arc.

■ EXERCISES 4.1

1. Using your intuition and experiences in geometry, decide what distance along the unit circle a point travels in moving from the first point to the second in each case (movement from the first point to the second to be taken in a counterclockwise direction).

   a. $(1, 0)$ to $(0, 1)$.   **d.** $(1, 0)$ to $(\sqrt{3}/2, \frac{1}{2})$.
   b. $(1, 0)$ to $(-1, 0)$.   **e.** $(1, 0)$ to $(1/\sqrt{2}, 1/\sqrt{2})$.
   c. $(1, 0)$ to $(0, -1)$.   **f.** $(1, 0)$ to $(\frac{1}{2}, \sqrt{3}/2)$.

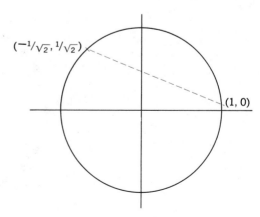

Figure 4.1.3
Determining another arc length

[Hint for (d): $d((1, 0), (\sqrt{3}/2, \frac{1}{2})) = d((\sqrt{3}/2, \frac{1}{2}), (\frac{1}{2}, \sqrt{3}/2)) = d((\frac{1}{2}, \sqrt{3}/2), (0, 1))$.

g. $(1, 0)$ to $(-\frac{1}{2}, \sqrt{3}/2)$.   **l.** $(1, 0)$ to $(-\frac{1}{2}, -\sqrt{3}/2)$.
h. $(1, 0)$ to $(-1/\sqrt{2}, 1/\sqrt{2})$.   **m.** $(1, 0)$ to $(\frac{1}{2}, -\sqrt{3}/2)$.
i. $(1, 0)$ to $(-\sqrt{3}/2, \frac{1}{2})$.   **n.** $(1, 0)$ to $(1/\sqrt{2}, -1/\sqrt{2})$.
j. $(1, 0)$ to $(-\sqrt{3}/2, -\frac{1}{2})$.   **o.** $(1, 0)$ to $(\sqrt{3}/2, -\frac{1}{2})$.
k. $(1, 0)$ to $(-1/\sqrt{2}, -1/\sqrt{2})$.
(See Fig. 4.1.4.)

2. Redo Exercise 1 by taking a clockwise direction of traversal.
3. Consider the circle $x^2 + y^2 = 25$. This circle is centered at the origin and has radius 5. Find the lengths of the arcs between the following pairs of points on the circle (counterclockwise traversal). See Fig. 4.1.4 for an analogous situation.

a. $(5, 0)$ to $(0, 5)$.   **i.** $(5, 0)$ to $(-5\sqrt{3}/2, \frac{5}{2})$.
b. $(5, 0)$ to $(-5, 0)$.   **j.** $(5, 0)$ to $(-5\sqrt{3}/2, -\frac{5}{2})$.
c. $(5, 0)$ to $(0, -5)$.   **k.** $(5, 0)$ to $(-5/\sqrt{2}, -5/\sqrt{2})$.
d. $(5, 0)$ to $(5\sqrt{3}/2, \frac{5}{2})$.   **l.** $(5, 0)$ to $(-\frac{5}{2}, -5\sqrt{3}/2)$.
e. $(5, 0)$ to $(5/\sqrt{2}, 5/\sqrt{2})$.   **m.** $(5, 0)$ to $(\frac{5}{2}, -5\sqrt{3}/2)$.
f. $(5, 0)$ to $(\frac{5}{2}, 5\sqrt{3}/2)$.   **n.** $(5, 0)$ to $(5/\sqrt{2}, -5/\sqrt{2})$.
g. $(5, 0)$ to $(-\frac{5}{2}, 5\sqrt{3}/2)$.   **o.** $(5, 0)$ to $(5\sqrt{3}/2, -\frac{5}{2})$.
h. $(5, 0)$ to $(-5/\sqrt{2}, 5/\sqrt{2})$.

How do these arc lengths compare with those in Exercise 1?
4. Redo Exercise 3 by taking clockwise traversals.
5. Do Exercises 3 and 4 have any relation to Exercises 1 and 2? If so, what relationship is noted?

Figure 4.1.4
Special points

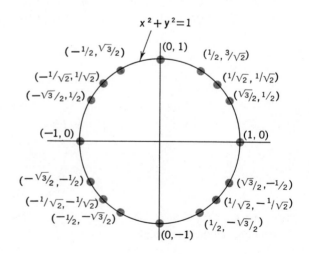

Figure 4.1.5
The four quadrants

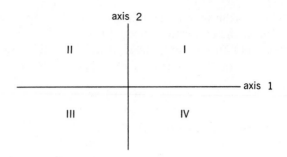

6. The axes split the plane into 4 natural quadrants as listed and numbered in Fig. 4.1.5. Determine the conditions (on the values of the first and second coordinate) that place a point in a particular quadrant (do this for each quadrant).

7. Suppose (a, b) and (c, d) lie on the same ray (Exercise 2.4.12). Then (a, b) belongs to the circle given by $x^2 + y^2 = a^2 + b^2$ and (c, d) lies on the circle $x^2 + y^2 = c^2 + d^2$. How does the arc length from $((a^2 + b^2)^{1/2}, 0)$ to (a, b) compare with the arc length from $((c^2 + d^2), 0)$ to (c, d)? Your answer may be formed by your answer to the last question of Exercise 3. Observe that each of the arc lengths above can be compared to that from $(1, 0)$ to $(a/(a^2 + b^2)^{1/2}, b/(a^2 + b^2)^{1/2})$ along the unit circle. Show that $(a/(a^2 + b^2)^{1/2}, b/(a^2 + b^2)^{1/2})$ is on the unit circle and on the ray through (a, b) and (c, d).

8. In Exercise 7, show that, if $a^2 + b^2 < 1$ and $c^2 + d^2 > 1$, then $(a/(a^2 + b^2)^{1/2}, b/(a^2 + b^2)^{1/2})$ lies between (a, b) and (c, d).

4.2 THE FUNCTION α

The following statement will be made without proof, the justification being beyond the level of this course. (See Exercise 4.9.2 at the end of this chapter.)

Let C be the unit circle. There is an onto function $\alpha: R \longrightarrow C$ with the following properties.

1. $\alpha(0) = (1, 0)$.
2. α restricted to the interval $[0, 2\pi)$ is $1:1$ and an onto function ($\alpha | [0, 2\pi)$ is $1:1$ and $\alpha[[0, 2\pi)] = C$).
3. $0 < a < e < 2\pi$ implies that $\alpha(a)$ is a point on the arc from $(1, 0)$ to $\alpha(e)$ where the arc is taken to be that one traversed by counterclockwise movement from $(1, 0)$ to $\alpha(e)$.
4. $\alpha(a + 2\pi) = \alpha(a)$.
5. $\alpha(-a) = (c, -d)$ where $\alpha(a) = (c, d)$.
6. $d(\alpha(a), \alpha(b)) = d(\alpha(a - b), \alpha(0))$.

Furthermore, the following values are some "special values" found under the correspondence. [The term "special value" will be used to refer to: (1) the matings listed below, (2) the special domain elements used below, or (3) the special image points below.]

$$\alpha(\pi/6) = (\sqrt{3}/2,\ \tfrac{1}{2})$$
$$\alpha(\pi/4) = (1/\sqrt{2},\ 1/\sqrt{2})$$
$$\alpha(\pi/3) = (\tfrac{1}{2},\ \sqrt{3}/2)$$
$$\alpha(\pi/2) = (0, 1)$$
$$\alpha(2\pi/3) = (-\tfrac{1}{2},\ \sqrt{3}/2)$$
$$\alpha(3\pi/4) = (-1/\sqrt{2},\ 1/\sqrt{2})$$
$$\alpha(5\pi/6) = (-\sqrt{3}/2,\ \tfrac{1}{2})$$
$$\alpha(\pi) = (-1, 0)$$
$$\alpha(7\pi/6) = (-\sqrt{3}/2,\ -\tfrac{1}{2})$$
$$\alpha(5\pi/4) = (-1/\sqrt{2},\ -1/\sqrt{2})$$
$$\alpha(4\pi/3) = (-\tfrac{1}{2},\ -\sqrt{3}/2)$$
$$\alpha(3\pi/2) = (0, -1)$$
$$\alpha(5\pi/3) = (\tfrac{1}{2},\ -\sqrt{3}/2)$$
$$\alpha(7\pi/4) = (1/\sqrt{2},\ -1/\sqrt{2})$$
$$\alpha(11\pi/6) = (\sqrt{3}/2,\ -\tfrac{1}{2})$$

Compare these with Exercise 4.1.1.

The function α is called the **arc length function:** intuitively, $\alpha(a)$ is the point on the unit circle with the property that the arc from $(1, 0)$ to $\alpha(a)$ has length $|a|$. (If a is positive, the arc is that traversed in a counterclockwise direction while, for negative a, the arc is described by a clockwise traversal.) Necessarily, our concept of an arc must be generalized to include those generated by moving in such a way as to traverse the circle more than once.

In some sense, property 1 shows that $(1, 0)$ is a "starting" point on the unit circle.

Property 2 declares that α is an onto function, that is, each point (x, y) on the unit circle is $\alpha(a)$ for some real number a. Moreover, we can find a real number $a \in [0, 2\pi)$ that has this property. That this is true is of significance. (Recall that $a \in [0, 2\pi)$ implies the inequality $0 \leqslant a < 2\pi$.) Furthermore, if $\alpha(a) = (x, y)$ and $a \in [0, 2\pi)$, a is the **only such** element from $[0, 2\pi)$ having this particular property. In other words, α is 1:1 on the given restricted domain.

Does α order the points on the unit circle in any way? Property 3 shows that the ordering of the real numbers by "$\leqslant$" orders these points in some manner. We could say that if $0 < a < e < 2\pi$, $\alpha(a)$ "precedes" $\alpha(e)$ in the sense that, upon traversing the unit circle in a counterclockwise direction from $(1, 0)$, we reach $\alpha(a)$ before we reach $\alpha(e)$. This is not necessarily true unless the condition $0 < a < e < 2\pi$ is fulfilled.

Figure 4.2.1

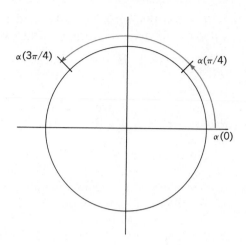

Example 4.2.1   As an example, $0 < \pi/4 < 3\pi/4 < 2\pi$.   Figure 4.2.1 shows that, upon traversing the unit circle C in a counterclockwise direction, $\alpha(\pi/4) = (1/\sqrt{2}, 1/\sqrt{2})$ is reached prior to $\alpha(3\pi/4) = (-1/\sqrt{2}, 1/\sqrt{2})$.

Property 4 says several things.   First, α is not a $1:1$ function on its entire domain.   Adding 2π to the argument of the function results in the same image point; that is, given the real number x, we can find $\alpha(x)$ (theoretically).   However, $\alpha(x) = \alpha(x + 2\pi)$, according to property 4.

Example 4.2.4   Since $\alpha(0) = (1, 0)$, we also know that $\alpha(2\pi) = (1, 0)$ $(2\pi = 0 + 2\pi)$.   Then we also know $(1, 0) = \alpha(6\pi) = \alpha(8\pi)$, and so forth.

Figure 4.2.2
An intuitive concept of $\alpha(5\pi)$

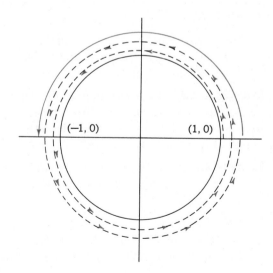

The fact that $\alpha|[0, 2\pi)$ is $1:1$ and an onto function is important (α itself does not share the former of the two properties).

Theorem 4.2.1 If n is any integer, $\alpha(a + 2n\pi) = \alpha(a)$.

PROOF. If $n = 0$, $\alpha(a + 2n\pi) = \alpha(a + 0) = \alpha(a)$ and the theorem is trivially satisfied.

For positive n, we need to use mathematical induction. If $n = 1$, $\alpha(a + 2n\pi) = \alpha(a + 2\pi) = \alpha(a)$ by property 4. Thus, the theorem holds for $n = 1$.

If the theorem holds for $n = k$ (that is, $\alpha(a + 2k\pi) = \alpha(a)$), then $\alpha(a + 2(n + 1)\pi) = \alpha(a + 2(k + 1)\pi) = \alpha((a + 2\pi) + 2k\pi) = \alpha(a + 2\pi) = \alpha(a)$. Consequently, the theorem holds for $n = k + 1$ and necessarily for all positive integers n.

If n is a negative integer, $-n$ is a positive integer and we have $\alpha(a + 2n\pi) = \alpha((a + 2n\pi) + 2(-n)\pi) = \alpha(a + 2n\pi - 2n\pi) = \alpha(a)$. The proof is complete.

The theorem states that adding an even multiple of π to the argument of α does not change the image point.

More than the simple examples above, property 4 discloses that to know $\alpha(a)$ for each $a \in [0, 2\pi)$ is sufficient to know the correspondence for the entire domain; that is, if $b \in R$, we can find $\alpha(b)$ knowing only the correspondence α on $[0, 2\pi)$.

PROOF. There is an integer n so that:

$$2n\pi \leqslant b < 2(n + 1)\pi$$

(b lies between two successive multiples of 2π).

Then $\qquad\qquad\qquad 0 \leqslant b - 2n\pi < 2\pi$

Alternatively, $\qquad\qquad (b - 2n\pi) \in [0, 2\pi)$

Since $\alpha(b) = \alpha((b - 2n\pi) + 2n\pi) = \alpha(b - 2n\pi)$, it is concluded that $\alpha(b)$ is known. (It was given that we knew the correspondence for the restricted domain $[0, 2\pi)$.)

Example 4.2.3 As an example of the calculation of an image point of α (using 4), find $\alpha(5\pi)$. We write 5π as $\pi + 4\pi$. Necessarily, $\alpha(5\pi) = \alpha(\pi + 4\pi) = \alpha(\pi) = (-1, 0)$. The intuitive concept of traversing an arc length of 5π is seen in Fig. 4.2.2.

Example 4.2.4 Look also at the example $\alpha(-11\pi/6)$. Now, $-11\pi/6 = \pi/6 - 2\pi$. Using property 4 we see that $\alpha(-11\pi/6) = \alpha(\pi/6 - 2\pi) = \alpha(\pi/6) = (\sqrt{3}/2, \frac{1}{2})$.

Example 4.2.5 As a third example, find $\alpha(13\pi/6)$. Since $13\pi/6 = \pi/6 + 2\pi$, $\alpha(13\pi/6) = \alpha(\pi/6) = (\sqrt{3}/2, \frac{1}{2})$.

In conjunction with property 4, a further note about functions is in order. Let $X \subset R$. If $f:X \longrightarrow Y$ is a function and there is a nonzero real number p so that $f(x + p) = f(x)$ for all x in the domain, we say that f is **periodic** and that p is a **period** of f. If a smallest positive period exists, it will be called the **primitive period.**

According to property 4 and what was discussed above, the arc length function α is of period (has a period) 2π. It is also true that this is the primitive period of α.

Theorem 4.1.2 Suppose f is a function with periods p and q. Show that $p + q$ is also a period of f.
PROOF. Let x be in the domain of f. Then $f(x + p) = f(x) = f(x + q)$, since each of p and q is a period. Now, $f(x + (p + q)) = f((x + p) + q) = f(x + p) = f(x)$ using first the fact that q is a period and then the fact that p is a period. The technique used in this proof is helpful in the proof of Exercise 4.9.4 at the end of this chapter.

Let $a \in R$. For purposes of simplicity, suppose $0 \leqslant a \leqslant \pi/2$. (See Fig. 4.2.3.) Now, $\alpha(a)$ is some point on the unit circle, say (c, d). By property 5, $\alpha(-a)$ is the point $(c, -d)$. It is intuitively clear that the length of the arc from $(1, 0)$ to $\alpha(a)$ taken in a counterclockwise direction is the same as the length of the arc from $(1, 0)$ to $\alpha(-a)$ taken in a clockwise direction $d(\alpha(a), \alpha(0)) = d(\alpha(-a), \alpha(0))$. The length of each arc is a, $a \geqslant 0$.

Example 4.2.6 As a concrete example of the message of 5, find $\alpha(-\pi/2)$. The special cases under α include $\alpha(\pi/2)$. Since $\alpha(\pi/2) = (0, 1)$, it follows that $\alpha(-\pi/2) = (0, -1)$. Intuitively, we should feel that the negative sign in $-\pi/2$ indicates a clockwise traversal (from $(1, 0)$) of arc length $\pi/2$. (See Fig. 4.2.4.) The "$-$" in $\alpha(-a)$ indicates traversal opposite to that taken in the case $\alpha(a)$.

Figure 4.2.3
$\alpha(a)$ versus $\alpha(-a)$

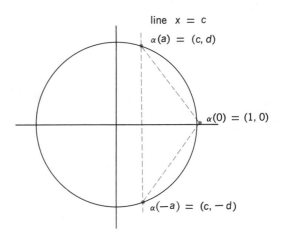

line x = c
$\alpha(a) = (c, d)$
$\alpha(0) = (1, 0)$
$\alpha(-a) = (c, -d)$

Figure 4.2.4
Arc of negative length

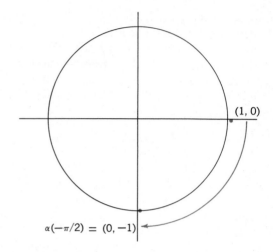

$$\alpha(-\pi/2) = (0, -1)$$

Property 6 may seem more complicated than the others. $\alpha(a)$ and $\alpha(b)$ are points on the unit circle, and, as such, we can calculate the distance between them. This distance is equal to the distance from $\alpha(0) = (1, 0)$ to $\alpha(a - b)$. An example will illustrate.

Example 4.2.7 Consider the real numbers π and $\pi/2$. $\alpha(\pi) = (-1, 0)$ and $\alpha(\pi/2) = (0, 1)$ so that $d(\alpha(\pi), \alpha(\pi/2)) = \sqrt{2}$. On the other hand, $\pi - \pi/2 = \pi/2$ giving $\alpha(\pi - \pi/2) = \alpha(\pi/2) = (0, 1)$. But, $d(\alpha(\pi - \pi/2), \alpha(0)) = d((0, 1), (1, 0)) = \sqrt{2}$ as anticipated from property 6.

Figure 4.2.5 illustrates property 6. The chords are of the same length. Geometrically, we feel that the "spanned" arcs are of equal length also.

Figure 4.2.5
A property of α

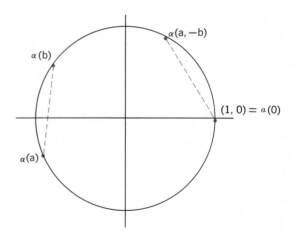

Example 4.2.8 Compute $\alpha(\pi/12)$. We know that since $\pi/12 = \pi/4 - \pi/6$,

$$
\begin{aligned}
d(\alpha(\pi/12), (1,0)) &= d(\alpha(\pi/4), \alpha(\pi/6)) \\
&= d((1/\sqrt{2}, 1/\sqrt{2}), (\sqrt{3}/2, \tfrac{1}{2})) \\
&= [(1/\sqrt{2} - \sqrt{3}/2)^2 + (1/\sqrt{2} - \tfrac{1}{2})^2]^{1/2} \\
&= [2 - (\sqrt{3} + 1)/\sqrt{2}]^{1/2}
\end{aligned}
$$

or $d^2(\alpha(\pi/12), (1,0)) = 2 - (\sqrt{3} + 1)/\sqrt{2}$

However, letting $\alpha(\pi/12) = (x, y)$, we see that $(x^2 + y^2 = 1)$

$$
\begin{aligned}
d^2(\alpha(\pi/12), (1,0)) &= (x - 1)^2 + (y - 0)^2 \\
&= x^2 - 2x + 1 + y^2 \\
&= x^2 - 2x + 1 + 1 - x^2 \\
&= 2 - 2x
\end{aligned}
$$

Consequently, $2 - (\sqrt{3} + 1)/\sqrt{2} = 2 - 2x$

or
$$
2x = \frac{\sqrt{3} + 1}{\sqrt{2}}
$$

$$
x = \frac{\sqrt{3} + 1}{2\sqrt{2}}
$$

Since $\alpha(\pi/12)$ is in the first quadrant,

$$
y = (1 - x^2)^{1/2} = \frac{\sqrt{3} - 1}{2\sqrt{2}}
$$

Whence
$$
\alpha(\pi/12) = \left(\frac{\sqrt{3} + 1}{2\sqrt{2}}, \frac{\sqrt{3} - 1}{2\sqrt{2}} \right)
$$

The student should familiarize himself with all the special values listed for the correspondence α. As we have mentioned, techniques more advanced than those appearing in this text are needed to compute all points $\alpha(a)$. Because of this, the special values listed will be used for the most part.

■ EXERCISES 4.2

1. Find each of the following. (Hint: Use properties 1 and 4 of α.)

a. $\alpha(0)$.
b. $\alpha(2\pi)$.
c. $\alpha(4\pi)$.
d. $\alpha(-2\pi)$.
e. $\alpha(-4\pi)$.

2. Suppose $a \in [0, 2\pi)$ and suppose $\alpha(a) = \alpha(b)$. Complete the following to make true statements (use property 2 of α).

a. If $b \in [0, 2\pi)$, then $\cdots$.
b. If $(x, y) \in C$, there is an $a \in [0, 2\pi)$ such that $\cdots$.

3. Use property 3 of α to determine the quadrant in which each of the following is located.

a. $\alpha(5\pi/24)$.　　　　**e.** $\alpha(23\pi/12)$.
b. $\alpha(18\pi/11)$.　　　　**f.** $\alpha(25\pi/51)$.
c. $\alpha(17\pi/16)$.　　　　**g.** $\alpha(24\pi/53)$.
d. $\alpha(18\pi/23)$.　　　　**h.** $\alpha(101\pi/100)$.

4. Find each point using property 5 of α.

a. $\alpha(-\pi/6)$.　　　　**e.** $\alpha(-3\pi/2)$.
b. $\alpha(-\pi/2)$.　　　　**f.** $\alpha(-4\pi/3)$.
c. $\alpha(-\pi)$.　　　　　**g.** $\alpha(-11\pi/6)$.
d. $\alpha(-3\pi/4)$.

5. Locate the quadrant of $\alpha(-a)$ if $\alpha(a)$ is in

a. quadrant I.
b. quadrant II.
c. quadrant III.
d. quadrant IV.

6. Give the point designated in each of the following cases.

a. $\alpha(6\pi)$.　　　　　**i.** $\alpha(37\pi/4)$.
b. $\alpha(-7\pi/4)$.　　　**j.** $\alpha(-\pi/3)$.
c. $\alpha(31\pi/3)$.　　　**k.** $\alpha(203\pi/2)$.
d. $\alpha(-3\pi/2)$.　　　**l.** $\alpha(721\pi/6)$.
e. $\alpha(32\pi/3)$.　　　**m.** $\alpha(15\pi/4)$.
f. $\alpha(-17\pi)$.　　　**n.** $\alpha(-3\pi/4)$.
g. $\alpha(29\pi/6)$.　　　**o.** $\alpha(35\pi/6)$.
h. $\alpha(-19\pi/6)$.　　**p.** $\alpha(26\pi/3)$.

7. Describe the following sets in terms of the points of the unit circle.

a. $\alpha[[0, \pi/2]]$.　　　　**f.** $\alpha[[\pi/2, 2\pi]]$.
b. $\alpha[[0, \pi]]$.　　　　　**g.** $\alpha[[\pi, 3\pi/2]]$.
c. $\alpha[[0, 3\pi/2]]$.　　　**h.** $\alpha[[\pi, 2\pi]]$.
d. $\alpha[[\pi/2, \pi]]$.　　　**i.** $\alpha[[3\pi/2, 2\pi]]$.
e. $\alpha[[\pi/2, 3\pi/2]]$.

8. Describe an arc from the first point to the second point in each case as the image of some interval under α. Find four such intervals. Do this for both counterclockwise and clockwise traversals.

a. $(1/\sqrt{2}, 1/\sqrt{2})$ to $(-1/\sqrt{2}, 1/\sqrt{2})$. [Hint: This arc can be given as $\alpha[[\pi/4, 3\pi/4]]$.] Find some other acceptable intervals.
b. $(\frac{1}{2}, \sqrt{3}/2)$ to $(-1/\sqrt{2}, 1/\sqrt{2})$.
c. $(-\frac{1}{2}, -\sqrt{3}/2)$ to $(\frac{1}{2}, \sqrt{3}/2)$.
d. $(-\sqrt{3}/2, \frac{1}{2})$ to $(\frac{1}{2}, -\sqrt{3}/2)$.

9.a. Show $\alpha(-a) = \alpha(2\pi - a)$.   (Hint: Use property 4.)
  b. Show $\alpha(a + \pi) = \alpha(a - \pi)$.
  c. $\alpha(a + (2n - 1)\pi) = \alpha(a + \pi)$, where $n \in N$.

10. For each domain element used in Exercise 1, find the value of the image under each function below.

a. $p_1 \circ \alpha$.

e. $\dfrac{p_1 \circ \alpha}{p_2 \circ \alpha}$.

b. $p_2 \circ \alpha$.

f. $\dfrac{p_2 \circ \alpha}{p_1 \circ \alpha}$.

c. $\dfrac{1}{p_1 \circ \alpha}$.

g. $(p_1 \circ \alpha)^2 + (p_2 \circ \alpha)^2$.

d. $\dfrac{1}{p_2 \circ \alpha}$.

11. Repeat Exercise 10 where the domain elements are those used in the following exercises: a. 4.   b. 6.
12. Repeat Exercise 10 where:

a. $\alpha(a) = (\frac{3}{5}, \frac{4}{5})$.

f. $\alpha(a) = (-\frac{8}{17}, \frac{15}{17})$.

b. $\alpha(a) = (-\frac{3}{5}, \frac{4}{5})$.

g. $\alpha(a) = (-\frac{7}{25}, \frac{24}{25})$.

c. $\alpha(a) = (-\frac{5}{13}, -\frac{12}{13})$.

h. $\alpha(a) = (\frac{24}{25}, -\frac{7}{25})$.

d. $\alpha(a) = (\frac{12}{13}, -\frac{5}{13})$.

i. $\alpha(a) = (-\frac{7}{25}, -\frac{24}{25})$.

e. $\alpha(a) = (\frac{8}{17}, -\frac{15}{17})$.

13. Compute the values of the six functions in (a) to (f) of Exercise 10 at $-a$, where a is as given in each part of Exercise 12.
14. Find a period for each of the functions of Exercise 10.   (It would probably be helpful to determine the natural domain in each case.)
15. Suppose $\alpha(a) = (x, y)$.   Show (using property 6) that $\alpha(\pi - a) = (-x, y)$, $\alpha(\pi + a) = (-x, -y)$, and $\alpha(2\pi - a) = (x, -y)$.
16. Show that, if $\alpha(a) = (x, y)$ and $b = \pi/2 - a$, then $\alpha(b) = (y, x)$.
17. Given numbers a and b, we can determine points on the unit given by

$$\left(\frac{a}{(a^2 + b^2)^{1/2}}, \frac{b}{(a^2 + b^2)^{1/2}}\right), \left(\frac{b}{(a^2 + b^2)^{1/2}}, \frac{a}{(a^2 + b^2)^{1/2}}\right)$$

Determine such a pair of points for the following pairs of numbers:

a. 1, 1.

d. 8, 15.

b. 1, 3.

e. 7, 24.

c. 5, 12.

18. Determine the following points.

a. $\alpha(7\pi/12)$.

c. $\alpha(143\pi/12)$.

b. $\alpha(11\pi/12)$.

d. $\alpha(13\pi/12)$.

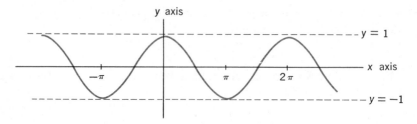

Figure 4.2.6
A sketch of $p_1 \circ \alpha$

19. In Exercise 10, you were asked to compute certain values given in terms of composition of functions. Attempt to sketch a graph of each composition in Exercise 7. We might help a bit by discussing the sketch of $p_1 \circ \alpha$.

We may envision our starting the sketch by looking at $p_1 \circ \alpha(0)$ and then proceeding through the positive domain elements. Since $\alpha(0) = (1, 0)$ and p_1 picks out the first coordinate, $p_1 \circ \alpha(0) = 0$. As a goes from 0 through $\pi/2$, $\alpha(a)$ goes from $(1, 0)$ to $(0, 1)$, whence $p_1 \circ \alpha(a)$ goes from 1 to 0, steadily decreasing. Similarly, as a goes from $\pi/2$ to π, $\alpha(a)$ traverses that part of the unit circle in the second quadrant and $p_1 \circ \alpha(a)$ decreases from 0 to -1. Continuing, we see that as a passes from π to $3\pi/2$, $p_1 \circ \alpha(a)$ increases from -1 to 0 and as a continues to 2π, $p_1 \circ \alpha(a)$ continues increasing to 1. If we retrace our steps from 2π back to 0, we see exactly what happens as a goes from 0 to -2π. (Why?) The graph we sketch appears in Fig. 4.2.6. The sketch indicates the effect of the periodicity of α.

Complete the other sketches.

20. Compute the equation of the line through the origin and

   a. the points in Exercise 1.
   b. the points in Exercise 3.
   c. the points in Exercise 4.
   d. the points in Exercise 6.
   e. the points in Exercise 12.
   f. the points in Exercise 17.

21. From Exercise 17 we find that, given a point $(a, b) \neq (0, 0)$, there is a unique point $(a/(a^2 + b^2)^{1/2}, b/(a^2 + b^2)^{1/2})$ on the unit circle. Show that this point is on the line passing through the origin and (a, b). Show further that the new point is on the ray through (a, b) emanating from $(0, 0)$. In this sense we are tying each point different from the origin to a single point on the unit circle. In fact, all points on a given ray emanating from the origin are related to the same point. From this idea, discover how each point (not the origin) is related to α. (Hint: Show that, for each $(x, y) \neq$

$(0, 0)$, $(x, y) = (rp, rq)$, where $(x^2 + y^2)^{1/2} = r$ and $(p, q) = \alpha(a)$ for some real number a.)

4.3 TWO TRIGONOMETRIC FUNCTIONS

Enough material has now been extended to allow the definition of the trigonometric functions. The definitions of two basic *trigonometric functions* will be given in this section.

$$\text{sine} = p_2 \circ \alpha : R \longrightarrow R$$
$$\text{cosine} = p_1 \circ \alpha : R \longrightarrow R$$

(Note: *The two functions just defined are real-valued functions of a real variable.*)

Sine (a) and cosine (a) will be abbreviated to $\sin a$ and $\cos a$, respectively.

What are some of the properties of these functions, and how does one set about evaluating them? The second question will be answered first.

Let $a \in R$. Then $\alpha(a)$ is a point (x, y) on the unit circle and it follows from the definitions above that:

$$\sin (a) = \sin a = p_2 \circ \alpha(a) = p_2(\alpha(a)) = p_2((x, y)) = y$$
$$\cos (a) = \cos a = p_1 \circ \alpha(a) = p_1((x, y)) = x$$

From the above analysis, we may say that the *value* of the sine function for any real number a is merely the second coordinate of the point that is the image of a under α. The cosine of a is the first coordinate of the same point. Thus, to determine the sine and cosine functional values, we must know the correspondence α.

What are the ranges of sine and cosine? We only note that $\alpha[R]$ is C and $p_1[C] = p_2[C] = [-1, 1]$. Necessarily, $\sin [R] = \cos [R] = [-1, 1]$. Hence, we may write:

$$\sin : R \longrightarrow [-1, 1]$$
$$\cos : R \longrightarrow [-1, 1]$$

These statements imply that, for each real number a,

$$-1 \leqslant \sin a \leqslant 1$$
$$-1 \leqslant \cos a \leqslant 1$$

Figures 4.3.1 and 4.3.2 show graphs of the sine and cosine functions (recall Exercise 4.2.14). The graph portrays the periodic nature of these functions. The periodicity is seen by the following (using the fact that $\alpha(a) = \alpha(a + 2\pi)$):

$$\sin a = p_2 \circ \alpha(a) = p_2(\alpha(a)) = p_2(\alpha(a + 2\pi)) = \sin (a + 2\pi)$$
$$\cos a = p_1 \circ \alpha(a) = p_1(\alpha(a)) = p_1(\alpha(a + 2\pi)) = \cos (a + 2\pi)$$

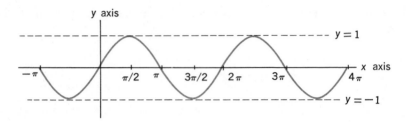

Figure 4.3.1
A sketch of sine

The graphs are then sketched by tracing $\alpha(a)$ once around the unit circle, starting at $(1, 0)$ (that is, following $\alpha(a)$ as a goes from 0 to 2π).

■ **EXERCISES 4.3**

1. Find the value of the image of each of the following under sine and cosine:

   a. All the special values of α listed in Section 4.2.
   b. The negative of all special values of α listed in Section 4.2.

2. Determine the polarity (positiveness or negativeness) of sin a and cos a, where $\alpha(a)$ is in quadrant I. II. III. IV.

3. Describe the following sets relative to the special values. For example, $\{x:\sin x = 0\} = \{x:\alpha(x) = (1, 0)\} = \{n\pi:n \in Z\}$.

   a. $\{x:\sin x = \frac{1}{2}\}$.
   b. $\{x:\sin x = -\frac{1}{2}\}$.
   c. $\{x:\sin x = \sqrt{3}/2\}$.
   d. $\{x:\sin x = -\sqrt{3}/2\}$.
   e. $\{x:\sin x = 1/\sqrt{2}\}$.
   f. $\{x:\sin x = -1/\sqrt{2}\}$.
   g. $\{x:\sin x = 1\}$.
   h. $\{x:\sin x = -1\}$.
   i. $\{x:\sin x = 2\}$.
   j. $\{x:\sin^2 x + \cos^2 x = 1\}$.
   k. Do Exercises (a) to (i) with cosine replacing sine.

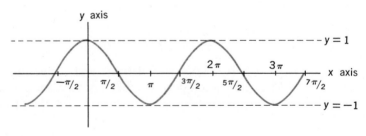

Figure 4.3.2
A sketch of cosine

4. If $\sin a = y$ and $\cos a = x$, compute $\alpha(a)$.
5. If $\sin a = x$ and $\cos a = y$, compute $\sin(-a)$ and $\cos(-a)$.
6. If $\sin a = x$ and $\cos a = y$, compute $\sin(\pi/2 - a)$ and $\cos(\pi/2 - a)$. (See Exercise 4.2.16.)
7. Show that, if $(x, y) \neq (0, 0)$, there is a real number a such that $(x, y) = (r \cos a, r \sin a)$, where $r = (x^2 + y^2)^{1/2}$. (See Exercise 4.2.21.)

4.4 SOME PROPERTIES OF SINE AND COSINE

In the next few paragraphs, a number of characteristics of the sine and cosine functions will be portrayed and developed algebraically through the use of: (1) properties of the α function and (2) some ingenuity in the use of manipulative techniques of algebra.

Let $\alpha(a) = (x, y)$. By property 5 of α, $\alpha(-a) = (x, -y)$.

Thus,
$$\sin(a) = y$$
$$\cos(a) = x$$
$$\sin(-a) = p_2 \circ \alpha(-a) = p_2(\alpha(-a))$$
$$= p_2((x, -y)) = -y = -\sin a$$
$$\cos(-a) = p_1 \circ \alpha(-a) = p_1(\alpha(-a))$$
$$= p_1((x, -y)) = x = \cos a$$

Restating these relationships, we have:

$$\sin(-a) = -\sin a$$
$$\cos(-a) = \cos a$$

It is seen here that changing the polarity (sign) of a changes the polarity of the value of the image under the sine function while causing no change in the value of the image under the cosine. This pair of results will be used several times in the coming work.

Example 4.4.1 Compute $\sin(-\pi/4)$ and $\cos(-7\pi/6)$.
Since $\sin \pi/4 = 1/\sqrt{2}$, $\sin(-\pi/4) = -1/\sqrt{2}$. Moreover, the statement $\cos 7\pi/6 = -\sqrt{3}/2$ shows that $\cos(-7\pi/6) = -\sqrt{3}/2$ also.
Note also that:

$$(\sin a)^2 + (\cos a)^2 = x^2 + y^2 = 1 \ ((x, y) \text{ is on } C)$$

In the future, $(\sin a)^2$ will be written $\sin^2 a$. In general, we shall write $f^2(x)$ for $(f(x))^2$. Thus, we have

$$\sin^2 a + \cos^2 a = 1$$

Are the two trigonometric functions periodic? Indeed they are.

$$\sin(a + 2\pi) = p_2(\alpha(a + 2\pi)) = p_2(\alpha(a)) = \sin a$$

using only the fact that α has period 2π. Similarly,

$$\cos (a + 2\pi) = p_1(\alpha(a + 2\pi)) = p_1(\alpha(a)) = \cos a$$

Hence, the sine and cosine functions have as a period the number 2π. It also happens to be the case that 2π is the *primitive period* for each of the two functions. Since induction shows that $2n\pi$ is a period for any $n \in Z$,

$$\sin (a + 2n\pi) = \sin a$$
$$\cos (a + 2n\pi) = \cos a$$

Necessarily, adding an even multiple of π to the argument a of the sine or cosine function does not change the image value.

Consider any point (x, y) on the unit circle. Since the image of α is all of the circle, it follows that $(x, y) = \alpha(a)$ for some real number a. (Note again that $\sin a = y$ and $\cos a = x$.) We now see,

$$(x, y) = (\cos a, \sin a), \qquad \text{where } (x, y) = \alpha(a)$$

That is, every point on the unit circle can be written in terms of the sine and cosine functions. (See Exercise 4.3.7 for a generalization of this idea.)

This can also be interpreted as: if x and y are real numbers so that $x^2 + y^2 = 1$, (x, y) is on the unit circle and $(x, y) = (\cos a, \sin a)$ for some real number a.

If we couple this fact with property 6 of α, we derive many interesting and important functional properties of sine and cosine. Again, property 6 states that:

$$d(\alpha(a), \alpha(b)) = d(\alpha(a - b), \alpha(0))$$

However, from what has just been said, $\alpha(a) = (\cos a, \sin a)$, while $\alpha(b) = (\cos b, \sin b)$.

Writing out the formula for $d(\alpha(a), \alpha(b))$, we see that

$$\begin{aligned}
d(\alpha(a), \alpha(b)) &= d((\cos a, \sin a)), ((\cos b, \sin b)) \\
&= [(\cos a - \cos b)^2 + (\sin a - \sin b)^2]^{1/2} \\
&= [(\cos^2 a + \sin^2 a) + (\cos^2 b + \sin^2 b) \\
&\qquad - 2(\cos a, \cos b + \sin a \sin b)]^{1/2} \\
&= [2 - 2(\cos a \cos b + \sin a \sin b)]^{1/2}
\end{aligned}$$

In like manner

$$d(\alpha(a - b), \alpha(0)) = [2 - 2 \cos (a - b)]^{1/2}$$

Equating the two expressions as related by property 6, and squaring both sides, we have

$$2 - 2(\cos a \cos b + \sin a \sin b) = 2 - 2 \cos (a - b)$$

Alternatively,

$$\cos (a - b) = \cos a \cos b + \sin a \sin b$$

We can now determine functional values at $(a - b)$ in terms of the functional values at a and b.

Example 4.4.2   For example, find $\cos \pi/12$.

The number $\pi/12$ is not one of the special values. However, since $\pi/12 = \pi/4 - \pi/6$:

$$\cos\left(\frac{\pi}{12}\right) = \cos\left(\frac{\pi}{4} - \frac{\pi}{6}\right)$$

$$= \cos\left(\frac{\pi}{4}\right)\cos\left(\frac{\pi}{6}\right) + \sin\left(\frac{\pi}{4}\right)\sin\left(\frac{\pi}{6}\right)$$

$$= \left(\frac{1}{\sqrt{2}}\right)\left(\frac{\sqrt{3}}{2}\right) + \left(\frac{1}{\sqrt{2}}\right)\left(\frac{1}{2}\right)$$

$$= \frac{\sqrt{3} + 1}{2\sqrt{2}}$$

If we make use of specific values for a (say, $\pi/2$, π, and so forth), various useful equalities result. For instance, if $a = \pi/2$:

$$\cos(a - b) = \cos\left(\frac{\pi}{2} - b\right) = \cos\frac{\pi}{2}\cos b + \sin\frac{\pi}{2}\sin b$$

$$\cos\left(\frac{\pi}{2} - b\right) = \sin b$$

Moreover, if in this equality we replace b by $\pi/2 - c$, we see that

$$\cos c = \sin\left(\frac{\pi}{2} - c\right)$$

The last two results above may be stated in the following manner: if two numbers a and b are such that $a + b = \pi/2$, $\sin a = \cos b$ and $\sin b = \cos a$. The numbers a and b with this characteristic (that is, $a + b = \pi/2$) are called **complementary numbers;** if functions f and g are such that, whenever a and b are complementary numbers, $f(a) = g(b)$ and $g(a) = f(b)$, f and g are called **cofunctions.** Examples of cofunctions other than sine and cosine will be given at a later time.

Example 4.4.3   Suppose $\cos \pi/24 = .9914$. Compute $\sin \pi/24$, $\cos 11\pi/24$, and $\sin 11\pi/24$.

The relation $\sin^2 a + \cos^2 a = 1$, together with the fact that $\alpha(\pi/24)$ is in the first quadrant, forces $\sin \pi/24 = [1 - \cos^2 \pi/24]^{1/2}$ or approximately $.1305$. Since $11\pi/24 = \pi/2 - \pi/24$, $\cos 11\pi/24 = \sin \pi/24 = .1305$ approximately and $\sin 11\pi/24 = \cos \pi/24 = .9914$.

Having been successful in computing $\cos(a - b)$, we might feel that it is possible to formulate such an expression for $\cos(a + b)$. This we can do, since we can write $a + b$ as $a - (-b)$. Making use of previous knowledge, we write:

$$\cos (a + b) = \cos (a - (-b))$$
$$\cos (a + b) = \cos a \cos (-b) + \sin a \sin (-b)$$
$$\cos (a + b) = \cos a \cos b - \sin a \sin b$$

The final form follows from $\cos (-b) = \cos b$, while $\sin (-b) = -\sin b$.

Example 4.4.4   Using the information from Example 4.4.3 and this new identity, we see that $\cos 13\pi/24 = \cos (\pi/2 + \pi/24) = \cos \pi/2 \cos \pi/24 - \sin \pi/2 \sin \pi/24 = -\sin \pi/24 = -.1305$.

Our previous knowledge, together with our last achievements, will now be used to develop "formulas" for $\sin (a + b)$ and $\sin (a - b)$.

$$\sin (a + b) = \cos \left(\frac{\pi}{2} - (a + b)\right) = \cos \left(\left(\frac{\pi}{2} - a\right) - b\right)$$
$$= \cos \left(\frac{\pi}{2} - a\right) \cos b + \sin \left(\frac{\pi}{2} - a\right) \sin b$$

$$\sin (a + b) = \sin a \cos b + \cos a \sin b$$

Similar to our tactics employed above, we write $a - b$ as $a + (-b)$ so that

$$\sin (a - b) = \sin a \cos (-b) + \cos a \sin (-b)$$
$$\sin (a - b) = \sin a \cos b - \cos a \sin b$$

These formulas, of course, are analogous to those for the cosine of the sum and difference of two numbers. It has been mentioned that we can now find sine and cosine values for some numbers not found among the special values given for α. In retrospect however, it also allows us to increase the list of values under α.

Example 4.4.5   We know that $\alpha(\pi/12) = (\cos \pi/12, \sin \pi/12) = ((\sqrt{3} + 1)/2\sqrt{2}, (\sqrt{3} - 1)/2\sqrt{2})$, since

$$\sin \frac{\pi}{12} = \sin \left(\frac{\pi}{4} - \frac{\pi}{6}\right) = \sin \frac{\pi}{4} \cos \frac{\pi}{6} - \cos \frac{\pi}{4} \sin \frac{\pi}{6}$$
$$= \frac{1}{\sqrt{2}} \cdot \frac{\sqrt{3}}{2} - \frac{1}{\sqrt{2}} \cdot \frac{1}{2} = \frac{\sqrt{3} - 1}{2\sqrt{2}}$$

Taking special values for a, we again are capable of deriving even more relationships.

$$\cos \left(\frac{\pi}{2} + b\right) = \cos \frac{\pi}{2} \cos b - \sin \frac{\pi}{2} \sin b = -\sin b$$
$$\cos (\pi + b) = \cos \pi \cos b - \sin \pi \sin b = -\cos b$$
$$\cos (b + (2n + 1)\pi) = \cos ((b + \pi) + 2n\pi) = \cos (b + \pi) = -\cos b$$

$$\sin (\pi + b) = \sin \pi \cos b + \cos \pi \sin b = -\sin b$$
$$\sin (b + (2n + 1)\pi) = \sin ((b + \pi) + 2n\pi) = \sin (b + \pi) = -\sin b$$

The above few lines show that adding an odd multiple of π to the argument of either the sine or cosine function merely changes the polarity of the image value. Coupled with the fact that adding an even multiple of π changes nothing so far as the image value is concerned, we may put the two statements together into a reduction formula for each of the functions. Let n be any integer.

$$\sin (b + n\pi) = (-1)^n \sin b$$
$$\cos (b + n\pi) = (-1)^n \cos b$$

Example 4.4.6   Calculate $\sin (25\pi/4)$ and $\cos (34\pi/3)$. Now, $25\pi/4 = 6\pi + \pi/4$ so that $\sin 25\pi/4 = \sin (6\pi + \pi/4) = (-1)^6 \sin \pi/4 = 1/\sqrt{2}$. Likewise, $34\pi/3 = 11\pi + \pi/3$ so that $\cos 34\pi/3 = \cos (\pi/3 + 11\pi) = (-1)^{11} \cos \pi/3 = -\frac{1}{2}$.

Example 4.4.7   Write $\cos 8\pi/3$ as $\pm\cos a$ or $\sin a$ for some a satisfying $0 \leqslant a \leqslant \pi/4$.

Since $8\pi/3 = 2\pi + 2\pi/3$, we may write (using the property of cofunctions) $\cos 8\pi/3 = \cos 2\pi/3 = \sin (\pi/2 - 2\pi/3) = \sin (-\pi/6) = -\sin \pi/6$. Alternatively, we could have written $\cos 8\pi/3 = \cos (3\pi - \pi/3) = -\cos \pi/3 = -\sin \pi/6$, using the reduction formula at another point.

The following is a list of the relationships developed in this section.

$$\alpha(a) = (\cos a, \sin a)$$
$$\sin (-a) = -\sin a$$
$$\cos (-a) = \cos a$$
$$\sin^2 a + \cos^2 a = 1$$
$$\sin (a + n\pi) = (-1)^n \sin a; \, n \in Z$$
$$\cos (a + n\pi) = (-1)^n \cos a; \, n \in Z$$
$$\cos (a - b) = \cos a \cos b + \sin a \sin b$$
$$\cos (a + b) = \cos a \cos b - \sin a \sin b$$
$$\cos (\pi/2 - b) = \sin b$$
$$\sin (\pi/2 - b) = \cos b$$
$$\sin (a + b) = \sin a \cos b + \cos a \sin b$$
$$\sin (a - b) = \sin a \cos b - \cos a \sin b$$

■ EXERCISES 4.4

1. Given that $\cos a = \frac{3}{5}$, $\sin b = \frac{8}{17}$, $\cos c = -\frac{5}{13}$, $\sin d = -\frac{24}{25}$, $\alpha(a) \in$ quadrant IV, $\alpha(b) \in$ quadrant I, $\alpha(c) \in$ quadrant II, and $\alpha(d) \in$ quadrant III, compute:

a. $\sin a$.
b. $\cos b$.
c. $\sin c$.
d. $\cos d$.
e. $\sin(-a)$.
f. $\cos(-b)$.
g. $\sin(-c)$.
h. $\cos(-d)$.

i. $\sin(36\pi + a)$.
j. $\cos(b - 15\pi)$.
k. $\cos(c + 4367\pi)$.
l. $\sin(d + 19\pi)$.
m. $\cos(397\pi/2 + a)$.
n. $\sin(43\pi/6 + b)$.
o. $\cos(c + 945\pi/4)$.
p. $\sin(1631\pi/3 - d)$.

2. Using a, b, c, and d as in Exercise 1, compute:

a. $\sin(a + b)$.
b. $\cos(a + b)$.
c. $\sin(a - b)$.
d. $\cos(a - b)$.
e. $\sin(a + c)$.
f. $\cos(a + c)$.
g. $\sin(-c)$.
h. $\cos(a - c)$.

i. $\sin(a + d)$.
j. $\cos(a + d)$.
k. $\sin(a - d)$.
l. $\cos(a - d)$.
m. $\sin(b + c)$.
n. $\cos(b + c)$.
o. $\sin(b - c)$.
p. $\cos(b - c)$.

q. $\sin(b + d)$.
r. $\cos(b + d)$.
s. $\sin(b - d)$.
t. $\cos(b - d)$.
u. $\sin(c + d)$.
v. $\cos(c + d)$.
w. $\sin(c - d)$.
x. $\cos(c - d)$.

3. Using a, b, c, and d as in Exercise 1, compute:

a. $\sin 2a = \sin(a + a)$.
b. $\cos 2a$.
c. $\sin 2b$.
d. $\cos 2b$.
e. $\sin 2c$.
f. $\cos 2c$.
g. $\sin 2d$.

h. $\cos 2d$.
i. $\alpha(a)$.
j. $\alpha(b)$.
k. $\alpha(c)$.
l. $\alpha(d)$.
m. $\alpha(a + b)$.

n. $\alpha(b + c)$.
o. $\alpha(c + d)$.
p. $\alpha(a - c)$.
q. $\alpha(b - c)$.
r. $\alpha(c - d)$.
s. $\alpha(b - d)$.

4. Given that $\sin \pi/5 = .5878$, $\sin \pi/8 = .3827$, and $\cos \pi/10 = .9511$, calculate each of the following:

a. $\cos \pi/5$.
b. $\cos \pi/8$.
c. $\sin \pi/10$.

d. $\cos(-\pi/5)$.
e. $\cos(-\pi/8)$.
f. $\sin(-\pi/10)$.

g. $\sin(-\pi/5)$.
h. $\sin(-\pi/8)$.
i. $\cos(-\pi/10)$.

5. Using the answers and information in Exercise 4, calculate:

a. $\cos 3\pi/10$. (Hint: $3\pi/10 = \pi/2 - (-\pi/5)$.)
b. $\sin 3\pi/10$.
c. $\cos 3\pi/8$.
d. $\sin 3\pi/8$.
e. $\cos 2\pi/5$.
f. $\sin 2\pi/5$.
g. $\cos(-3\pi/10)$.

h. $\sin(-3\pi/10)$.
i. $\cos(-3\pi/8)$.
j. $\sin(-3\pi/8)$.
k. $\cos(-2\pi/5)$.
l. $\sin(-2\pi/5)$.

6. Given that a and b are complementary numbers, find $\sin a$, $\cos a$, $\sin b$, $\cos b$, and $\alpha(b)$ in each of the following.

a. $\alpha(a) = (\frac{3}{5}, \frac{4}{5})$.
b. $\alpha(a) = (\frac{5}{13}, \frac{12}{13})$.
c. $\alpha(a) = (1/\sqrt{2}, -1/\sqrt{2})$.

d. $\alpha(a) = (-\frac{3}{5}, \frac{4}{5})$.
e. $\alpha(a) = (\frac{8}{17}, \frac{15}{17})$.
f. $\alpha(a) = (x, y)$.

7. Using the answers and information from Exercises 4 and 5, find the values for:

a. $\cos 23\pi/10$.
b. $\sin 53\pi/10$.
c. $\cos 25\pi/8$.
d. $\sin 33\pi/8$.
e. $\cos 21\pi/5$.
f. $\sin 46\pi/5$.
g. $\cos 83\pi/8$.

h. $\sin 502\pi/5$.
i. $\cos 13\pi/40$.
j. $\sin 3\pi/40$.
k. $\cos \pi/40$.
l. $\sin 9\pi/40$.
m. $\cos 3\pi/5$.
n. $\sin 3\pi/5$.

8. Find each of the following values using some of the identities developed.   (See Exercise 1 for some values needed.)

a. $\sin 5\pi/12$.
b. $\cos 5\pi/12$.
c. $\sin 7\pi/12$.
d. $\cos 7\pi/12$.
e. $\sin (-7\pi/12)$.
f. $\cos (-7\pi/12)$.

g. $\sin (-5\pi/12)$.
h. $\cos (-5\pi/12)$.
i. $\sin 17\pi/60$.
j. $\cos 3\pi/20$.
k. $\sin \pi/8$.
l. $\cos 13\pi/10$.

9. Compute $\alpha(a)$ for each a below.   (Hint: Use Exercise 8.)

a. $a = 5\pi/12$.
b. $a = 7\pi/12$.

c. $a = -7\pi/12$.
d. $a = -5\pi/12$.

10. Show that $\sin (a + \pi/2) = \cos a$.   Then calculate $\alpha(a + \pi/2)$ in each of the following.

a. $\alpha(a) = (\frac{3}{5}, \frac{4}{5})$.
b. $\alpha(a) = (\frac{5}{13}, \frac{12}{13})$.
c. $\alpha(a) = (-\frac{8}{17}, \frac{15}{17})$.

d. $\alpha(a) = (\frac{7}{25}, -\frac{24}{25})$.
e. $\alpha(a) = (x, y)$.

11. Write each of the following as $\pm\sin a$ or $\pm\cos a$ for a satisfying $0 \leqslant a \leqslant \pi/4$.

a. $\cos 21\pi/4$.
b. $\sin 23\pi/6$.
c. $\sin 5\pi/6$.
d. $\cos 89\pi/4$.

e. $\sin 72\pi/5$.
f. $\cos 23\pi/5$.
g. $\sin 58\pi/3$.
h. $\cos 3163\pi/16$.

12. Write the equation of the line passing through the origin and the following. (See Exercise 4.3.7.)

a. $(\cos \pi/4, \sin \pi/4)$.
b. $(\cos \pi/5, \sin \pi/5)$.
c. $(\cos \pi/10, \sin \pi/10)$.
d. $(\cos 3\pi/4, \sin 3\pi/4)$.
e. $(\cos 5\pi/3, \sin 5\pi/3)$.
f. $(\cos 7\pi/6, \sin 7\pi/6)$.

If A is a ray emanating from the origin, what do all the points $(x, y) \neq (0, 0)$ on A have in common when written in the form $(r \cos a, r \sin a)$ for $0 \leqslant a \leqslant 2\pi$?

13. Why do we know what the entire graph of $y = \sin x$ looks like if we only graph the curve for $0 \leqslant x < 2\pi$?

14. Which of the sine and cosine is an even function? Odd function? See Exercise 3.4.13.

4.5 SOME MORE RELATIONSHIPS INVOLVING SINE AND COSINE

In this section, we shall develop more identities using properties derived in the preceding section. Let us develop some *double argument* formulas.

$$\sin 2a = \sin (a + a) = \sin a \cos a + \cos a \sin a$$
$$\sin 2a = 2 \sin a \cos a$$
$$\cos 2a = \cos (a + a) = \cos a \cos a - \sin a \sin a$$
$$\cos 2a = \cos^2 a - \sin^2 a$$
$$\cos 2a = 2 \cos^2 a - 1$$
$$\cos 2a = 1 - 2 \sin^2 a$$

The latter two forms of the double argument formulas for the cosine come as a result of writing the identity $\sin^2 a + \cos^2 a = 1$ as:

$$\cos^2 a = 1 - \sin^2 a,$$
or
$$\sin^2 a = 1 - \cos^2 a$$

Having double argument formulas, we might feel that *half argument* formulas should arise in a quite natural way. (We can view a as half of $2a$ as well as think of $2a$ as being twice a.)

Replacing $2a$ by b (and hence a by $b/2$) in the formulas for the cosine, we see that they become (looking at the latter forms only):

$$\cos b = 2 \cos^2 \frac{b}{2} - 1$$

$$\cos b = 1 - 2 \sin^2 \frac{b}{2}$$

Solving each equation for the expression involving $b/2$, we successfully arrive at the half argument formulas:

$$\cos b/2 = \pm[(1 + \cos b)/2]^{1/2}$$
$$\sin b/2 = \pm[(1 - \cos b)/2]^{1/2}$$

The ambiguous signs in front of the radical must be determined by observing the quadrant location of $\alpha(b/2)$.

Example 4.5.1   Compute $\cos \pi/8$ and $\sin 8\pi/5$, knowing that $\cos 16\pi/5 = -.81$.

Since $\pi/8 = \frac{1}{2}(\pi/4)$ and $\alpha(\pi/8)$ is in quadrant I,

$$\cos \pi/8 = \left[\frac{1 + \cos \pi/4}{2}\right]^{1/2} = \left[\frac{1 + 1/\sqrt{2}}{2}\right]^{1/2}$$

or approximately .92.   Now, $8\pi/5 = \frac{1}{2}(16\pi/5)$ and $3\pi/2 < 8\pi/5 < 2\pi$, whence $\alpha(8\pi/5)$ is in quadrant IV and

$$\sin 8\pi/5 = -\left[\frac{1 - \cos 16\pi/5}{2}\right]^{1/2} = -\left[\frac{1 + .81}{2}\right]^{1/2}$$

or approximately $-.95$.

Expressions involving products of functional values for the sine and cosine are easily generated.   For example, if we add the identities for $\cos (a - b)$ and $\cos (a + b)$, we see:

$$\cos (a - b) = \cos a \cos b + \sin a \sin b$$
$$\underline{\cos (a + b) = \cos a \cos b - \sin a \sin b}$$
$$\cos (a - b) + \cos (a + b) = 2 \cos a \cos b$$

The expression may be rewritten as:

$$\cos a \cos b = \frac{1}{2}(\cos (a - b) + \cos (a + b))$$

If, instead of adding (in the above derivation) we subtract, the result is:

$$\sin a \sin b = \frac{1}{2}(\cos (a - b) - \cos (a + b))$$

Similar results are obtained by using $\sin (a + b)$ and $\sin (a - b)$.
Add the two equations below.

$$\sin (a + b) = \sin a \cos b + \cos a \sin b$$
$$\underline{\sin (a - b) = \sin a \cos b - \cos a \sin b}$$
$$\sin (a + b) + \sin (a - b) = 2 \sin a \cos b$$

Solving as done above, we see that:

$$\sin a \cos b = \frac{1}{2}(\sin (a + b) + \sin (a - b))$$

Interchanging a and b, we find

$$\sin b \cos a = \tfrac{1}{2}(\sin (a + b) - \sin (a - b))$$

Example 4.5.2 Compute $\cos 5\pi/12 \cos \pi/12$.

By the first of the identities developed above, $\cos 5\pi/12 \cos \pi/12 = \tfrac{1}{2}(\cos (5\pi/12 - \pi/12) + \cos (5\pi/12 + \pi/12)) = \tfrac{1}{2}(\cos \pi/3 + \cos \pi/2) = \tfrac{1}{4}$.

The form of the above identities is easily changed by replacing $(a - b)$ by y and $(a + b)$ by x. We solve for a and b to observe that

$$a = \tfrac{1}{2}(x + y)$$
$$b = \tfrac{1}{2}(x - y)$$

Substituting these expressions into the aforementioned equations, we have

$$\cos y + \cos x = 2 \cos \tfrac{1}{2}(x + y) \cos \tfrac{1}{2}(x - y)$$
$$\cos y - \cos x = 2 \sin \tfrac{1}{2}(x + y) \sin \tfrac{1}{2}(x - y)$$
$$\sin x + \sin y = 2 \sin \tfrac{1}{2}(x + y) \cos \tfrac{1}{2}(x - y)$$
$$\sin x - \sin y = 2 \sin \tfrac{1}{2}(x - y) \cos \tfrac{1}{2}(x + y)$$

Example 4.5.3 Compute $\sin 3\pi/10 - \sin \pi/5$ knowing that $\sin \pi/20 = .156$.

Since $\tfrac{1}{2}(3\pi/10 + \pi/5) = \pi/4$ and $\tfrac{1}{2}(3\pi/10 - \pi/5) = \pi/20$, we can calculate the quantity desired by using the special values and the given information.

$$\sin 3\pi/10 - \sin \pi/5 = 2 \sin \tfrac{1}{2}(3\pi/10 - \pi/5) \sin \tfrac{1}{2}(3\pi/10 + \pi/5)$$
$$= 2 \sin \pi/20 \sin \pi/4$$
$$= 2(.156)(1/\sqrt{2})$$

or approximately .221.

The following is a list of the identities derived in this section.

$$\sin 2a = 2 \sin a \cos a$$
$$\cos 2a = \cos^2 a - \sin^2 a$$
$$= 2 \cos^2 a - 1$$
$$= 1 - 2 \sin^2 a$$
$$\sin \frac{a}{2} = \pm \left[\frac{1 - \cos a}{2} \right]^{1/2}$$
$$\cos \frac{a}{2} = \pm \left[\frac{1 + \cos a}{2} \right]^{1/2}$$

$$\cos a \cos b = \tfrac{1}{2}(\cos (a - b) + \cos (a + b))$$
$$\sin a \sin b = \tfrac{1}{2}(\cos (a - b) - \cos (a + b))$$
$$\sin a \cos b = \tfrac{1}{2}(\sin (a - b) + \sin (a + b))$$
$$\cos a + \cos b = 2 \cos \tfrac{1}{2}(a + b) \cos \tfrac{1}{2}(a - b)$$
$$\cos a - \cos b = 2 \sin \tfrac{1}{2}(a + b) \sin \tfrac{1}{2}(b - a)$$
$$\sin a + \sin b = 2 \sin \tfrac{1}{2}(a + b) \cos \tfrac{1}{2}(a - b)$$
$$\sin a - \sin b = 2 \sin \tfrac{1}{2}(a - b) \cos \tfrac{1}{2}(a + b)$$

■ **EXERCISES 4.5**

1. Compute $\sin 2a$ and $\cos 2a$ in each of the following cases.

  a. $\alpha(a) = (\tfrac{3}{5}, \tfrac{4}{5})$.
  b. $\alpha(a) = (-\tfrac{3}{5}, \tfrac{4}{5})$.
  c. $\sin a = \tfrac{15}{17}$, $\cos a = \tfrac{8}{17}$.
  d. $\sin a = -\tfrac{15}{17}$, $\alpha(a)$ is in quadrant III.
  e. $\cos a = \tfrac{5}{13}$, $\alpha(a)$ is in quadrant IV.
  f. $\cos a = \tfrac{12}{13}$, $\alpha(a)$ is in quadrant I.

2. Using half angle formulas, compute $\sin a/2$ and $\cos a/2$ in each case below.

  a. $\alpha(a) = (\tfrac{3}{5}, \tfrac{4}{5})$, $a \in [0, 2\pi)$.   [Hint: $\alpha(a) \in$ quadrant I, and $a \in [0, 2\pi)$ implies $a \in [0, \pi/2)$. Whence $a/2 \in [0, \pi/4)$.]
  b. $\alpha(a) = (-\tfrac{3}{5}, \tfrac{4}{5})$, $a \in [0, 2\pi)$.
  c. $\alpha(a) = (\tfrac{8}{17}, \tfrac{15}{17})$, $a \in [2\pi, 4\pi)$.
  d. $\sin a = -\tfrac{15}{17}$, $a \in [\pi, 3\pi/2)$.
  e. $\sin a = -\tfrac{5}{13}$, $a \in [3\pi/2, 2\pi)$.
  f. $\cos a = \tfrac{12}{13}$, $a \in [50\pi, 101\pi/2)$.
  g. $\sin a = \tfrac{7}{25}$, $a \in (0, 2\pi)$, $\alpha(a) \notin$ quadrant I.

3. Write each of the following products as a sum.

  a. $\sin 2a \cos 3a$.   **d.** $\sin 2a \cos 5a$.
  b. $\sin 3a \cos 3a$.   **e.** $\sin 3a \sin 2a$.
  c. $\cos 5a \cos 2a$.   **f.** $2 \sin 7a \cos 2a$.

4. Compute each of the following. (Write each product as a sum.)

  a. $\cos \pi/8 \cos 3\pi/8$.
  b. $\sin \pi/12 \sin 7\pi/12$.
  c. $\sin \pi/16 \cos 3\pi/16$.
  d. $\cos 7\pi/5 \sin 2\pi/5$. (Use $\sin \pi/5 = .59$.)

5. Write each of the following sums as a product.

  a. $\sin a + \sin 2a$.   **d.** $\sin 2a - \sin a$.
  b. $\cos 3a + \cos 5a$.   **e.** $\cos 3a - \cos 5a$.
  c. $\sin 5a/3 - \sin 5a/6$.   **f.** $\cos 6a + \cos 7a$.

6. Compute each of the following. (Write each of the sums as a product.)

  a. $\cos 7\pi/12 + \cos \pi/12$.
  b. $\cos \pi/8 - \cos 5\pi/8$.
  c. $\sin 13\pi/16 + \sin 3\pi/16$.
  d. $\sin 3\pi/5 - \sin 2\pi/5$. (Use $\sin \pi/10 = .31$.)

7. Use the sine and cosine to describe the slope of a line through $(0, 0)$ and $(x, y) \neq (0, 0)$. [Hint: Write (x, y) in the form $(r \cos a, r \sin a)$.]

4.6 WORKING WITH IDENTITIES

An equation is a (not necessarily true) mathematical statement of equality. Whenever variable terms are involved in such statements, there is, a priori, a set called the *substitution set* (that is, the set from which values of the variable are chosen). In our encounters, the substitution set will be $R - A$, where A is the *restriction set* (that is, the set of values for which some expression in the statement of the equality is not well defined).

Example 4.6.1 In the statement $1/x = 1/x$, the restriction set is $\{0\}$, for only zero renders any expression undefined. The substitution set is then $R - \{0\}$, the set of all nonzero real numbers.

Any equation that is a valid statement for every value of the substitution set is said to be an **identity**. All other equations are said to be **conditional**. The statement $1/x = 1/x$ examined above is an identity, since $1/x = 1/x$ is valid for all x in $R - \{0\}$, that is, if $x \neq 0$, $1/x = 1/x$. (We cannot say that $1/0 = 1/0$, since $1/0$ is not defined.)

Example 4.6.2 The equation $x = 2x$ must be conditional. The substitution set is R, but, if $x = 1$, $x \neq 2x$. In fact, only $x = 0$ gives validity to the equation.

Example 4.6.3 The equation $\sqrt{x} = -2$ is also a conditional equation. By definition, $\sqrt{x} \geqslant 0$ and, necessarily, $\sqrt{x}$ cannot be negative. Consequently, the equation fails to be valid for *any* element of the substitution set $\{x \in R : x \geqslant 0\}$.

The relationships derived in Sections 4.2 through 4.5 were derived *independent of the choices of values for the variables.* (The variables were usually a, b, c, x, or y.) Thus, these relationships are indeed *trigonometric identities.*

You might suspect that countless new identities can be developed from those already given. In order to do so, however, familiarity with the identities already given is essential. The following is a complete list of our previous results.

$$\sin^2 a + \cos^2 a = 1$$
$$\cos(-a) = \cos a$$
$$\sin(-a) = -\sin a$$
$$\cos(a + n\pi) = (-1)^n \cos a$$
$$\sin(a + n\pi) = (-1)^n \sin a$$
$$\cos a = \sin(\pi/2 - a)$$

$\sin a = \cos (\pi/2 - a)$
$\cos (a + b) = \cos a \cos b - \sin a \sin b$
$\cos (a - b) = \cos a \cos b + \sin a \sin b$
$\sin (a + b) = \sin a \cos b + \cos a \sin b$
$\sin (a - b) = \sin a \cos b - \cos a \sin b$
$\cos 2a = \cos^2 a - \sin^2 a = 2 \cos^2 a - 1 = 1 - 2 \sin^2 a$
$\sin 2a = 2 \sin a \cos a$

$$\cos \left(\frac{a}{2}\right) = \pm(\tfrac{1}{2}(1 + \cos a))^{1/2}$$

$$\sin \left(\frac{a}{2}\right) = \pm(\tfrac{1}{2}(1 - \cos a))^{1/2}$$

$\cos a \cos b = \tfrac{1}{2}(\cos (a - b) + \cos (a + b))$
$\sin a \sin b = \tfrac{1}{2}(\cos (a - b) - \cos (a + b))$
$\sin a \cos b = \tfrac{1}{2}(\sin (a - b) + \sin (a + b))$
$\cos a \sin b = \tfrac{1}{2}(\sin (a + b) - \sin (a - b))$
$\cos a + \cos b = 2 \cos \tfrac{1}{2}(a + b) \cos \tfrac{1}{2}(a - b)$
$\cos a - \cos b = 2 \sin \tfrac{1}{2}(a + b) \sin \tfrac{1}{2}(b - a)$
$\sin a + \sin b = 2 \sin \tfrac{1}{2}(a + b) \cos \tfrac{1}{2}(a - b)$
$\sin a - \sin b = 2 \sin \tfrac{1}{2}(a - b) \cos \tfrac{1}{2}(a + b)$

Several problems will be investigated in order to exhibit some uses of a few of the derived identities. Some of the work is repetitious and will serve as a review.

Example 4.6.4   Suppose $\alpha(a) = (\tfrac{3}{5}, \tfrac{4}{5})$ and $\alpha(b) = (\tfrac{15}{17}, \tfrac{8}{17})$.   Find each of the following:

1. $\cos a$.	9. $\cos 2a$.
2. $\cos b$.	10. $\sin 2a$.
3. $\sin a$.	11. $\cos a/2$.
4. $\sin b$.	12. $\sin a/2$.
5. $\cos (a + b)$.	13. $\cos (a + 3\pi)$.
6. $\cos (a - b)$.	14. $\sin (a + 4\pi)$.
7. $\sin (a + b)$.	15. $(a + b)$.
8. $\sin (a - b)$.	16. $(a - b)$.

SOLUTIONS:

1. $\cos a = \tfrac{3}{5}$.
2. $\cos b = \tfrac{15}{17}$.
3. $\sin a = \tfrac{4}{5}$.
4. $\sin b = \tfrac{8}{17}$.
5. $\cos (a + b) = \cos a \cos b - \sin a \sin b = \tfrac{3}{5} \cdot \tfrac{15}{17} - \tfrac{4}{5} \cdot \tfrac{8}{17} = \tfrac{17}{85}$.
6. $\cos (a - b) = \cos a \cos b + \sin a \sin b = \tfrac{3}{5} \cdot \tfrac{15}{17} + \tfrac{4}{5} \cdot \tfrac{8}{17} = \tfrac{77}{85}$.
7. $\sin (a + b) = \sin a \cos b + \cos a \sin b = \tfrac{4}{5} \cdot \tfrac{15}{15} + \tfrac{3}{5} \cdot \tfrac{8}{17} = \tfrac{84}{85}$.

8. $\sin (a - b) = \sin a \cos b - \cos a \sin b = \frac{4}{5} \cdot \frac{15}{17} - \frac{3}{5} \cdot \frac{8}{17} = \frac{36}{85}$.

9. $\cos 2a = \cos^2 a - \sin^2 a = (\frac{3}{5})^2 - (\frac{4}{5})^2 = \frac{9}{25} - \frac{16}{25} = -\frac{7}{25}$.

10. $\sin 2a = 2 \sin a \cos a = 2(\frac{4}{5})(\frac{3}{5}) = \frac{24}{25}$.

11. $\cos a/2 = \pm[(1 + \cos a)/2]^{1/2} = \pm[(1 + \frac{3}{5})/2]^{1/2} = \pm(\frac{8}{10})^{1/2} = \pm(\frac{4}{5})^{1/2} = \pm 2/\sqrt{5}$. There is ambiguity as to the $+$ and $-$, since the value of a is not known.

12. $\sin a/2 = \pm[(1 - \cos a)/2]^{1/2} = \pm[(1 - \frac{3}{5})/2]^{1/2} = \pm(\frac{2}{10})^{1/2} = \pm(\frac{1}{5})^{1/2} = \pm 1/\sqrt{5}$. The $\pm$ is ambiguous for the same reason as in (11).

13. $\cos (a + 3\pi) = (-1)^3 \cos a = -\cos a$.

14. $\sin (a + 4\pi) = (-1)^4 \sin a = \sin a$.

15. $\alpha(a + b) = (\cos (a + b), \sin (a + b)) = (\frac{17}{85}, \frac{84}{85})$.

16. $\alpha(a - b) = (\cos (a - b), \sin (a - b)) = (\frac{77}{85}, \frac{36}{85})$.

The following example illustrates a technique for justifying identities. The process used is that of showing that one of the two expressions can be transformed into the other by (1) substitution of identical expressions and (2) the use of reversible algebraic manipulations.

Example 4.6.5   Show that:

$$\cos a + \frac{\sin^2 a}{\cos a} = \frac{1}{\cos a}$$

is an identity.

The more complicated expression will be simplified. The statement of equality rules out the possibility that $\cos a = 0$ since $1/\cos a$ would in that situation be meaningless. Since $\cos a \neq 0$ implies that $a \neq (2n + 1)\pi/2$ for some integer n, a will not be allowed to take on any such value during this discussion.

To add the two expressions on the left, we need a common denominator. Thus,

$$\cos a + \frac{\sin^2 a}{\cos a} = \frac{\cos^2 a}{\cos a} + \frac{\sin^2 a}{\cos a} = \frac{\cos^2 a + \sin^2 a}{\cos a}$$

However, one of the basic identities developed shows that

$$\frac{\cos^2 a + \sin^2 a}{\cos a} = \frac{1}{\cos a}$$

Thus, the left expression (in the original statement of the problem) has been simplified to look exactly like the right. This procedure could have been reversed. In fact, if the identity is valid, the process *must* be reversible.

$$\frac{1}{\cos a} = \frac{\cos^2 a + \sin^2 a}{\cos a} = \frac{\cos^2 a}{\cos a} + \frac{\sin^2 a}{\cos a} = \cos a + \frac{\sin^2 a}{\cos a}$$

If reversible steps are used, we *need not* show both procedures.

Example 4.6.6   Examine the expression $\sin a + \sqrt{3} \cos a$, which may be rewritten $1 \cdot \sin a + \sqrt{3} \cos a$. Since $1^2 + (\sqrt{3})^2 = 4$, the point $(\frac{1}{2}, \sqrt{3}/2)$ is on the unit circle. In particular, $(\frac{1}{2}, \sqrt{3}/2) = \alpha(\pi/3)$. Thus, $\sin a + \sqrt{3} \cos a = 2(\frac{1}{2} \sin a + \sqrt{3}/2 \cos a) = 2(\sin a \cos \pi/3 + \cos a \sin \pi/3) = 2 \sin (a + \pi/3)$.

Alternatively, the point $(\sqrt{3}/2, \frac{1}{2}) = \alpha(\pi/6)$ is on the unit circle and $\sin a + \sqrt{3} \cos a = 2(\frac{1}{2} \sin a + \frac{3}{2} \cos b) = 2(\sin a \sin \pi/6 + \cos a \cos \pi/6) = 2 \cos (a - \pi/6)$. Observe that the two numbers $\pi/6$ and $\pi/3$ brought into the discussion are complementary numbers.

■ EXERCISES 4.6

Using established identities, algebraic maneuvering, and substitution, verify that the following are trigonometric identities, giving the substitution set and restriction set for each statement.

1. $\sin (\pi/2 + a) = \sin (\pi/2 - a)$.
2. $1 + \cos^2 a/\sin^2 a = 1/\sin^2 a$.   (Note: $\sin a$ cannot be zero, whence $a \neq n\pi$ for $n \in Z$.)
3. $1 + \sin^2 a/\cos^2 a = 1/\cos^2 a$.   What values of a are to be eliminated?
4. $\sin a/\cos a + \cos a/\sin a = 1/(\cos a \sin a)$.

5. $\dfrac{\cos a/\sin a}{1 + (\cos^2 a/\sin^2 a)} = \sin a \cos a$.

6. $\sin a/(1 + \cos a) = 1/\sin a - \cos a/\sin a$.
7. $1/\sin a - \cos^2 a/\sin a = \sin a$.
8. $1 - \cos^2 a/(1 + \sin a) = \sin a$.
9. $1/(\cos^2 a \sin^2 a) = 1/\cos^2 a + 1/\sin^2 a$.
10. $\cos a = \sin (a + \pi/6) + \cos (a + \pi/3)$.
11. $\sin (a + \pi/6) + \cos (\pi/3 - a) = 2 \sin (a + \pi/6)$.
12. $\sin (a + \pi/6) + \cos (\pi/3 - a) = 2 \cos (\pi/3 - a)$.
13. $\sin (a + \pi/6) = \cos (\pi/3 - a)$.
14. $\sin (\pi/4 - a) \sin (\pi/4 + a) = \frac{1}{2} \cos 2a$.
15. $3 \sin a + 4 \cos a = 5 \sin (a + b)$, where $\alpha(b) = (\frac{3}{5}, \frac{4}{5})$.
16. $5 \sin a + 12 \cos a = 13 \sin (a + b)$, where $\alpha(b) = (\frac{5}{13}, \frac{12}{13})$.
17. $5 \sin a + 12 \cos a = 13 \cos (a - b)$, where $\alpha(b) = (\frac{12}{13}, \frac{5}{13})$.
18. $(1/\sqrt{2}) \sin a + (1/\sqrt{2}) \cos a = \sin (a + \pi/4)$.
19. $x \sin a + y \cos a = (x^2 + y^2)^{1/2} \sin (a + b)$, where $\alpha(b) = (x/(x^2 + y^2)^{1/2}, y/(x^2 + y^2)^{1/2})$.
20. $(\sin^3 a + \cos^3 a)/(\sin a + \cos a) = 1 - \sin a \cos a$.
21. $(1 + \sin a)/(1 - \sin a) - (1 - \sin a)/(1 + \sin a) = 4 \sin a/\cos^2 a$.
22. $\sin 2a/\cos a = 2 \sin a$.
23. $2 \cos 2a/\sin 2a = \cos a/\sin a - \sin a/\cos a$.

24. $\sin (a + b) \sin (a - b) = \sin^2 a - \sin^2 b.$
25. $\cos (a + b) \cos (a - b) = \cos^2 a - \sin^2 b.$
26. $\sin a = 2 \sin a/2 \cos a/2.$
27. $1/(1 + \sin a) + 1/(1 - \sin a) = 2/\cos^2 a.$
28. $(1 - \cos a)/\sin a = \sin a/(1 + \cos a).$
29. $\sin a = \sin (a - b) \cos b + \cos (a - b) \sin b.$
30. $\sin a/(1 + \cos a) + (1 + \cos a)/\sin a = 2/\sin a.$
31. $\cos^4 a - \sin^4 a = \cos^2 a - \sin^2 a.$
32. $\sin 3a = 3 \sin a - 4 \sin^3 a.$ [Hint: $\sin 3a = \sin (2a + a)$.]
33. $(\cos a - \sin a)^2 + 2 \sin a \cos a = 1.$
34. $(\sin^3 a + \cos^3 a)/(\sin a + \cos a) = 1 - \tfrac{1}{2} \sin 2a.$
35. $\sin^3 a/\sin a - \cos 3a/\cos a = 2.$

4.7 FOUR MORE TRIGONOMETRIC FUNCTIONS

Now that the sine and cosine functions have been discussed, it is possible, and profitable, to look at four other trigonometric functions. They are:

$$\text{tangent} = \frac{\text{sine}}{\text{cosine}} = \frac{p_2 \circ \alpha}{p_1 \circ \alpha}$$

$$\text{cotangent} = \frac{\text{cosine}}{\text{sine}} = \frac{p_1 \circ \alpha}{p_2 \circ \alpha}$$

$$\text{secant} = \frac{1}{\text{cosine}} = \frac{1}{p_1 \circ \alpha}$$

$$\text{cosecant} = \frac{1}{\text{sine}} = \frac{1}{p_2 \circ \alpha}$$

The first problem at hand is to determine a domain and a range for each of the four. Furthermore, it is of interest to know the mechanics of obtaining the image of any given value from the domain.

Let $\alpha(a) = (x, y)$. Then $\cos a = x$ and $\sin a = y$. We conclude:

$$\tan a = \frac{y}{x} \qquad \cot a = \frac{x}{y}$$

$$\sec a = \frac{1}{x} \qquad \csc a = \frac{1}{y}$$

The abbreviations are obvious.

Now, when do each of the above make sense (when are they well defined)? Clearly, y/x makes sense as long as x is different from zero. The next logical question is: when does $x = 0$? The only points (on the unit circle) having first coordinate zero are the points $(0, 1)$ and $(0, -1)$. The points form the images of α under all odd multiples of $\pi/2$. That is to say, the first coordinate value x of a point $(x, y) \in C$ is zero if and only if (x, y) is $\alpha(a)$ for a some odd multiple of $\pi/2$. Thus, we have

$$\text{tangent}: R - \{a : a \text{ is an odd multiple of } \pi/2\} \longrightarrow R$$
$$\text{secant}: R - \{a : a \text{ is an odd multiple of } \pi/2\} \longrightarrow R$$

By a like analysis, $y = 0$ implies that $\alpha(a)$ is $(1, 0)$ or $(-1, 0)$. But $\alpha(a)$ is one of these points if and only if $a = n\pi$ for some integer. Necessarily, the cotangent and cosecant functions are not well defined for such values a.

$$\text{cotangent}: R - \{a : A = n\pi \text{ for some } n \in Z\} \longrightarrow R$$
$$\text{cosecant}: R - \{a : A = n\pi \text{ for some } n \in Z\} \longrightarrow R$$

It can be shown that the images of the domains of the tangent, cotangent, secant, and cosecant are, respectively, R, R, $R - A$, and $R - A$, where $A = \{x \in R : -1 < x < 1\}$.

Example 4.7.1 An illustration that the image of the secant function is $R - A$ will be given.

First, $\sec a$ cannot be in A (that is, $-1 < \sec a < 1$ is impossible). Let $\alpha(a) = (x, y)$, $x = 0$. Then $\sec a = 1/x$. Since $-1 \leqslant x \leqslant 1$, $|x| \leqslant 1$. Then, $|\sec a| = |1/x| = 1/|x| \geqslant 1$. The inequality $-1 < \sec a < 1$ is equivalent to the inequality $|\sec a| < 1$, and, since the two inequalities are contradictory, $R - A$ certainly contains the range of the secant.

Let $a \in R - A$. Then $|a| \geqslant 1$ and $|1/a| \leqslant 1$. We can show (and the reader should show) that $((1/a, 1 - 1/a^2)^{1/2})$ is a point on the unit circle. Necessarily, there is an $x \in R$ with $\alpha(x)$ this given point. Consequently, $\sec x = 1/(1/a) = a$, and a is an element of the image of the secant. This argument illustrates that all elements of R except those in A are image elements under this trigonometric function. Thus, the range of the secant is $R - A$.

The mechanics of finding values of images under each of the new functions has actually been pointed out. For instance, if $\alpha(a) = (\tfrac{3}{5}, \tfrac{4}{5})$, then $\sin(a) = \tfrac{4}{5}$ and $\cos a = \tfrac{3}{4}$ giving:

$$\tan a = (\tfrac{4}{5})/(\tfrac{3}{5}) = \tfrac{4}{3}$$
$$\cot a = (\tfrac{3}{5})/(\tfrac{4}{5}) = \tfrac{3}{4}$$
$$\sec a = 1/(\tfrac{3}{5}) = \tfrac{5}{3}$$
$$\csc a = 1/(\tfrac{4}{5}) = \tfrac{5}{4}$$

The secant and cosecant are obviously seen as the reciprocals of the cosine and sine functions, respectively. In other words, when each makes sense

$$\cos a \sec a = 1$$

and

$$\sin a \csc a = 1$$

Furthermore, it is noticeable that the tangent and cotangent share this same relationship. That is,

$$\tan a \cot a = 1$$

Many other relationships exist in the form of identities. (Again, it must be kept in mind that identities must be valid for all values of the variables where each individual expression makes sense.) For instance,

$$\tan (a + b) = \frac{\sin (a + b)}{\cos (a + b)} = \frac{\sin a \cos b + \cos a \sin b}{\cos a \cos b - \sin a \sin b}$$

$$= \frac{(\sin a \cos b)/(\cos a \cos b) + (\cos a \sin b)/(\cos a \cos b)}{(\cos a \cos b)/(\cos a \cos b) - (\sin a \sin b)/(\cos a \cos b)}$$

$$\tan (a + b) = \frac{\tan a + \tan b}{1 - \tan a \tan b}$$

In general, division is not a reversible operation. How do we know we did not divide the numerator and the denominator by zero? The answer is, we do not know.

If $\cos a \cos b$ is zero, then either $\cos a = 0$ or $\cos b = 0$, which is to say that either a or b is an odd multiple of $\pi/2$. Necessarily, either $\tan a$ or $\tan b$ fails to be defined. The above relationship holds for all a and b *as long as neither is an odd multiple of $\pi/2$.* The expression is seen to be an identity.

It should not take you long to convince yourself that 2π is a period for each of the new trigonometric functions. It is also the primitive period for the secant and cosecant. However,

$$\tan (a + \pi) = \frac{\sin (a + \pi)}{\cos (a + \pi)} = \frac{-\sin a}{-\cos a}$$

$$= \frac{\sin a}{\cos a} = \tan a$$

Likewise, $\cot (a + \pi) = \cot a$.

Because of the property just displayed, π is seen to be a period for the tangent and cotangent functions. The number π is also the primitive period for both. No proof of this is offered.

Graphs of the four new trigonometric functions are given in Figs. 4.7.1, 4.7.2, 4.7.3, and 4.7.4 with a comparison of the mutually reciprocal functions shown in Figs. 4.7.5, 4.7.6, and 4.7.7. The fact that the sine and cosine functions "differ by $\pi/2$" [that is, $\sin (a + \pi/2) = \cos a$ and $\cos (a - \pi/2) = \sin a$] is seen in Fig. 4.7.8.

Still other identities may be developed. The identity for $\tan (a + b)$ yields an identity for $\tan 2a$ (a *double argument formula*).

$$\tan 2a = \tan (a + a) = \frac{\tan a + \tan a}{1 - \tan a \tan a}$$

$$\tan 2a = \frac{2 \tan a}{1 - \tan^2 a}$$

Furthermore,

$$\tan \frac{a}{2} = \frac{\sin a/2}{\cos a/2}$$

$$= \pm\{[(1 - \cos a)/2]/[(1 + \cos a)/2]\}^{1/2}$$

$$= \pm[(1 - \cos a)/(1 + \cos a)]^{1/2}$$

As implied by their names, tangent and cotangent are cofunctions. Let a and b be complementary numbers. Then $\sin a = \cos b$ and $\cos a = \sin b$. Consequently,

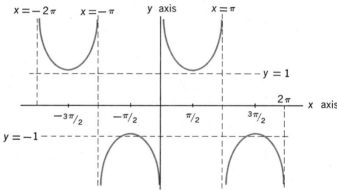

Figure 4.7.1
A graph of cosecant

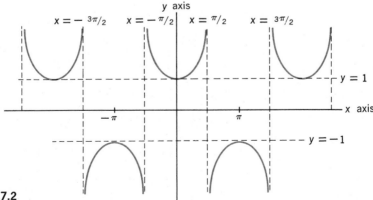

Figure 4.7.2
A graph of secant

Figure 4.7.3
A graph of tangent

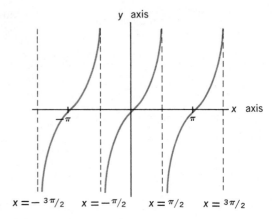

Figure 4.7.4
A graph of cotangent

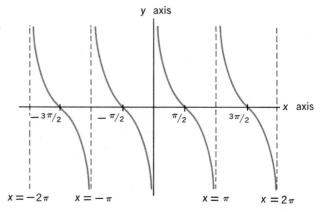

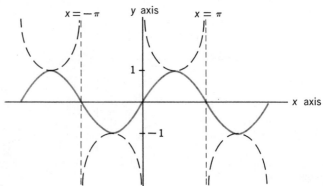

Figure 4.7.5
Sine versus cosecant

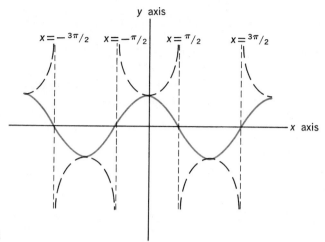

Figure 4.7.6
Cosine versus secant

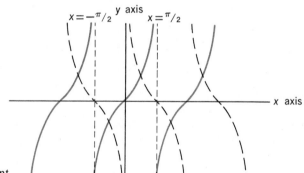

Figure 4.7.7
Tangent versus cotangent

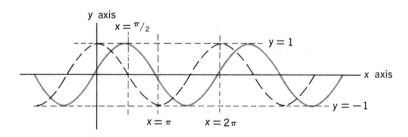

Figure 4.7.8
Sine versus cosine

$$\tan a = \frac{\sin a}{\cos a} = \frac{\cos b}{\sin b} = \cot b$$

and
$$\tan b = \frac{\sin b}{\cos b} = \frac{\cos a}{\sin a} = \cot a$$

Similarly, the secant and cosecant are cofunctions. The exercises ask you to show many more relationships.

■ EXERCISES 4.7

1. Fill in the following table.

	0	$\frac{\pi}{6}$	$\frac{\pi}{4}$	$\frac{\pi}{3}$	$\frac{\pi}{2}$	$\frac{2\pi}{3}$	$\frac{3\pi}{4}$	$\frac{5\pi}{6}$	π	$\frac{7\pi}{6}$	$\frac{5\pi}{4}$	$\frac{4\pi}{3}$	$\frac{3\pi}{2}$	$\frac{5\pi}{3}$	$\frac{7\pi}{4}$	$\frac{11\pi}{6}$	2π
sin																	
cos																		
tan																		
cot																		
sec																		
csc																		

2. Give the polarity of $\tan a$, $\cot a$, $\sec a$, and $\csc a$, where $\alpha(a)$ is in quadrant I, II, III, and IV.

3. Develop an identity for $\cot (a + b)$ in a manner analogous to that for $\tan (a + b)$.

4. Give a "double argument formula" for the cotangent.

5. Develop a "half argument formula" for the cotangent.

6. Show that the secant and cosecant are cofunctions.

7. Verify that the following are identities.

    a. $\tan (-a) = -\tan a$.   **c.** $\sec (-a) = \sec a$.

    b. $\cot (-a) = -\cot a$.   **d.** $\csc (-a) = -\csc a$.

8. Develop identity for $\tan (a - b)$ in terms of $\tan a$ and $\tan b$.

9. Let n be in Z and prove that the following are identities.

    a. $\tan (a + n\pi) = \tan a$.

    b. $\cot (a + n\pi) = \cot a$.

    c. $\tan (\pi/2 + a) = -\cot a$.

    d. $\cot (\pi/2 + a) = -\tan a$.

10. Verify the following identities.

   a. $\sec(\pi/2 + a) = -\csc a$.
   b. $\csc(\pi/2 + a) = -\sec a$.
   c. $\tan a + \cot a = \sec a \csc a$.
   d. $1 + \tan^2 a = \sec^2 a$.
   e. $1 + \cot^2 a = \csc^2 a$.
   f. $(\tan x + 1)/(\sin x + \cos x) = \sec x$.
   g. $\csc a - \cos a \cot a = \sin a$.
   h. $\cot a/(1 + \cot^2 a) = \sin a \cos a$.
   i. $(\sin t + \sin q)/(\cot t - \cos q) = -\cot \tfrac{1}{2}(t - q)$.
   j. $(\sin x - \sin y)/(\cos x + \cos y) = \tan \tfrac{1}{2}(x - y)$.
   k. $(\cos 6x + \cos 4x)/(\sin 6x + \sin 4x) = \cot x$.
   l. $[\sin(2x - y) + \sin y]/[\cos(2x - y) + \cos y] = \tan x$.
   m. $\sin t - \sin 3t = 2 \sin^3 t - 2 \sin t \cos^2 t$.
   n. $(\tan a - \cot a)/(\tan a + \cot a) = 2 \sin^2 a - 1$.
   o. $\cos^2 a - \sin^2 a = (1 - \tan^2 a)/(1 + \tan^2 a)$.
   p. $\cot a + \tan a = \cot a \sec^2 a$.
   q. $(\sec a - \tan a)^2 = (1 - \sin a)/(1 + \sin a)$.
   r. $(1 + \csc a)/(\csc a - 1) = (1 + \sin a)/(1 - \sin a)$.

11. Prove the following identities.

   a. $\sin a/(1 + \cos a) = \csc a - \cot a$.
   b. $2 \cot 2a = \cot a - \tan a$.
   c. $\sin a/(1 + \cos a) + (1 + \cos a)/\sin a = 2 \csc a$.

4.8 REVIEW EXERCISES

   1. If a circle has radius r and one of its arcs is ⅛ the circle, what is the associated length of the arc? If the arc is ⅙ of the circle? If the arc is ¹⁄₁₂ of the circle?

   2. Answer the questions of Exercise 1 where $r = 1$. Where $r = 5$.

   3. What is the length of the arc (counterclockwise traversal) from $(1, 0)$ to each indicated point?

   a. $(\sqrt{3}/2, \frac{1}{2})$.
   b. $(1/\sqrt{2}, 1/\sqrt{2})$.
   c. $(\frac{1}{2}, \sqrt{3}/2)$.
   d. $(0, 1)$.
   e. $(-\frac{1}{2}, \sqrt{3}/2)$.
   f. $(-1/\sqrt{2}, 1/\sqrt{2})$.
   g. $(-\sqrt{3}/2, \frac{1}{2})$.
   h. $(-1, 0)$.

   i. $(-\sqrt{3}/2, -\frac{1}{2})$.
   j. $(-1/\sqrt{2}, -1/\sqrt{2})$.
   k. $(-\frac{1}{2}, -\sqrt{3}/2)$.
   l. $(0, -1)$.
   m. $(\frac{1}{2}, -\sqrt{3}/2)$.
   n. $(1/\sqrt{2}, -1/\sqrt{2})$.
   o. $(\sqrt{3}/2, -\frac{1}{2})$.

4. In which quadrant is each of the following?

a. $\alpha(1)$.

b. $\alpha(3)$.

c. $\alpha(5)$.

d. $\alpha(7)$.

e. $\alpha(9)$.

f. $\alpha(-1)$.

g. $\alpha(-3)$.

h. $\alpha(-5)$.

i. $\alpha(-7)$.

j. $\alpha(-9)$.

k. $\alpha(16)$.

l. $\alpha(-3\pi/8)$.

m. $\alpha(27\pi/10)$.

n. $\alpha(-3\pi/5)$.

o. $\alpha(\frac{1}{2})$.

p. $\alpha(-\frac{1}{3})$.

q. $\alpha(2095\pi/3)$.

r. $\alpha(34\pi/7)$.

s. $\alpha(-2000)$.

5. Find each of the following points.

a. $\alpha(2700\pi)$.

b. $\alpha(-51\pi)$.

c. $\alpha(\pi/3)$.

d. $\alpha(2702\pi/3)$.

e. $\alpha(195\pi/4)$.

f. $\alpha(-85\pi/5)$.

g. $\alpha(95\pi/6)$.

h. $\alpha(1191\pi/6)$.

i. $\alpha(75\pi/2)$.

j. $\alpha(-3491\pi/2)$.

6. Find the values of sine and cosine at the values of Exercise 5.

7. Find the values of secant, cosecant, tangent, and cotangent of each value in Exercise 5.

8. Given that $\sin 1 = .84$, find the following.

a. cos 1.

b. sec 1.

c. csc 1.

d. tan 1.

e. cot 1.

f. sin 2.

g. cos 2.

h. tan 2.

i. sin ½.

j. cos ½.

k. tan ½.

l. sin 3.

m. cos 3.

n. tan 3.

o. cot 3.

p. sin ³⁄₂.

q. cos ⁵⁄₂.

r. tan ⁷⁄₂.

s. cos (-1).

t. sin (-1).

u. tan (-1).

v. $\alpha(1)$.

w. $\alpha(-1)$.

x. $\alpha(2)$.

y. $\alpha(\frac{1}{2})$.

9. Show that for $0 < a < \pi/2$, $0 < \sin a < \tan a$.

10. Show that the following are identities.

a. $\sin a \cot a = \cos a$.

b. $\cos a \tan a = \sin a$.

c. $\tan a \csc a = \sec a$.

d. $\cot a \sec a = \csc a$.

e. $\sec a / \csc a = \tan a$.

f. $\csc a / \sec a = \cot a$.

g. $(\sin a + \cos a)/(\sec a + \csc a) = \sin a \cos a$.

h. $\cot a + \csc a + 2 = (1 + \sin a) + \cot a(1 + \cos a)$.

i. $2 \sin a + \sec a = \sec a(\sin a + \cos a)^2$.

j. $\frac{1}{2} \csc a \sec a = \csc 2a$.

k. $\sin 4a = 8 \cos^3 a \sin a - 4 \cos a \sin a$.

l. $\cos 4a = \cos^4 a - 8 \cos^2 a + 1$.

m. $(\tan a - \sin a)/(\cos a) + (\sin a \cos a)/(1 + \cos a)$
$$= (\tan a)/(1 + \cos a).$$

n. $\dfrac{\sec a + \tan a}{\cos a - \tan a - \sec a} = -\csc a$.

o. $\sin a + \cos a + \sin a/\cot a = \sec a + \csc a - \cos a/\tan a$.

p. $\tan a/2 = \sin a/(1 + \cos a)$.

q. $\tan a/2 = \csc a - \cot a$.

r. $(1 - \cos 2a)/(\sin 2a) = \tan a$.

s. $\sin 2a/\sin a = \sec a + (\cos 2a/\cos a)$.

t. $(\sin 3a)/(\sin a) - (\cos 3a)/(\cos a) = 2$.

u. $(\sin 5a - \sin 3a)/(\cos 5a - \cos 3a) = \tan a$.

11. Write the following products as sums and the sums as products.

a. $\sin 6a + \sin 10a$.

b. $\sin 6a + \cos 10a$.

c. $\cos 6a + \cos 10a$.

d. $\sin (x + h) - \sin x$.

e. $\sin 10a \sin 6a$.

f. $\sin 10a \cos 6a$.

g. $\cos 10a \cos 6a$.

h. $\cos (x - h) - \cos x$.

12. Write each of the following in the form $(r \cos a, r \sin a)$. (See Exercise 4.3.7.)

a. All those points of Exercise 4.1.1.

b. All those points of Exercise 4.1.3.

c. $(1, 1)$.

d. $(-1, -1)$.

e. (x, x), where $x > 0$.

f. (x, x), where $x < 0$.

g. $(x, -x)$, where $x > 0$.

h. $(x, -x)$, where $x < 0$.

i. $(x, \sqrt{3}x)$, $x > 0$.

j. $(x, \sqrt{3}x)$, $x < 0$.

k. $(\sqrt{3}x, x)$, $x > 0$.

l. $(x, -\sqrt{3}x)$, $x < 0$.

m. $(x, -\sqrt{3}x)$, $x > 0$.

n. $(\sqrt{3}x, -x)$, $x < 0$.

o. $(\sqrt{3}x, -x)$, $x > 0$.

p. $(x, 0)$, $x < 0$.

q. $(x, 0)$, $x > 0$.

r. $(0, x)$, $x < 0$.

s. $(0, x)$, $x > 0$.

13. Show that $x \sin a + y \cos a = (x^2 + y^2)^{1/2} \sin (a + b)$, where $\alpha(b) = [x/(x^2 + y^2)^{1/2}, y/(x^2 + y^2)^{1/2}]$.

4.9 ADVANCED EXERCISES

1. $\alpha(\pi/8)$ can be calculated from half angle formulas for the sine and cosine functions. However, knowing that equal chords determine equal angles, find $\alpha(\pi/8)$ by the following methods.

   a. Write the equation of the line ℓ passing through $(1, 0)$ and $(1/\sqrt{2}, 1/\sqrt{2})$.

   b. Find the equation of ℓ', the perpendicular bisector of that segment of ℓ from $(1, 0)$ to $(1/\sqrt{2}, 1/\sqrt{2})$.

   c. Find the point of intersection of ℓ' and C.

Calculate $\alpha(\pi/8)$ by this method.

2. In Fig. 4.9.1, we see that a finite number of points of the circle $x^2 + y^2 = r^2$ are chosen and that the line segment connecting "successive" points form an inscribed polygon. The perimeter of the polygon is the sum of the lengths of its sides. Define:

$$S = \{p \in R : p \text{ is the perimeter of a polygon}$$
$$\text{inscribed in the circle } x^2 + y^2 = r^2\}$$

Now, S is the set formed of the perimeters of all polygons inscribed in our circle. Show that S is bounded above (Exercise 1.5.5) by:

   a. showing that each polygon as in Fig. 4.9.1 has perimeter not greater than the associated polygon in Fig. 4.9.2, and

   b. each polygon in Fig. 4.9.2 has perimeter less than the square in Fig. 4.9.3.

Let C be the l.u.b. of S and define $C/2r = \pi$, whence we have the equation $C = 2\pi r$. Derive an analogous technique for determining arc length and the equation $s = r\theta$.

3. If $f : R \longrightarrow Y$ has periods p and q, then $p + q$, np, and $np + mq$ are also periods for f with $n, m \in Z$. $p + q$ was shown to be a period. If we

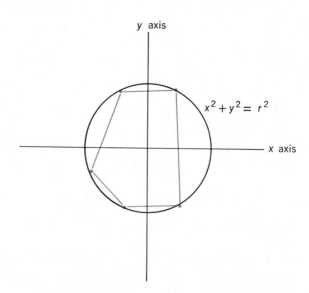

Figure 4.9.1

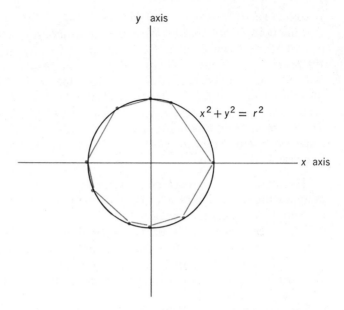

Figure 4.9.2

can show np to be a period also, we know that mq is a period as well, and $np + mq$ is necessarily a period. Show, then, that np is a period for f. [Hint: The theorem is obvious for $n = 0$. Use induction to show that np is a period for f where $n \in N$. It follows that $-np$ is a period, since $f(x - np) = f((x - np) + np) = f(x)$. In particular, since 2π is a period, the exercise shows that $\alpha(x + 2n\pi) = \alpha(x)$ (that is, $2n\pi$ is a period for α if $n \in Z$).]

4. Exercise 4.3.7 showed that all points could be written in the form

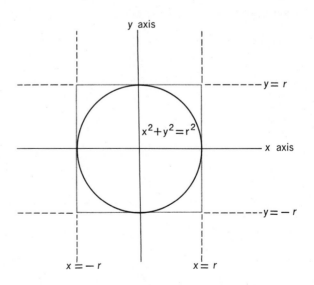

Figure 4.9.3

$(r \cos a, r \sin a)$ for some $r > 0$ and some $a \in [0, 2\pi)$, provided, of course, that the point is not $(0, 0)$. We can then symbolize the "location" of such a point by specifying r and a. Such a manner of depicting points is known as the *polar form*. That is, $(r \cos a, r \sin a)$ can be written as $(r, a)^*$. The ordered pair notation here is different from the original *rectangular system* introduced in Chapter 2. The first element r denotes distance from the origin, while a determines a ray starting at $(0, 0)$ and passing through $\alpha(a)$. The intersection of this half line and the circle $x^2 + y^2 = r^2$ determines a single point (the point in question). Write in polar form those points in Exercise 4.8.12.

[Note: Polar form may be more general than this description. For example, the restriction of a to $[0, 2\pi)$ need not hold. Some do not prefer to restrict r to being nonnegative.]

5. Let (a, b) and (c, d) be two open intervals. Show that $f(x) = ((d - c)/(b - a))(x - a) + c$ is a $1:1$ and an onto function $f: (a, b) \longrightarrow (c, d)$. Figure 4.9.4 graphically illustrates the geometric interpretation. This result shows that any two open intervals of real numbers are equivalent, that is, $(a, b) \sim (c, d)$.

6. Show that $\tan[(-\pi/2, \pi/2)] = R$ and that $\tan|(-\pi/2, \pi/2)$ is $1:1$. In particular, this shows that $(-\pi/2, \pi/2) \sim R$. [Hint: Let r be any real number. Then, let $x = 1/(1 + r^2)$ and $y = \pm(1 - x^2)^{1/2}$, the $+$ or $-$ agreeing with whether $r > 0$ or $r < 0$. Show that $(x, y) \in C$ and is on the arc from $(0, -1)$ to $(0, 1)$ when traveling in a counterclockwise direction. Then show that, if $\alpha(a) = (x, y)$, $-\pi/2 < a < \pi/2$ and $\tan a = r$, then

Figure 4.9.4

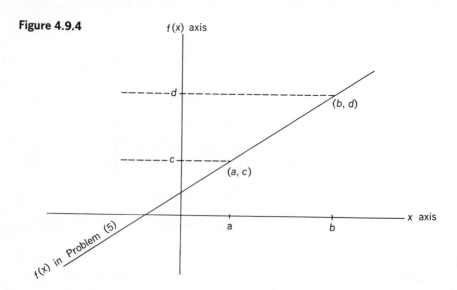

$x = [1/(1 + r^2)]^{1/2}$ and $y = \pm(1 - x^2)^{1/2}$. This latter shows $\tan|(-\pi/2, \pi/2)$ is $1:1$, while the former argument shows that $\tan[(-\pi/2, \pi/2)] = R$.]

7. Putting Exercises 5 and 6 together, show that, if (a, b) is any open interval (a, b) of real numbers, $(a, b) \sim R$.

8. Verify the following.

    a. $\sin|[-\pi/2, \pi/2]:[-\pi/2, \pi/2] \longrightarrow [-1, 1]$ is $1:1$ and an onto function.

    b. $\cos|[0, \pi]:[0, \pi] \longrightarrow [-1, 1]$ is $1:1$ and an onto function.

    c. $\cot|(0, \pi):(0, \pi) \longrightarrow R$ is $1:1$ and an onto function.

[Hint: It is necessary to show only $\tan(a + \pi/2) = \cot a$.]

Chapter 5
Inverse
Trigonometric
Functions

5.1 SETS DETERMINED BY TRIGONOMETRIC FUNCTIONS

Various equations involving trigonometric functions have been encountered throughout the material presented in the text. These equations have, for the most part, been identities rather than conditional equations. Since conditional trigonometric equations have much value in mathematics, we will take time to examine a few.

One simple example of a conditional trigonometric equation is given by:

$$\sin x = k, \, k \in R$$

Example 5.1.1 Find the solution set for:

$$\sin x = \tfrac{1}{2}$$

You should recognize by inspection that $x = \pi/6$ is a possible solution to the latter equation. The value $5\pi/6$ is another candidate. These two numbers represent the only values in $[0, 2\pi)$ satisfying the conditions. (Why?) However, $\pi/6 + 2n\pi$ and $5\pi/6 + 2n\pi$ are also solutions for n

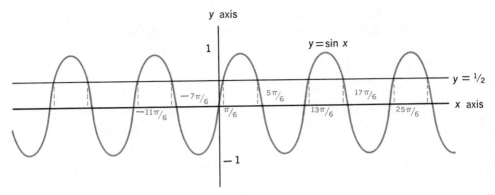

Figure 5.1.1
$\{x : \sin x = \frac{1}{2}\}$

any integer (2π is a period for the sine function). The set of solutions
(solution set) for the equation $\sin x = \frac{1}{2}$ is

$$\begin{aligned}
\{x : \sin x = \tfrac{1}{2}\} \\
= \{x : x = \pi/6 + 2n\pi \text{ or } 5\pi/6 + 2n\pi, n \in Z\} \\
= \{x : x = n\pi + (-1)^n \pi/6 \text{ for } n \in Z\}
\end{aligned}$$

Figure 5.1.1 shows a geometric interpretation of this solution. See
Exercise 4.3.3 for more of this kind of analysis.

Example 5.1.2   Find the set of solutions for the equation $\cos x = 3$.
This problem is easily answered. We recall from a previous discussion
the fact that $-1 \leqslant \cos x \leqslant 1$. Thus, $\cos x = 3$ is impossible and there
exist no solutions. Consequently, the solution set is seen to be $\varnothing$.
We shall introduce at this time some new symbolism. The set **arcsin** k
($k \in R$) is given by

$$\textbf{arcsin } k = \{x \in R : \sin x = k\}$$

The symbol **arcsin** k is actually an abbreviation of arcsine k. The pronun-
ciation is obvious. We will read the statement "$x \in$ **arcsin** k" in the following way: x is a number whose image under the sine function is k. This
may be shortened (by an abuse of the language) to "x is a number whose
sine is k."
In a similar manner, we define:

$$\begin{aligned}
\text{arccosine } k &= \textbf{arccos } k = \{x : \cos x = k\} \\
\text{arctangent } k &= \textbf{arctan } k = \{x : \tan x = k\} \\
\text{arcsecant } k &= \textbf{arcsec } k = \{x : \sec x = k\} \\
\text{arccosecant } k &= \textbf{arccsc } k = \{x : \csc x = k\} \\
\text{arccotangent } k &= \textbf{arccot } k = \{x : \cot x = k\}
\end{aligned}$$

We could have defined arcsine instead of **arcsin** k. This idea lends itself to the functional notation (for example, compare the meaning of "sine" as opposed to "sin a"). Our concept of **arcsin** k would not be greatly altered if we were to consider

$$\textbf{arcsin } k = \{(k, x): \sin x = k\}$$

We simply are taking our original set **arcsin** k and making it into a collection of ordered pairs, each having first element k. Then define:

$$\text{arcsine} = \bigcup_{k \in R} (\text{arcsin } k)$$

the union of all sets **arcsin** k. The set arcsine is a collection of ordered pairs of real numbers. That is, arcsine $\subset R \times R$. Moreover, arcsine can be formed by reversing the order of the elements of the sine function. We might say that arcsine is the *inverse* of the sine relation (function). It would be nice if arcsine were a function, but unfortunately it is not.

Now examine some further trigonometric equations.

Example 5.1.3   Solve the equation

$$2 + 3 \sin x = 6 - \sin x$$

Adding $\sin x$ to both sides and subtracting 2, we reveal an alternative form of the equation to be

$$4 \sin x = 4$$

Division of both sides by 4 yields

$$\sin x = 1$$

The indicated solution set is therefore **arcsin** 1. The (expected) solution set can also be described as $\{x : x = \pi/2 + 2n\pi, n \in Z\}$. Whether or not this is the solution set must be checked against the original equation.

To verify that each element of **arcsin** 1 is a solution of the given equation, let us suppose that $x \in$ **arcsin** 1. Does x satisfy

$$2 + 3 \sin x = 6 - \sin x?$$

Seeing that $\sin x = 1$, we realize that the left expression of the original equation becomes $2 + 3 = 5$ while the right is $6 - 1 = 5$. Necessarily, each element of the set **arcsin** 1 is a solution of (satisfies) the stated equation.

Example 5.1.4   Find the solution set for

$$\cos^2 x - \sin x = -1$$

Since $\cos^2 x = 1 - \sin^2 x$, the original equation becomes

$$1 - \sin^2 x - \sin x = -1$$

or
$$\sin^2 x + \sin x - 2 = 0$$

The left-hand side of the latter form is a *quadratic* equation in the variable $\sin x$. Substituting $y = \sin x$, we observe that the equation takes on the appearance of the algebraic equation $y^2 + y - 2 = 0$, which in turn factors into the form

$$(y + 2)(y - 1) = 0$$

The trigonometric equation then must factor likewise into

$$(\sin x + 2)(\sin x - 1) = 0$$

Now, if $a \cdot b = 0$, $a = 0$ or $b = 0$. As a result, we conclude from above that $\sin x + 2 = 0$ or $\sin x - 1 = 0$. This is to say, $\sin x = -2$ or $\sin x = 1$. However, $\sin x = -2$ is not possible, since $-1 \leqslant \sin x \leqslant 1$. Thus, the only possible solutions arise from the equation $\sin x = 1$. Consequently, the supposed solution set is **arcsin 1**.

To investigate this supposed solution, note that, if $\sin x = 1$, then $\cos x = \pm(1 - \sin^2 x)^{1/2} = 0$. Substituting these values into the original form of the equation, we have

$$\cos^2 x - \sin x = 0 - 1 = -1$$

Hence, the solution set is as predicted.

Example 5.1.5 Solve the equation

$$\tan x \sin x - \tan x = 0$$

Since $\tan x$ is a factor of each expression on the left-hand side, we can "factor it out" to obtain:

$$\tan x(\sin x - 1) = 0$$

What conclusion remains?

$$\tan x = 0 \quad \text{or} \quad \sin x - 1 = 0$$

The latter result is to say that $\sin x = 1$. Remembering that **arcsin 1** $= \{x : x = \pi/2 + 2n\pi$ for $n \in Z\}$, we remark that for $x \in$ **arcsin 1**, $\tan x$ is not defined (review again the domain of the tangent function). It is then clear that *no* member of **arcsin 1** can be a solution for $\tan x \sin x - \tan x = 0$.

Consider **arctan 0**. If $x \in$ **arctan 0**, is x a solution of our equation? Given that $\tan x = 0$, $\tan x = \sin x/\cos x$ implies that $\sin x = 0$ ($\cos x \neq 0$). Thus, $\tan x \sin x - \tan x = 0 \cdot 0 - 0 = 0$. Clearly, the solution set is **arctan 0**.

The last problem emphasizes the need for checking tentative results to see whether or not they represent genuine solutions.

Example 5.1.6   Compute the solution set for $5 \sin x + 12 \cos x = 13$. By a previous identity, $5 \sin x + 12 \cos x = 13 \sin (x + a)$ where $\sin a = \frac{12}{13}$ and $\cos a = \frac{5}{13}$. The equation now becomes,

$$13 \sin (x + a) = 13$$

or

$$\sin (x + a) = 1$$

giving

$$x + a \in \textbf{arcsin } 1$$

The latter statement implies that $x + a = \pi/2 + 2n\pi$ for some $n \in Z$. Alternatively, $x = (\pi/2 - a) + 2n\pi$. The proposed solution set $\{x:x = (\pi/2 - a) + 2n\pi, n \in Z, \sin a = \frac{12}{13}, \cos a = \frac{5}{13}\}$ should be verified by check.

Example 5.1.7   As a final example, find the solution set for

$$\sin^2 2x = 1$$

The equation is similar to the algebraic equation $y^2 = 1$, having roots 1 and -1. Thus, $2x \in \textbf{arcsin } 1 \cup \textbf{arcsin } (-1) = \{a:a = (2n + 1)\pi/2 \text{ for } n \in Z\}$. Since $2x = (2n + 1)\pi/2$, $x = (2n + 1)\pi/4$. This is merely a means of describing the set of all *odd multiples* of $\pi/4$. Alternatively, the set is given as $(\textbf{arcsin } 1/\sqrt{2}) \cup (\textbf{arcsin } (-1/\sqrt{2}))$.

The solution set, like that of the previous example, needs verification.

■ EXERCISES 5.1

1. Describe each of the following sets in terms of the special values $0, \pi/6, \pi/4$, and so forth.

   a. arcsin 0.
   b. arcsin $\frac{1}{2}$.
   c. arcsin $(-\frac{1}{2})$.
   d. arcsin $(\sqrt{3}/2)$.
   e. arcsin $(-\sqrt{3}/2)$.

   f. arcsin 1.
   g. arcsin (-1).
   h. arcsin $(1/\sqrt{2})$.
   i. arcsin $(-1/\sqrt{2})$.

2. Redo Exercise 1 with arccos replacing arcsin.
3. Follow the instruction of Exercise 1.

   a. arctan 0.
   b. arctan 1.
   c. arctan (-1).
   d. arctan $(\sqrt{3})$.

   e. arctan $(-\sqrt{3})$.
   f. arctan $(1/\sqrt{3})$.
   g. arctan $(-1/\sqrt{3})$.

4. Do Exercise 3 with arccot replacing arctan.
5. Follow the instruction of Exercise 1.

   a. arcsec 1.
   b. arcsec (-1).

   c. arcsec $(\sqrt{2})$.
   d. arcsec $(-\sqrt{2})$.

e. arcsec $(2/\sqrt{3})$.   **g.** arcsec 2.
f. arcsec $(-2/\sqrt{3})$.   **h.** arcsec (-2).

6. Do Exercise 5 with arccsc replacing arcsec.
7. Describe each of the sets below (in terms of special values).

  a. arcsin 1 $\cap$ arccos 0.
  b. arcsin ½ $\cap$ arccos $\sqrt{3}/2$.
  c. arcsin 0 $\cap$ arccot 0.
  d. arccos $(-1/\sqrt{2})$ $\cap$ arctan (-1).
  e. arctan $(-\sqrt{3})$ $\cap$ arcsin $\sqrt{3}/2$.
  f. arcsin $\sqrt{3}/2$ $\cup$ arcsin $(-\sqrt{3}/2)$.
  g. arccos $1/\sqrt{2}$ $\cup$ arccos $(-1/\sqrt{2})$.
  h. arctan 1 $\cup$ arctan (-1).
  i. arctan 0 $\cup$ arccos 0.

8. List the elements in each set.

  a. arcsin 0 $\cap$ $[-\pi/2, \pi/2]$.
  b. arcsin $1/\sqrt{2}$ $\cap$ $[-\pi/2, \pi/2]$.
  c. arcsin ½ $\cap$ $[-\pi/2, \pi/2]$.
  d. arcsin $\sqrt{3}/2$ $\cap$ $[-\pi/2, \pi/2]$.
  e. arcsin 1 $\cap$ $[-\pi/2, \pi/2]$.
  f. arcsin (-1) $\cap$ $[-\pi/2, \pi/2]$.
  g. arcsin $(-½)$ $\cap$ $[-\pi/2, \pi/2]$.
  h. arcsin $(-\sqrt{3}/2)$ $\cap$ $[-\pi/2, \pi/2]$.
  i. arcsin $(-1/\sqrt{2})$ $\cap$ $[-\pi/2, \pi/2]$.
  j. arccos 0 $\cap$ $[0, \pi]$.
  k. arccos $1/\sqrt{2}$ $\cap$ $[0, \pi]$.
  l. arccos ½ $\cap$ $[0, \pi]$.
  m. arccos $\sqrt{3}/2$ $\cap$ $[0, \pi]$.
  n. arccos 1 $\cap$ $[0, \pi]$.
  o. arccos (-1) $\cap$ $[0, \pi]$.
  p. arccos $(-½)$ $\cap$ $[0, \pi]$.
  q. arccos $(-\sqrt{3}/2)$ $\cap$ $[0, \pi]$.
  r. arccos $(-1/\sqrt{2})$ $\cap$ $[0, \pi]$.
  s. arctan 0 $\cap$ $(-\pi/2, \pi/2)$.
  t. arctan $1/\sqrt{3}$ $\cap$ $(-\pi/2, \pi/2)$.
  u. arctan 1 $\cap$ $(-\pi/2, \pi/2)$.
  v. arctan $\sqrt{3}$ $\cap$ $(-\pi/2, \pi/2)$.
  w. arctan $(-\sqrt{3})$ $\cap$ $(-\pi/2, \pi/2)$.
  x. arctan (-1) $\cap$ $(-\pi/2, \pi/2)$.
  y. arctan $(-1/\sqrt{3})$ $\cap$ $(-\pi/2, \pi/2)$.

9. Evaluate each of the following.

   a. $\tan a$, $a \in$ **arctan** 3.
   b. $\sin a$, $a \in$ **arcsin** ⅓.
   c. $\cos a$, $a \in$ **arccos** ¼.
   d. $\sin a$, $a \in$ **arccos** ⅗.
   e. $\cos a$, $a \in$ **arcsin** ⁸⁄₁₇.
   f. $\tan a$, $a \in$ **arcsin** ⁵⁄₁₃.
   g. $\cos a$, $a \in$ **arctan** ¾.
   h. $\tan a$, $a \in$ **arccos** 0.

10. Describe each set in terms of special values.

   a. **arcsin** $(\sin \pi/6)$.
   b. **arccos** $(\cos \pi/4)$.
   c. **arctan** $(\tan \pi/3)$.
   d. **arccos** $(\tan \pi/4)$.

11. Solve the following trigonometric equations using the "arc" notation. Where possible, also write the solution set in terms of the special values 0, $\pi/6$, $\pi/4$, and so forth.

   a. $2 \sin \theta = \sqrt{2}$.
   b. $3 \tan \theta = \sqrt{3}$.
   c. $5 \cos \theta = 3\sqrt{3} - \cos \theta$.
   d. $\sin^2 a = 1$.
   e. $\cos^2 a = 1$.
   f. $\tan^2 a = 3$.
   g. $\tan^2 a = 1$.
   h. $\sin a + (\sin^2 a)/(\cos a) = 3/\sqrt{2}$.
   i. $2 \sin^2 a - \sin a - 1 = 0$.
   j. $2 \cos^2 a - \cos a = 0$.
   k. $\sin^2 a - \cos^2 a = 0$.
   l. $\cos a(\sin^2 a - 1) = 0$.
   m. $\tan a = \sin a$.
   n. $\tan a = \cos a$.
   o. $\cos a = \sin a$.
   p. $2 \sin^2 a + \cos a = 2$.
   q. $2 \cos^2 a + \sin a = 1$.
   r. $\sin a \tan a = 0$.
   s. $\sin a \cos a = 2$.
   t. $\tan^2 a + \sin^2 a = \sec a - \cos^2 a$.
   u. $\tan a - \sec^2 a = 1$.
   v. $\sin^2 a + \cos a = 0$.
   w. $\sin^2 a - \sin a - 2 = 1$.
   x. $\sin 2\theta = \cos \theta$.

y. $\tan \theta = \sin 2\theta$.
z. $\sin 2\theta = 1$.
a'. $2 \cos 3\theta = 2$.
b'. $\theta \cos \theta - \theta - \cos \theta + 1 = 0$.
c'. $3 \sin a + 4 \cos a = 5$.

12. Find the possible choices for $\alpha(a)$ where

a. $a \in$ **arcsin** 1.
b. $a \in$ **arccos** $\sqrt{3}/2$.
c. $a \in$ **arctan** $(-\sqrt{3})$.
d. $a \in$ **arccsc** $(-\sqrt{2})$.
e. $a \in$ **arcsec** $(2/\sqrt{3})$.

f. $a \in$ **arcsin** ⅗.
g. $a \in$ **arccos** $-5/13$.
h. $a \in$ **arctan** 8/15.
i. $a \in$ **arcsin** $-7/25$.

5.2 INVERSE TRIGONOMETRIC FUNCTIONS

The "arc" sets given in Section 5.1 show an important characteristic of the trigonometric functions, a property that should have been evident as a result of their periodicity.

Consider, for instance, the sine function. Reiterating, $\sin : R \longrightarrow [-1, 1]$ is an onto function. The sine function relates to each x, the number $\sin x$. Reversing the correspondence does *not* give rise to a function (the sine function is not 1:1). For each $y \in [-1, 1]$, the inverse correspondence relates to y every element of **arcsin** y. Thus, the inverse correspondence is not a function; **arcsin** y is an "infinite" set. Figure 5.1.1 shows that ½, for example, is mated to many real numbers x.

The colored portion of the curve in Fig. 5.2.1 shows that section of the sine function whose domain is restricted to the closed interval $[-\pi/2, \pi/2]$. This portion of the curve is seen to be a graph of $\sin | [-\pi/2, \pi/2]$. Alternatively, this curve segment is a graph of $y = \sin x$, where $x \in [-\pi/2, \pi/2]$.

Every horizontal line (of the form $y = k$) strikes the solid part of the curve *at most once*. If $-1 \leqslant k \leqslant 1$, the line strikes this segment *exactly once*. Such a situation is a geometric indication that the restriction is 1:1.

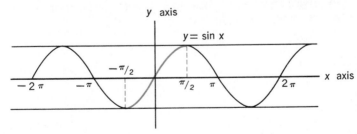

Figure 5.2.1
Sine $| [-\pi/2, \pi/2]$

In addition, the graph also points to the fact that, for the colored portion, $\sin x$ makes an excursion through *all* the values of $[-1, 1]$. Hence, the restriction (restricted function) has an inverse. The inverse will be symbolized by

$$\sin^{-1}: [-1, 1] \longrightarrow [-\pi/2, \pi/2]$$

The correspondence is given by $y \longrightarrow a \in [-\pi/2, \pi/2]$, where $\sin a = y$ (that is, $\sin^{-1} y = a$, where $\sin a = y$).

The inverse mating may also be displayed in the following manner.

$$\sin^{-1} y \in [-\pi/2, \pi/2] \qquad \text{and} \qquad \sin(\sin^{-1} y) = y$$

Furthermore, $\sin^{-1} y \in \textbf{arcsin } y \cap [-\pi/2, \pi/2]$.

The symbolism $\sin^{-1} k$ is not the (-1) power (reciprocal) of the $\sin k$. The symbolism to be used for an exponent of (-1) is $1/\sin k = (\sin k)^{-1}$.

The **inverse sine function** $(\sin^{-1})$ is really the inverse of the given restriction and not of the sine function itself.

Figure 5.2.2 shows a graph of $x = \sin y$ with $y \in [-\pi/2, \pi/2]$. Figure 5.2.1 shows $y = \sin x$. The two figures differ only because x and y are interchanged. Figure 5.2.3 illustrates a graph of $y = \sin^{-1} x$. We note that Figs. 5.2.2 and 5.2.3 are identical. This can be shown to be correct from a set theoretic analysis.

Example 5.2.1 As examples of the $\sin^{-1}$ correspondence, compute: (1) $\sin^{-1} \frac{1}{2}$, (2) $\sin^{-1} (-\frac{1}{2})$, and (3) $\sin^{-1} 1$.

By definition, $\sin^{-1} \frac{1}{2} \in \textbf{arcsin } \frac{1}{2}$ and $\sin^{-1} \frac{1}{2} \in [-\pi/2, \pi/2]$. Elements of **arcsin** $\frac{1}{2}$ are of the form $n\pi + (-1)^n \pi/6$. (See Fig. 5.1.1.) The only element from this collection belonging to $[-\pi/2, \pi/2]$ is $\pi/6$. Thus, $\sin^{-1} \frac{1}{2} = \pi/6$.

We cannot overemphasize the fact that $\sin^{-1} \frac{1}{2}$ is a real number, not a set.

Now, **arcsin** $(-\frac{1}{2})$ includes the numbers $7\pi/6$, $11\pi/6$, $19\pi/6$, $\cdots$ and none of these is in $[-\pi/2, \pi/2]$. However, $-\pi/6$, $-5\pi/6$, $-13\pi/6$, $-17\pi/6$, $\cdots$ are also in **arcsin** $(-\frac{1}{2})$. Only the element $-\pi/6$ is in the necessary interval. Problem (2) is then answered by $\sin^{-1} (-\frac{1}{2}) = -\pi/6$.

To solve (3) we only need note that **arcsin** 1 contains the number $\pi/2$. Necessarily, we conclude that $\sin^{-1} 1 = \pi/2$.

Figure 5.2.2

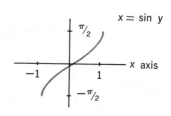

Figure 5.2.3

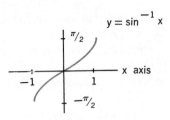

For each of the trigonometric functions there are many restrictions giving rise to an inverse. It is important that the reader be aware that ranges of the restrictions and the ranges of the corresponding unrestricted functions coincide.

AGREEMENT. .The restrictions to be used (when discussing inverse trigonometric functions) are listed below.

$$\text{cosine} | [0, \pi]$$
$$\text{tangent} | (-\pi/2, \pi/2)$$
$$\text{cosecant} | (0, \pi/2) \cup (-\pi, -\pi/2)$$
$$\text{secant} | (0, \pi/2) \cup (\pi, 3\pi/2)$$
$$\text{cotangent} | (0, \pi)$$

The corresponding inverse functions are:

$$\cos^{-1} : [-1, 1] \longrightarrow [0, \pi]$$
$$\tan^{-1} : R \longrightarrow (-\pi/2, \pi/2)$$
$$\csc^{-1} : R - (-1, 1) \longrightarrow (0, \pi/2) \cup (-\pi, -\pi/2)$$
$$\sec^{-1} : R - (-1, 1) \longrightarrow (0, \pi/2) \cup (\pi, 3\pi/2)$$
$$\cot^{-1} : R \longrightarrow (0, \pi)$$

These functions are called, respectively, the **inverse cosine, inverse tangent, inverse cosecant, inverse secant,** and **inverse cotangent.** In each case, the domain of the inverse function agrees with the image of the trigonometric function for which it is named. The range of each inverse function is the restricted domain of the corresponding function.

Examples will be used to further illustrate the new inverse functions.

Example 5.2.2   Find (1) $\cos^{-1} \frac{1}{2}$, (2) $\cos^{-1} (-\frac{1}{2})$, (3) $\tan^{-1} 1$, (4) $\csc^{-1} 2$, (5) $\sec^{-1} (-2)$, and (6) $\cot^{-1} (-1)$.

As in the case of $\sin^{-1} \frac{1}{2}$, there are two items to note upon examination of $\cos^{-1} \frac{1}{2}$. First, $\cos^{-1} \frac{1}{2}$ belongs to **arccos** $\frac{1}{2}$. Secondly, $\cos^{-1} \frac{1}{2} \in [0, \pi]$. The set **arccos** $\frac{1}{2} = \{x : \cos x = \frac{1}{2}\} = \{x : x = \pm\pi/3 + 2n\pi, n \in Z\}$. The one element from this set that is also in $[0, \pi]$ is $\pi/3$. In set symbolism, **arccos** $\frac{1}{2} \cap [0, \pi] = \{\pi/3\}$. By definition, $\cos^{-1} \frac{1}{2} = \pi/3$.

Now, **arccos** $(-\frac{1}{2}) = \{x : x = 2\pi/3 + 2n\pi \text{ or } 4\pi/3 + 2n\pi, n \in Z\} = \{x : x = (2n + 1)\pi \pm \pi/3\}$. Then, since **arccos** $(-\frac{1}{2}) \cap [0, \pi] = \{2\pi/3\}$, $\cos^{-1} (-\frac{1}{2}) = 2\pi/3$.

In order to answer (3) we must see that **arctan** $1 = \{x : x = \pi/4 + n\pi$ for $n \in Z\}$. Since $\pi/4 \in (-\pi/2, \pi/2)$, $\tan^{-1} 1 = \pi/4$.

Problem (4) can be solved by observing that **arccsc** $2 =$ **arcsin** $\frac{1}{2} = \{x : x = n\pi + (-1)^n \pi/6\}$ ($x \in$ **arccsc** 2 implies csc $x = 2$ and sin $x = \frac{1}{2}$ whence $x \in$ **arcsin** $\frac{1}{2}$ and conversely). Since $\pi/6$ belongs to the set, $\pi/6 =$ csc^{-1} 2.

Similarly, **arcsec** $(-2) =$ **arccos** $(-\frac{1}{2})$. Because $4\pi/3 \in (\pi, 3\pi/2)$, sec^{-1} $(-2) = 4\pi/3$.

The set **arccot** $(-1) =$ **arctan** $(-1) = \{x : x = 5\pi/4 + n\pi, n \in Z\}$. It follows by definition that cot^{-1} $(-1) = 3\pi/4$.

The following list affords a check as to whether or not an inverse function has been correctly evaluated. The list makes use of (1) the "arc" sets and (2) the ranges of the individual inverse functions. The intersections will contain either a single element or will be empty (depending on k).

$$\sin^{-1} k \in (\textbf{arcsin } k) \cap [-\pi/2, \pi/2]$$
$$\cos^{-1} k \in (\textbf{arccos } k) \cap [0, \pi]$$
$$\tan^{-1} k \in (\textbf{arctan } k) \cap (-\pi/2, \pi/2)$$
$$\csc^{-1} k \in (\textbf{arccsc } k) \cap [(0, \pi/2) \cup (-\pi, -\pi/2)]$$
$$\sec^{-1} k \in (\textbf{arcsec } k) \cap [(0, \pi/2) \cup (\pi, 3\pi/2)]$$
$$\cot^{-1} k \in (\textbf{arccot } k) \cap (0, \pi)$$

Example 5.2.3   As one more example, compute $\tan^{-1} 0$. Now,

$$(\textbf{arctan } 0) \cap \left(\frac{-\pi}{2}, \frac{\pi}{2}\right) = \{0\}$$

Hence, $\tan^{-1} 0 = 0$.

■ **EXERCISES 5.2**

1. Find the value of each of the following.

a. $\sin^{-1} (0)$.

b. $\sin^{-1} (\sqrt{3}/2)$.

c. $\sin^{-1} (1/\sqrt{2})$.

d. $\sin^{-1} (-1)$.

e. $\cos^{-1} (\sqrt{3}/2)$.

f. $\cos^{-1} (1/\sqrt{2})$.

g. $\cos^{-1} (1)$.

h. $\cos^{-1} (-1)$.

i. $\tan^{-1} (1/\sqrt{3})$.

j. $\tan^{-1} (\sqrt{3})$.

k. $\tan^{-1} (-1)$.

l. $\tan^{-1} (\sin 0)$.

m. $\sin^{-1} (-\sqrt{3}/2)$.

n. $\sin^{-1} (-1/\sqrt{2})$.

o. $\cos^{-1} (-\sqrt{3}/2)$.

p. $\cos^{-1} (-1/\sqrt{2})$.

q. $\cos^{-1} (-1/\sqrt{2})$.

r. $\tan^{-1} (-1/\sqrt{3})$.

s. $\tan^{-1} (-\sqrt{3})$.

2. Evaluate each of the following:

a. $\sin(\sin^{-1} \frac{1}{4})$.
b. $\cos(\cos^{-1} \frac{1}{5})$.
c. $\tan(\tan^{-1} 6)$.
d. $\sin(\tan^{-1} \frac{3}{4})$.
e. $\cos(\sin^{-1} (-\frac{4}{5}))$.
f. $\sin^{-1}(\sin \pi/10)$.

g. $\sin^{-1}(\sin(-\pi/7))$.
h. $\cos^{-1}(\cos 3\pi/4)$.
i. $\cos^{-1}(\cos(-\pi/5))$.
j. $\tan^{-1}(\tan \pi/7)$.
k. $\tan^{-1}(\tan 11\pi/10)$.
l. $\tan^{-1}(\tan 9\pi/10)$.

3. Sketch a graph of each inverse trigonometric function.
4. Determine the following points.

a. $\alpha(\sin^{-1} \frac{1}{2})$.
b. $\alpha(\cos^{-1}(-\frac{1}{2}))$.
c. $\alpha(\tan^{-1}(-1))$.
d. $\alpha(\sec^{-1} \frac{5}{4})$.
e. $\alpha(\csc^{-1}(-\frac{17}{8}))$.

f. $\alpha(\sin^{-1}(\sin 5\pi/4))$.
g. $\alpha(\sin^{-1}(\cos(-\pi/6)))$.
h. $\alpha(\cos^{-1}(\cos 5\pi/6))$.
i. $\alpha(\cos^{-1}(\sin \pi/3))$.
j. $\alpha(\tan^{-1}(\cot \pi/4))$.

5.3 HANDLING INVERSE TRIGONOMETRIC FUNCTIONS

Certain problems of evaluation (see Exercise 5.2.2) become interesting whenever the trigonometric functions and the inverse functions appear together.

Example 5.3.1 Find the values of: (1) $\sin(\sin^{-1} a)$, (2) $\sin^{-1}(\sin 5\pi/6)$, (3) $\tan(\sin^{-1} \frac{3}{5})$, and (4) $\cos(\tan^{-1} \frac{5}{12})$.

Problem (1) has been examined previously. It was seen that $\sin(\sin^{-1} a) = a$ if $-1 \leqslant a \leqslant 1$. This is really the age-old question, "What is the name of the man whose name is Jones?"

On the other hand, $\sin^{-1}(\sin a)$ is *not necessarily a*. Computation of a solution to (2) shows this truth: $\sin^{-1}(\sin 5\pi/6) = \sin^{-1}(-\frac{1}{2}) = -\pi/6$. The answer $\sin^{-1}(\sin 5\pi/6) = 5\pi/6$ is *invalid*.

The solution for (3) requires further thought. Since $\frac{3}{5} > 0$, the value of $\sin^{-1} \frac{3}{5}$, say x, must be such that $0 < x < \pi/2$. The identity $\sin^2 x + \cos^2 x = 1$ implies that $\cos x = \pm\frac{4}{5}$ ($\sin x = \frac{3}{5}$ by definition). The fact that $0 < x < \pi/2$ demands that $\cos x = \frac{4}{5}$. Rewriting $\sin^{-1} \frac{3}{5}$ in place of x, we see:

$$\cos(\sin^{-1} \tfrac{3}{5}) = \tfrac{4}{5}$$

Then,
$$\tan(\sin^{-1} \tfrac{3}{5}) = \frac{\sin(\sin^{-1} \tfrac{3}{5})}{\cos(\sin^{-1} \tfrac{3}{5})}$$

$$= \frac{\tfrac{3}{5}}{\tfrac{4}{5}} = \tfrac{3}{4}$$

Problem (4) is somewhat like (3). However, the calculation may not be so clear. Two techniques will be used, each having a special significance.

Since we wish to know the value of cos $(\tan^{-1} 5/12)$, we may just as well examine sec $(\tan^{-1} 5/12)$ and use the reciprocal. Let $\tan^{-1} 5/12 = x$. It is seen that $0 < x < \pi/2$ and $\tan x = 5/12$. Using the identity $\tan^2 x + 1 = \sec^2 x$, we quickly observe that sec $x = 13/12$, and cos $(\tan^{-1} 5/12) = \cos x = 1/\sec x = 12/13$.

An alternative approach utilizes the identity $\tan x = \sin x/\cos x$. Now, $\tan x = 5/12$ and it is clear that $\sin x \neq 5$ and $\cos x \neq 12$. (Why?) However, it is a known algebraic truth that there is a number r so that $5/r = \sin x$ and $12/r = \cos x$. Whence, $5r/12r = 5/12 = \sin x/\cos x$ as before. Recalling that $\sin^2 x + \cos^2 x = 1$, $1 = (5/r)^2 + (12/r)^2 = 169/r^2$, which is to say that $r^2 = 169$ and $r = \pm 13$. Since $\sin x > 0$, $r = 13$, whence $\sin x = 5/13$. Clearly, $\cos x = 12/13$.

The latter also extends a technique for finding $\sin x$ and $\cos x$ (up to $\pm$) whenever $\tan x$ is known. The r is a square root of the sum of the squares of the numerator and denominator of the expression for the value of the tangent.

As we might expect, equations arise during the use of the inverse trigonometric functions. The following illustrates how such an equation might be solved (using known identities).

Example 5.3.2   Show that $\tan^{-1} \frac{1}{2} + \tan^{-1} \frac{1}{3} = \pi/4$. Since $0 < \frac{1}{3} < \frac{1}{2} < 1$, it follows that $0 < \tan^{-1} \frac{1}{3} < \tan^{-1} \frac{1}{2} < \pi/4$. (Why?) Furthermore, $0 < \tan^{-1} \frac{1}{2} + \tan^{-1} \frac{1}{3} < \pi/2$. Thus, $\pi/4$ and $\tan^{-1} \frac{1}{2} + \tan^{-1} \frac{1}{3}$ are both in $(0, \pi/2)$. If we can show that $\tan (\tan^{-1} \frac{1}{3}) = \tan \pi/4 = 1$, we are finished. (Note: $\tan a = \tan b$ does not imply that $a = b$, unless both a and b are between the same two multiples of π):

$$\tan (\tan^{-1} \tfrac{1}{2} + \tan^{-1} \tfrac{1}{3}) = \frac{\tan (\tan^{-1} \tfrac{1}{2}) + \tan (\tan^{-1} \tfrac{1}{3})}{1 - \tan (\tan^{-1} \tfrac{1}{2}) \tan (\tan^{-1} \tfrac{1}{3})}$$

$$= \frac{\tfrac{1}{2} + \tfrac{1}{3}}{1 - \tfrac{1}{2} \tfrac{1}{3}} = \frac{5/6}{5/6} = 1$$

The equation is seen to be valid.

■ EXERCISES 5.3

1. Find the value of each of the following.

 a. $\sin (\sin^{-1} (\frac{1}{2}))$.
 b. $\cos (\cos^{-1} (\frac{1}{3}))$.
 c. $\tan (\tan^{-1} (65))$.
 d. $\sin^{-1} (\sin 17\pi/6)$.
 e. $\cos^{-1} (\cos (-\pi/12))$.

 f. $\tan^{-1} (\tan 5\pi/6)$.
 g. $\sin (\sin^{-1} 2)$.
 h. $\cos (\cos^{-1} (-\frac{3}{2}))$.
 i. $\tan^{-1} (\tan \pi/2)$.

2. Find the value of each of the following.

a. $\tan^{-1}(\sin \pi/2)$.

b. $\cos^{-1}(\sin 9\pi/4)$.

c. $\sin^{-1}(\cos \pi/6)$.

d. $\sin^{-1}(\tan 26\pi)$.

e. $\sin(\cos^{-1} \frac{1}{2})$.

f. $\tan(\sin^{-1}(-\frac{1}{2}))$.

g. $\cos(\tan^{-1} 1)$.

h. $\csc(\sin^{-1} \frac{1}{2})$.

i. $\sec(\sin^{-1}(-\frac{1}{2}))$.

j. $\cos(\sin^{-1} \frac{3}{5})$.

k. $\sin(\cos^{-1} \frac{3}{5})$.

l. $\tan(\cos^{-1} \frac{3}{5})$.

m. $\sin(\tan^{-1} \frac{12}{5})$.

n. $\cos(\tan^{-1} \frac{8}{15})$.

o. $\cos(\cot^{-1}(-\frac{7}{24}))$.

p. $\sin(\cos^{-1}(-\frac{7}{25}))$.

q. $\cos(\sin^{-1}(-\frac{5}{13}))$.

r. $\tan(\cot^{-1}(-\frac{4}{3}))$.

3. Use identities where necessary to find the value of each of the following.

a. $\sin(\tan^{-1}(1) + \cos^{-1}(-\frac{1}{2}))$.

b. $\sin(\sin^{-1}(\frac{3}{4}) + \cos^{-1}(\frac{7}{25}))$.

c. $\cos(\tan^{-1}(\frac{3}{4}) + \sin^{-1}(\frac{8}{17}))$.

d. $\cos(\cos^{-1}(\frac{1}{2}) - \sin^{-1}(1))$.

e. $\tan(\sin^{-1}(\frac{5}{13}) + \cos^{-1}(\frac{7}{25}))$.

f. $\tan(\sin^{-1}(\frac{5}{13}) = \sin^{-1}(\frac{4}{5}))$.

g. $\sin(2\tan^{-1}(\frac{4}{3}))$.

h. $\cos(2\sin^{-1}(\frac{3}{5}))$.

i. $\cos(\frac{1}{2}\tan^{-1}(1))$.

j. $\sin(\frac{1}{2}\cos^{-1}(\frac{3}{5}))$.

k. $\tan(2\sin^{-1}(\frac{24}{25}))$.

l. $\sin(2\sin^{-1}(\frac{1}{2}) + \cos^{-1}(\frac{4}{5}))$.

m. $\sin(\pi/2 - \cos^{-1}(\frac{1}{3}))$.

n. $\cos(\pi/2 - \sin^{-1}(-\frac{2}{3}))$.

4. Verify that the following equations hold.

a. $\tan^{-1}(6) + \tan^{-1}(\frac{1}{6}) = \pi/2$.

b. $\sin^{-1}(\frac{1}{2}) + \sin^{-1}(\frac{3}{2}) = \pi/2$.

c. $\sin^{-1}(a) + \sin^{-1}(-a) = 0$ if $a = 1$.

d. $\cos^{-1}(\frac{1}{2}) + \cos^{-1}(-\frac{1}{2}) = \pi$.

e. $\cos^{-1}(\frac{12}{13}) + \cos^{-1}(\frac{24}{25}) = \cos^{-1}(\frac{253}{325})$.

f. $\sin^{-1}(\frac{3}{5}) + \sin^{-1}(\frac{4}{5}) = \pi/2$.

g. $\tan^{-1}(\frac{3}{4}) + \tan^{-1}(\frac{1}{3}) = \tan^{-1}(\frac{13}{9})$.

h. $\tan^{-1}(\frac{1}{5}) + \tan^{-1}(\frac{1}{6}) = \tan^{-1}(\frac{11}{30})$.

i. $\tan^{-1}(\frac{1}{6}) - \tan^{-1}(\frac{1}{7}) = \tan^{-1}(\frac{1}{41})$.

j. $\tan^{-1}(\frac{1}{4}) + 2\tan^{-1}(\frac{1}{5}) = \tan^{-1}(\frac{32}{43})$.

k. $\tan^{-1}(\frac{1}{5}) + \tan^{-1}(\frac{3}{2}) = \pi/4$.

l. $\tan^{-1}(\frac{1}{8}) + \tan^{-1}(\frac{1}{2}) = \tan^{-1}(\frac{2}{3})$.

5.4 GEOMETRIC PROPERTIES OF THE TRIGONOMETRIC FUNCTIONS

We have, on previous occasions, seen graphs of $y = \sin x$. The outstanding characteristics of this curve are:

1. a maximum value of $+1$.
2. a minimum value of -1.
3. periodicity.
4. $(0, 0)$ a point on the curve.

The minimum and maximum values have the same absolute value, namely 1. This number is called the **amplitude** or peak value of the curve $y = \sin x$. Figure 5.4.1 illustrates.

Examine the curve $y = a \sin x$ with $a > 0$ ($a > 0$ is not a necessary condition but will be employed). Since the excursions of the expression $\sin x$ carry it between -1 and $+1$, it follows that $a \sin x$ must have minimum and maximum values $-a$ and a, respectively. In a manner analogous to the above, we define the amplitude of $y = a \sin x$ to be a.

Example 5.4.1   The example $y = 3 \sin x$ is shown in Fig. 5.4.2. The reader should compare Figs. 5.4.1 and 5.4.2 to see that the curves are identical except for minimum and maximum excursions. If we are not concerned with whether or not both axes have the same scale of measurement, both $\sin x$ and $a \sin x$ can be given by the same curve (by simply relabeling the graduations on the y axis).

The curve $y = \sin (x + b)$ (in Fig. 5.4.3) resembles $y = \sin x$ as given in Fig. 5.4.1. However, one difference appears. The point $(0, 0)$ is not on the curve $y = \sin (x + b)$. The argument of the sine curve given here is $x + b$, not x. The value $\sin 0 = 0$ still holds, but, when $x + b = 0$, $x = -b$. Thus, the point $(-b, 0)$ on $y = \sin (x + b)$ is analogous to the point $(0, 0)$ for the curve $y = \sin x$. The curve $y = \sin (x + b)$ appears to be the curve $y = \sin x$ after a slight "shift" to the left. The only visible difference between Figs. 5.4.1 and 5.4.3 is the placement of the y axis. The value $|b|$ is called the **phase displacement.**

Let $k > 0$ be any real number. The expression $\sin kx$ is periodic of primitive period 2π, that is, if kx_0 and kx_1 differ by 2π, $\sin kx_0 = \sin kx_1$. Furthermore, $kx_0 - kx_1 = 2\pi$ implies that $x_0 - x_1 = 2\pi/k$. Thus, the graph (in Fig. 5.4.4) of $\sin kx$ shows the apparent periodicity of the expression to be $2\pi/k$. We shall call $2\pi/k$ the **period** of $\sin kx$ **relative to** x.

Example 5.4.2   Figure 5.4.5 illustrates the curve $y = \sin 2x$. In this figure, $k = 2$, whence $2\pi/k = 2\pi/2 = \pi$. The period of $\sin 2x$ relative to x is π.

In the expression $\sin kx$, $k/2\pi$ is sometimes called the **frequency.** That part of a sine curve from some point x to $x + 2\pi/k$ (the difference of

Figure 5.4.1
$y = \sin x$

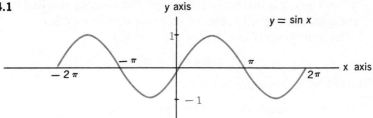

$y = \sin x$

Figure 5.4.2
$y = 3 \sin x$

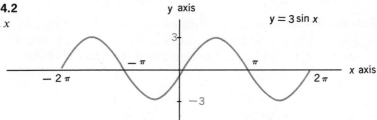

$y = 3 \sin x$

Figure 5.4.3
$y = \sin (x + b)$

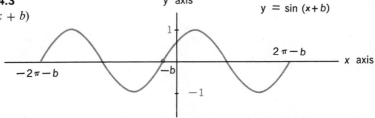

$y = \sin (x + b)$

Figure 5.4.4
$y = \sin kx$

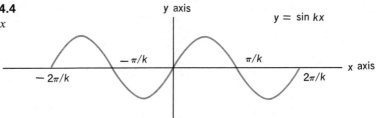

$y = \sin kx$

Figure 5.4.5
$y = \sin 2x$

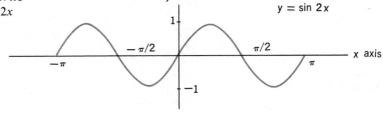

$y = \sin 2x$

$x + 2\pi/k$ and x is one period of sin kx relative to x) is called a **cycle.** The curve goes through all values of its range during a cycle.

The most general form of the sine curve is:

$$y = a \sin k(x + b), \qquad k > 0$$

The following items are to be noted:

1. The *amplitude* of $y = a \sin k(x + b)$ is a.
2. The *phase of* $y = a \sin k(x + b)$ is $|b|$ ($(-b, 0)$ is on the curve).
3. The *period* of $y = a \sin k(x + b)$ relative to x is $2\pi/k$.

Example 5.4.3   Figure 5.4.6 shows a step by step construction of a sketch of $y = a \sin k(x + b)$ for the special case $y = 3 \sin 2(x + \pi/4)$.

An example will be used to summarize the material of this section.

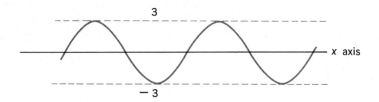

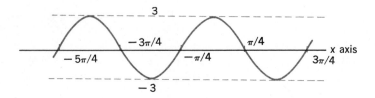

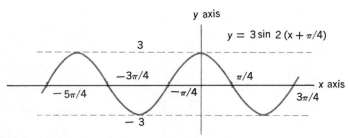

Figure 5.4.6
Construction of a graph of $y = 3 \sin 2 (x + \pi/4)$

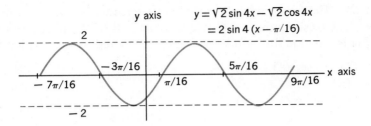

Figure 5.4.7
$y = \sqrt{2} \sin 4x - \sqrt{2} \cos 4x$

Example 5.4.4 Write $\sqrt{2} \sin 4x - \sqrt{2} \cos 4x$ in the form $a \sin k(x + b)$ and state the amplitude, period, and phase.

Using a previous identity,

$$\sqrt{2} \sin 4x - \sqrt{2} \cos 4x = 2 \sin \left(4x - \frac{\pi}{4}\right)$$

$$= 2 \sin 4\left(x - \frac{\pi}{16}\right)$$

The amplitude is 2, the period relative to x is $2\pi/4 = \pi/2$, and the phase is $\pi/16$. Figure 5.4.7 shows the sketch of this function.

A similar analysis exists for $a \cos k(x + b)$.

■ EXERCISES 5.4

1. Give the period relative to x, the phase, and the amplitude of each of the following. [Hint: Use identities to rewrite (h) to (j).]

   a. $3 \sin 5x$.
   b. $2 \sin (3x + 1)$.
   c. $\frac{1}{2} \cos 2(x - \pi/6)$.
   d. $\frac{3}{4} \cos (\frac{2}{3}x - \pi/8)$.
   e. $\sin (x - \pi/2)$.
   f. $6 \cos (x - \pi/3)$.
   g. $\frac{5}{3} \sin (5x/3 + \pi)$.
   h. $\sin x + \cos x$.
   i. $5 \sin 2x + 12 \cos 2x$.
   j. $\sin 3x + \sqrt{3} \cos 3x$.

2. Give the frequency in each part of Exercise 1.
3. Sketch each curve (3 cycles) from Exercise 1.
4. Find a period for f in each case. Example: Let $f(x) = \sin (x/7) +$

sin $(x/5)$. The component parts of $f(x)$ have period 14π and 10π, respectively. The least common multiple of 14 and 10 is 70. It can be shown that 70π is a period for both $\sin x/7$ and $\sin x/5$ and hence a period for f.

a. $f(x) = \sin x/3 + \sin x/4$.
b. $f(x) = \sin 3x/4 + \sin x$.
c. $f(x) = \cos 2x/3 + \cos x/6$.

5.5 REVIEW EXERCISES

1. Describe the elements of each of the following sets.

  a. arcsin $0 \cap$ arctan 0.
  b. arcsin $0 \cap$ arccos 0.
  c. arcsin $1/\sqrt{2} \cap$ arccos $1/\sqrt{2}$.
  d. arcsin $(-1/\sqrt{2}) \cap$ arccos $1/\sqrt{2}$.
  e. arccos $0 \cap$ arctan 0.
  f. arccos $0 \cap$ arccot 0.
  g. arcsin $3 \cap$ arccos $\frac{1}{2}$.
  h. arcsin $1 \cap$ arccsc 1.
  i. arctan $(-1) \cap$ arccos $1/\sqrt{2}$.
  j. arcsin $(-\sqrt{3}/2) \cap$ arctan $(-\sqrt{3})$.
  k. arctan $(\sin \pi/2)$.
  l. arcsin $1 \cup$ arccos 1.
  m. arcsin $0 \cup$ arccos 0.
  n. arcsin $1/\sqrt{2} \cup$ arccos $1/\sqrt{2}$.
  o. arctan $1 \cup$ arcsin $1/\sqrt{2}$.
  p. arctan $1 \cup$ arctan (-1).
  q. arcsin $1/\sqrt{2} -$ arccos $(-1/\sqrt{2})$.
  r. arcsin $1/\sqrt{2} -$ arccos $1/\sqrt{2}$.
  s. arctan $(-1) -$ arccos $(-1/\sqrt{2})$.

2. List the elements in each set.

  a. arcsin $0 \cap [\pi/2, 3\pi/2]$.
  b. arcsin $1/\sqrt{2} \cap [\pi/2, 3\pi/2]$.
  c. arcsin $\frac{1}{2} \cap [\pi/2, 3\pi/2]$.
  d. arcsin $\sqrt{3}/2 \cap [\pi/2, 3\pi/2]$.
  e. arcsin $1 \cap [\pi/2, 3\pi/2]$.
  f. arcsin $(-1) \cap [\pi/2, 3\pi/2]$.
  g. arcsin $(-\frac{1}{2}) \cap [\pi/2, 3\pi/2]$.
  h. arcsin $(-\sqrt{3}/2) \cap [\pi/2, 3\pi/2]$.
  i. arcsin $(-1/\sqrt{2}) \cap [\pi/2, 3\pi/2]$.
  j. arccos $0 \cap [-\pi, 0]$.
  k. arccos $1/\sqrt{2} \cap [-\pi, 0]$.

l. arccos ½ ∩ [−π, 0].
m. arccos √3/2 ∩ [−π, 0].
n. arccos 1 ∩ [−π, 0].
o. arccos (−1) ∩ [−π, 0].
p. arccos (−½) ∩ [−π, 0].
q. arccos (−√3/2) ∩ [−π, 0].
r. arccos (−1/√2) ∩ [−π, 0].
s. arctan 0 ∩ (π/2, 3π/2).
t. arctan 1/√3 ∩ (π/2, 3π/2).
u. arctan 1 ∩ (π/2, 3π/2).
v. arctan √3 ∩ (π/2, 3π/2).
w. arctan (−√3) ∩ (π/2, 3π/2).
x. arctan (−1) ∩ (π/2, 3π/2).
y. arctan (−1/√3) ∩ (π/2, 3π/2).

3. Each of the elements in the sets resulting in Exercise 5.1.8 can be described in terms of inverse functions. Do so.

4. Solve the following trigonometric equations.

a. $\sin x \cos x = 0$.
b. $\sin x \cos x = \sin x$.
c. $\tan x + \cot x = 2$.
d. $\sin x \cos x = 1$.
e. $\sin^2 x - \cos x + 4 = -1$.
f. $3 \csc^2 2x - 5 = \cot 2x$.

5. Evaluate the following:

a. $\sin (\sin^{-1} a)$.
b. $\cos (\cos^{-1} a)$.
c. $\tan (\tan^{-1} a)$.
d. $\sin^{-1} (\sin 5\pi/6)$.
e. $\sin^{-1} (\sin 11\pi/6)$.
f. $\cos^{-1} (\cos 2\pi/3)$.
g. $\cos^{-1} (\cos (-\pi/4))$.
h. $\tan^{-1} (\tan \pi/7)$.
i. $\tan^{-1} (\tan 8\pi/7)$.

j. $\sin (\cos^{-1} (-\tfrac{5}{12}))$.
k. $\cos (\sin^{-1} (-\tfrac{8}{17}))$.
l. $\tan (\sin^{-1} (\tfrac{3}{5}))$.
m. $\tan (\cos^{-1} (\tfrac{12}{13}))$.
n. $\sin (\tan^{-1} (\tfrac{3}{4}))$.
o. $\cos (\tan^{-1} (-\tfrac{8}{15}))$.
p. $\sin^{-1} (\tan 95\pi/4)$.
q. $\cos^{-1} (\tan 95\pi/4)$.
r. $\tan^{-1} (\sin 3\pi/4)$.

6. Prove or disprove the validity of the following statements.

a. $\sin^{-1} \tfrac{3}{5} + \cos^{-1} \tfrac{5}{13} = \cos^{-1} (-\tfrac{16}{65})$.
b. $\sin^{-1} \tfrac{8}{17} + \cos^{-1} \tfrac{12}{13} = \sin^{-1} \tfrac{21}{221}$.
c. $\sin^{-1} \tfrac{3}{5} + \cos^{-1} \tfrac{5}{13} + \sin^{-1} \tfrac{63}{65}$.
d. $\tan^{-1} \tfrac{1}{8} + \tan^{-1} \tfrac{1}{7} = \tan^{-1} \tfrac{3}{11}$.
e. $\tan^{-1} \tfrac{1}{7} - \tan^{-1} \tfrac{1}{8} = \tan^{-1} \tfrac{1}{55}$.
f. $\tan^{-1} \tfrac{4}{3} + \tan^{-1} \tfrac{4}{5} = \tan^{-1} (-32)$.

7. Give the amplitude, phase, and period with respect to x in each example below.

    a. $3 \sin (7x + 1)$.
    b. $\frac{1}{5} \cos (3x - 2)$.
    c. $\frac{3}{2} \sin 5(x - \frac{1}{3})$.
    d. $5 \cos 3(x - \pi/7)$.
    e. $3 \sin x + \cos x$.
    f. $5 \sin x + 12 \cos x$.
    g. $3 \sin 2x + 4 \cos 2x$.

8. Give the frequency of each expression in 7.
9. Sketch 4 cycles of each expression in 7.

5.6 ADVANCED EXERCISES

1. Find 3 closed intervals other than $[-\pi/2, \pi/2]$ which would make appropriate restrictions for the sine function in order that an inverse might be defined.
2. Find 3 closed intervals other than $[0, \pi]$ which would make appropriate restrictions for the cosine function in order that an inverse might be defined.
3. Find 3 open intervals other than $(-\pi/2, \pi/2)$ which would make appropriate restrictions for the tangent function in order that an inverse might be defined.
4. Why is the restriction for the tangent function the open interval $(-\pi/2, \pi/2)$ instead of $[-\pi/2, \pi/2]$?
5. Why does the restriction $\cos | [-\pi/2, \pi/2]$ prove unsatisfactory for purposes of defining an inverse cosine function?
6. Let m, n, p, and q be positive integers. What is the primitive period of $f(x) = \sin (m/n)x + \sin (p/q)x$ with respect to x?
7. Write $\cos x$ in the form $\sin (x + a)$.
8. Consider $a \sin k(x + b)$. Let f be the frequency and p the period relative to x. Show that $pf = 1$.
9. Write the equation of the line through $(0, 0)$ and the point:

    a. $(\sin^{-1} \frac{1}{2})$.
    b. $(\sin^{-1} (-\sqrt{3}/2))$.
    c. $(\cos^{-1} 1/\sqrt{2})$.
    d. $(\cos^{-1} (-1))$.
    e. $(\tan^{-1} 0)$.
    f. $(\tan^{-1} (-1/\sqrt{3}))$.

10. Note that $\sin^{-1} = \alpha^{-1} \circ p_2^{-1}$ with appropriate restrictions. Suppose

$f:X \longrightarrow Y$ and $g:Y \longrightarrow Z$ are $1:1$ and onto functions. Show that $(g \circ f)^{-1} = f^{-1} \circ g^{-1}$.

11. Let b **arcsin** $k = \{x : \sin x/b = k\}$. Show that if $cx \in$ **arcsin** k, $x \in 1/c$ **arcsin** k; $c \neq 0$.

12. Let $G \subset R \times R$ be given by

$$G = \{(x, y) : x, y \in R, \sin x = \sin y\}$$

Show that G is an equivalence relation on $R \times R$.

13. Let $G \subset [-\pi/2, \pi/2] \times [-\pi/2, \pi/2]$ be given as in Exercise 12. Describe G completely in terms of the types of ordered pairs it has.

14. Describe the relations on R determined by each of the 6 "arc" sets (that is, determine arccos, arctan, and so forth). Describe also the relations for each "arc" set determined analogous to the determination of G in Exercise 12. G in Exercise 13. What Cartesian product is used in each case?

Chapter 6
Extensions
of
Trigonometry

6.1 POLAR COORDINATES

Previous discussions and exercises have drawn attention to the fact that each point $(p, q) \neq (0, 0)$ of the plane can be written in the form $(r \cos a, r \sin a)$ where:

1. a is a real number, and
2. $r = \sqrt{p^2 + q^2} = d((p, q), (0, 0))$.

Since the line ℓ passing through (p, q) and $(0, 0)$ has the equation $py = qx$ (why?), the unit circle C and ℓ intersect at $(p/r, q/r)$ and $(-p/r, -q/r)$. The point $(p/r, q/r)$, lying on the unit circle, has the representation $\alpha(a) = (\cos a, \sin a)$ for some real number a $(0 \leqslant a < 2\pi$ if desired). Thus, $(p, q) = (r \cos a, r \sin a)$.

The point $(-p, -q)$ is given by $(-r \cos a, -r \sin a)$ or by $(r \cos (a+\pi), r \sin (a + \pi))$. Moreover, (p, q) itself may be given by $(-r \cos (a + \pi), -r \sin (a + \pi))$. That is, since r need not be positive, more than one such representation (for each point) exists. In fact, an infinite variety of forms exists since a may be replaced by $a + 2n\pi$ for any $n \in Z$.

Since the r and a, in this sense, completely describe (or "locate") the point (p, q), it seems reasonable that we might develop a system of identification for points of the plane using this type of information. If (p, q) is a point of the plane with $(p, q) = (r \cos a, r \sin a)$, the **polar** (coordinate) **form** for (p, q) is given by the ordered pair notation $(r, a)*$. To insure that no confusion arises because of the similarity between the notations of the rectangular coordinate and polar coordinate systems, the polar coordinates will always be denoted by an asterisk.

Example 6.1.1   Sketch a graph showing the location of the points $(6, \pi/3)*$ and $(-6, \pi/3)*$.

From the special value list, we determine that the line through the **pole** (origin) and $(6, \pi/3)*$ intersects the unit circle at $(\frac{1}{2}, \sqrt{3}/2)$. (The rectangular coordinates of $(6, \pi/3)*$ are $(3, 3\sqrt{3}/2)$.)

The location of $(6, \pi/3)*$ is shown on Fig. 6.1.1. The point $(-6, \pi/3)*$ is the same point as $(6, \pi/3 + \pi)*$ or $(6, 4\pi/3)*$ (verify this) so that $(-6, \pi/3)*$ can be located as in Fig. 6.1.2.

Example 6.1.2   Give four polar (coordinate) representations for $(3, 4)$.

Since $5^2 = 3^2 + 4^2$, $(3, 4)$ may be given by:

$$(5, \sin^{-1} \tfrac{4}{5})*$$
$$(-5, \pi + \sin^{-1} \tfrac{4}{5})*$$
$$(5, 2\pi + \sin^{-1} \tfrac{4}{5})*$$
$$(5, -2\pi + \sin^{-1} \tfrac{4}{5})*$$

This follows since $\alpha(\sin^{-1} \tfrac{4}{5}) = \alpha(2\pi + \sin^{-1} \tfrac{4}{5}) = \alpha(-2\pi + \sin^{-1} \tfrac{4}{5})$ and if $\alpha(\sin^{-1} \tfrac{4}{5}) = (x, y)$, $\alpha(\pi + \sin^{-1} \tfrac{4}{5}) = (-x, -y)$. [What is $\alpha(\sin^{-1} \tfrac{4}{5})$? $\alpha(\sin^{-1} a)$?]   (See Fig. 6.1.3.)

Figure 6.1.1
Location of $(6, \pi/3)*$

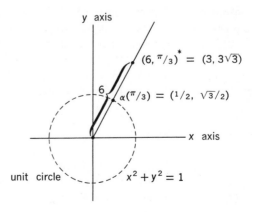

y axis

$(6, {}^{\pi}/_3)^* = (3, 3\sqrt{3})$

6

$\alpha({}^{\pi}/_3) = (1/2, \sqrt{3}/2)$

x axis

unit circle

$x^2 + y^2 = 1$

Figure 6.1.2
Location of $(-6, \pi/3)^*$

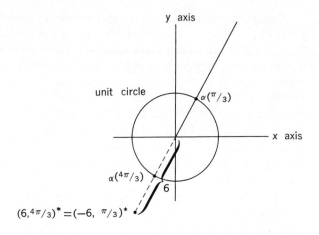

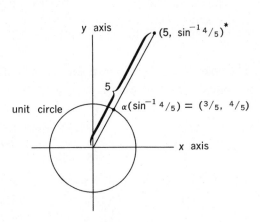

Figure 6.1.3

Example 6.1.3 Suppose $(3, \pi/2)^*$ is a polar form representation of a point. What is its rectangular representation?

The rectangular form for $(3, \pi/2)^*$ is given by $(0, 3)$; $(0, 3)$ is 3 units from the pole and $(0, 1)$ is $\alpha(\pi/2)$. (See Fig. 6.1.4.) We observe that $(3, 5\pi/2)^*$ and $(3, -3\pi/2)^*$ also represent polar forms of the same point. The representation of a point in polar form is not unique.

■ **EXERCISES 6.1**

1. Each of the following points is given in rectangular form. Write each in polar form $(r, a)^*$ for two different values of a, $0 \leqslant a < 2\pi$. (Hint: Use a negative value for r.)

Figure 6.1.4

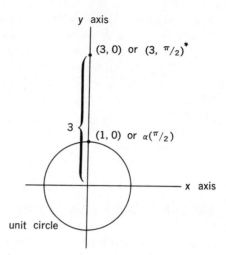

a. $(3, 0)$.   **i.** $(-5, 12)$.
b. $(0, 1)$.   **j.** $(-8, -15)$.
c. $(-1, 0)$.   **k.** $(6, 5)$.
d. $(0, -2)$.   **l.** $(-2, 3)$.
e. $(3, 4)$.   **m.** $(-2, 3)$.
f. $(5, 12)$.   **n.** $(-1, \sqrt{3})$.
g. $(8, 15)$.   **o.** $(-5\sqrt{3}, 5)$.
h. $(3, -4)$.   **p.** (x, y).

2. Do Exercise 1 with $-2\pi \leqslant a < 0$.

3. Let $(r, a)^*$ be a point written in polar form. Show that:

    a. $p_1((r, a)^*) = r \cos a$.   **b.** $p_2((r, a)^*) = r \sin a$.

4. If $(x, y) = (r \cos a, r \sin a)$, $x \neq 0$, $r > 0$, and $a \in \mathbf{arctan}\ k$; find k.

5. Convert each polar form to rectangular form.

    a. $(3, 0)^*$.   **i.** $(-2, \pi)^*$.
    b. $(2, \pi)^*$.   **j.** $(-3, \pi/2)^*$.
    c. $(3, -\pi/2)^*$.   **k.** $(-5, 5\pi/3)^*$.
    d. $(5, -\pi)^*$.   **l.** $(20, \tan^{-1} 8/15)^*$.
    e. $(1, \pi/2)^*$.   **m.** $(10, \pi + \tan^{-1} 5/12)^*$.
    f. $(6, 5\pi/6)^*$.   **n.** $(8, \pi - \sin^{-1} 1/2)^*$.
    g. $(2, 3\pi/4)^*$.   **o.** $(3, \pi + \cos^{-1}(-1/2))^*$.
    h. $(10, \tan^{-1} 3/4)^*$.

6. Let $y = mx + b$ represent the equation of a straight line in rectangular coordinates (that is, (x, y) is a point of the line, (x, y) a rectangular representation). Show that this becomes $a = \tan^{-1} m$ if $b = 0$ ($\tan^{-1} m$

is a constant). [Hint: Write (x, y) in polar form (in terms of r and a).]

7. Let $x^2 + y^2 = a^2$ represent a circle in rectangular form. Show that this equation becomes $r = a$ when the equation is written in terms of r and a $((r, a)^*$ a polar representation for $(x, y))$.

8. Let $(x - h)^2 + y^2 = h^2$ be a circle (in rectangular form). Show that, in polar form $((x, y)$ converted to the polar form $(r, a)^*)$, the equation has the appearance $r = 2h \cos a$.

9. Show $x^2 + (y - k)^2 = k^2$ $((x, y)$ rectangular coordinates) becomes $r = 2k \sin a$ when (x, y) is written in the polar form $(r, a)^*$.

6.2 THE PLANE AS AN ALGEBRAIC SYSTEM

If we consider the plane and define operations of addition, multiplication, and so forth, relative to its points, the resulting system has a very interesting and important algebraic structure.

Equality between points has already been defined by $(a, b) = (c, d)$ if and only if $a = c$ and $b = d$. Now, we define:

$$(a, b) + (c, d) = (a + c, b + d)$$
$$(a, b)(c, d) = (ac - ad, bc)$$
$$(a, b) - (c, d) = (e, f) \qquad \text{where } (c, d) + (e, f) = (a, b)$$

and if $c^2 + d^2 \neq 0$,

$$(a, b) \div (c, d) = (g, h) \qquad \text{where } (c, d)(g, h) = (a, b)$$

Addition is perhaps the easiest operation to perform. We merely add the respective first elements and then the respective second elements.

Example 6.2.1 From the above, we see: $(2, 1) + (3, 5) = (2+3, 1+5) = (5, 6)$.

Multiplication is somewhat more complicated, as the following example illustrates.

Example 6.2.2 $(2, 1)(3, 5) = (2 \cdot 3 - 1 \cdot 5, 2 \cdot 5 + 1 \cdot 3) = (6 - 5, 10 + 3) = (1, 13)$.

Our definition of subtraction does not provide a procedure for calculating a difference; it merely gives a means for verifying a correct answer. Now, it is claimed that $(a - c, b - d)$ is a solution to $(a, b) - (c, d)$, that is, the answer is found in a manner analogous to the process for addition. That the answer is valid is seen by

$$(c, d) + (a - c, b - d) = (c + a - c, d + b - d) = (a, b)$$

Example 6.2.3 From the above,

$$(2, 1) - (3, 5) = (2 - 3, 1 - 5) = (-1, -4)$$

As in the case of subtraction, a proposed answer will be given for the division shown above. (Note: If both c and d are 0, the division is not defined.) This proposed answer for $(a, b) \div (c, d)$ is

$$\left(\frac{ac + bd}{c^2 + d^2}, \frac{bc - ad}{c^2 + d^2}\right)$$

We verify this proposal via the following.

$$(c, d)\left(\frac{ac + bd}{c^2 + d^2}, \frac{bc - ad}{c^2 + d^2}\right)$$
$$= \left(\frac{ac^2 + bcd - bcd + ad^2}{c^2 + d^2}, \frac{dac + bd^2 + bc^2 - cad}{c^2 + d^2}\right)$$
$$= (a, b)$$

Example 6.2.4 The following shows a computation of a solution to a division process.

$$(2, 1)/(3, 5) = (2, 1) \div (3, 5) = \left(\frac{2 \cdot 3 + 1 \cdot 5}{3^2 + 5^2}, \frac{1 \cdot 3 - 2 \cdot 5}{3^2 + 5^2}\right)$$
$$= (^{11}\!\!/_{34}, -^{7}\!\!/_{34})$$

Let us now introduce another multiplication called **scalar multiplication**. Let k be any real number. We define:

$$k(a, b) = (ka, kb)$$

Scalar multiplication is simply a multiplication of each coordinate by a common real number.

Example 6.2.5 $6(2, 1) = (12, 6)$.

Examining this definition in a sort of reverse manner, we see that we may in effect **factor out** any real number appearing as a factor common to both elements of the ordered pair.

A geometric interpretation can be given to the operations of addition, subtraction, and scalar multiplication.

Example 6.2.6 The point (a, b) (Fig. 6.2.1) can be used to describe a "line segment" between $(0, 0)$ and (a, b). By placing an arrow head on the line segment at (a, b), we give rise to the intuitive feeling that we are considering a "vector" from $(0, 0)$ to (a, b). The line segment has become a directed line segment in some sense. (For the concept of a line segment see Review Exercise 2.4.14.)

Example 6.2.7 The point $(-a, -b) = -(a, b)$ is seen to determine a vector of the same length as that determined by (a, b) but the direction indicated is opposite that of Fig. 6.2.1. (See Fig. 6.2.2.)

Example 6.2.8 The point $6(a, b) = (6a, 6b)$ denotes a vector in the same direction as that of Fig. 6.2.1 but of length six times as great. (See

Figure 6.2.1
The vector (a, b)

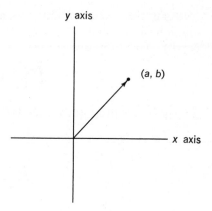

Figure 6.2.2
(a, b) versus $-(a, b)$

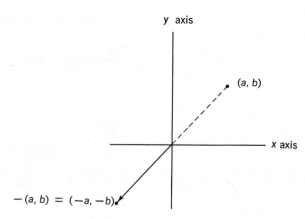

Figure 6.2.3
(a, b) versus (a, b)

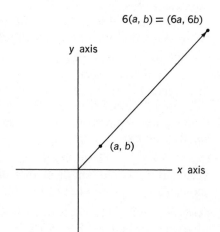

Fig. 6.2.3.) On the other hand, $-6(a, b) = (-6a, -6b) = -(6a, 6b)$ gives a vector point in the opposite direction with similar length (Fig. 6.2.4).

Example 6.2.9 Since $(a, b) + (c, d) = (a + c, b + d)$, the addition of the two is seen geometrically as the "head to tail" addition of the two vectors involved (Fig. 6.2.5). The vector for (c, d) is moved so that its initial point (originally $(0, 0)$) is superimposed with (a, b) and so that the vector is parallel (parallel in the sense that straight lines are parallel and the oriented direction coincides with that of the original vector). The resulting vector is that from $(0, 0)$ to the terminal point of the displaced vector. The vector sum is often called the *resultant* vector.

Figure 6.2.4
$-6(a, b)$

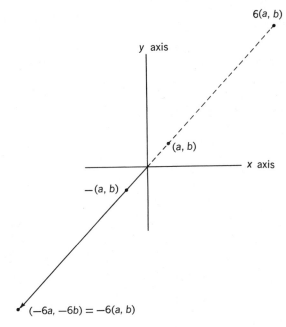

Figure 6.2.5
Vector addition

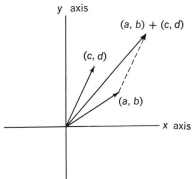

Example 6.2.10 The geometric interpretation of $(a, b) - (c, d) = (a, b) + (-(c, d))$ follows from Example 6.2.9 and is given in Fig. 6.2.6.

Figure 6.2.6
Vector subtraction

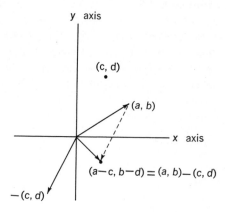

■ EXERCISES 6.2

1. Compute each of the following:

a. $(3, 2) + (1, 5)$.
b. $(2, -1) + (3, 2)$.
c. $(-3, -1) + (-2, 1)$.
d. $(3, 5) + (-1, -1)$.
e. $(3, 4) + (0, -2)$.
f. $(6, 2) - (1, 4)$.
g. $(3, 1) - (2, -3)$.
h. $(-2, 5) - (-3, 2)$.
i. $(-2, -1) - (-1, -2)$.
j. $(5, -6) - (5, 6)$.
k. $(1, 2)(2, 1)$.
l. $(1, 3)(3, 5)$.
m. $(1, 2)(6, 3)$.
n. $(2, 5) + (-1, -1)$.
o. $(2, 4)(-1, -1)$.

p. $(12, 2)(\frac{1}{2}, \frac{1}{2})$.
q. $(3, 0)(0, 1)$.
r. $(5, 6)(-1, -2)$.
s. $(-2, 1)(1, -2)$.
t. $(2, 0)(0, 3)$.
u. $(1, 2) \div (3, 1)$.
v. $(4, 4) \div (3, 5)$.
w. $(6, 2) \div (1, 2)$.
x. $(6, 2) \div (2, 0)$.
y. $(2, -3) \div (-4, 1)$.
z. $(5, 2) \div (-3, -2)$.
a'. $(16, 0) \div (0, 1)$.
b'. $(3, -2) \div (1, 1)$.
c'. $(5, -6) \div (2, 3)$.
d'. $(0, 5) \div (5, 0)$.

2. Find each of the following $((a, b)^n = (a, b)(a, b) \cdots (a, b)$, where n factors (a, b) appear).

a. $(a, b) + (0, 0)$.
b. $(a, b)(1, 0)$.
c. $(a, b) \div (1, 0)$.
d. $(a, b) - (0, 0)$.

e. $(a, b)(0, 0)$.
f. $(a, b)^2$.
g. $(a, b)(a, -b)$.
h. $(a, b)(0, 1)$.

i. $(0, 1)^2$.

j. $(0, 1)(0, -1)$.

k. $(0, 1)^3$.

l. $(0, 1)^4$.

m. $(0, 1)^{4n}, n \in N$.

n. $(0, 1)^{4n+1}, n \in N$.

o. $(0, 1)^{4n+2}, n \in N$.

p. $(0, 1)^{4n+3}, n \in N$.

3. Show that "=" is an equivalence relation on the plane [that is, $(a, b) = (a, b)$, if $(a, b) = (c, d)$, $(c, d) = (a, b)$, and if $(a, b) = (c, d)$, $(c, d) = (e, f)$, then $(a, b) = (e, f)$].

4. Show that, if $(a, b) + (c, d)$ can be given as both (e, f) and (g, h), $(e, f) = (g, h)$ (that is, show that there is "only one" sum). Answer the same question for $(a, b)(c, d)$, $(a, b) - (c, d)$, and $(a, b) \div (c, d)$, $(c, d) \neq (0, 0)$.

5. If for each (a, b), $(a, b) + (c, d) = (a, b)$, what is (c, d)?

6. If (c, d) is such that $(a, b)(c, d) = (a, b)$ for all points (a, b), what is (c, d)?

7. If $(a, b)(c, d) = (-b, a)$ for all (a, b), what is (c, d)?

8. Show that $(a, b)(c, 0) = c(a, b)$ and $(a, b) \div (c, 0) = (1/c)(a, b)$ if $c \neq 0$.

9. Compute:

a. $(0, k)(a, b)$.

b. $(a, b) \div (0, k)$.

c. $(a, b) + (a, -b)$.

d. $(a, b) - (a, -b)$.

e. $(a, b)(a, -b)$.

10. What is the geometric relation between (a, b) and $(a, -b)$?

11. Compute:

a. $(r \cos a, r \sin a) \cdot (s \cos b, s \sin b)$.

b. $(r \cos a, r \sin a) \div (s \cos b, s \sin b)$.

c. $(r \cos a, r \sin a) \cdot (r \cos a, r \sin a)$.

d. $(r \cos a, r \sin a)(r \cos a, r \sin a)(r \cos a, r \sin a)$.

6.3 THE COMPLEX PLANE (NUMBERS)

In the last section, we stated that the plane, with the operations defined, is an important algebraic structure. It is, in fact, a *field* (see Exercise 6.3.4).

Let us consider the function $i : R \longrightarrow C$, C the (complex) plane with these operations, given by $i(a) = (a, 0)$. For example, i takes 6 to $(6, 0)$ and $-\frac{1}{2}$ to $(-\frac{1}{2}, 0)$. Geometrically, i is the mapping that takes the real numbers onto the x axis in a $1:1$ fashion. Thus, we say that the complex plane contains a "copy" of the real numbers. In fact, from this point we shall not differentiate between a and $(a, 0)$.

The complex number (a, b) can be written in the form $(a, 0) + (0, b)$.

Thus, we say that a (or $(a, 0)$) is the **real part** of (a, b). The term b [or $(b, 0)$] is called the **imaginary part** of (a, b). The real and imaginary parts can be interpreted geometrically in terms of projections. (See Fig. 6.3.1.)

Figure 6.3.2 shows that (a, b) and $(a, -b)$ have the geometric relationship of being "reflections about the x axis." The point $\alpha(p)$ is on the vector (or its extension) determined by (a, b) while $\alpha(-p)$ is on that vector (or its extension) determined by its (complex) **conjugate** $(a, -b)$. (See Fig. 6.3.3.)

The complex field has one property not held by the field of real numbers. Each polynomial over the field of complex numbers has a zero in the field.

Example 6.3.1   The expression $x^2 + 1$ is a polynomial over the real numbers (that is, the coefficients are real numbers) but does not have a real

Figure 6.3.1

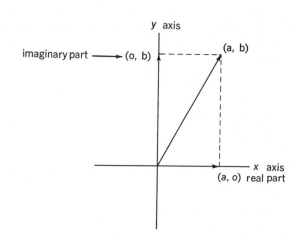

Figure 6.3.2
Complex conjugates

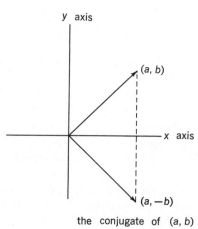

the conjugate of (a, b)

Figure 6.3.3

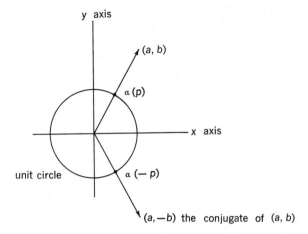

zero. In other words, $x^2 + 1 = 0$ is impossible if x must be a real number. However, if we allow x to be a complex number, there is a solution.

Example 6.3.2   The equation $x^2 + 1 = 0$ or $x^2 + (1,0) = (0,0)$ has a solution in the complex plane.

This follows since, letting $x = (0, 1)$, we have $(0, 1)(0, 1) + (1, 0) = (-1, 0) + (1, 0) = (0, 0)$.

■ EXERCISES 6.3

1. Show that the mapping $i: R \longrightarrow C$ by $i(a) = (a, 0)$ is $1:1$.

2. Show that i from 1 satisfies:

$$i(a + b) = i(a) + i(b)$$
$$i(ab) = a(i(b)) = i(a)i(b)$$
$$i(a) = (0, 0) \quad \text{if and only if } a = 0$$

A function i satisfying these properties is called an **isomorphism** from R into C.

3. Show (use Exercise 2) that $i(a) = ai(1)$.

4. Show that C is a field (that is, satisfies the first 9 conditions from Chapter 1).

5. a. Find the real part of each number and answer in Exercise 6.2.2.
   b. Find the imaginary part of each number and answer in Exercise 6.2.2.

6. In terms of real and imaginary parts, what is $p_1(a, b)$? $p_2(a, b)$?

7. What is the length of the vector associated with (a, b)?

8. It is possible that you have seen the complex numbers in another form. Writing (a, b) as $a + bi$, we observe the classic designation of a complex number. The i is endowed with the property that $i^2 = -1$ or $i = \sqrt{-1}$. Now it is agreed that $a + 0i = a$ and $0 + bi = bi$. The reader can then rewrite all the definitions of this section in terms of the $a + bi$ form. Multiplication, for instance, can be performed by the same mechanical pattern as a polynomial multiplication $[(a + bx)(c + dx) = ac + adx + bxc + bdx^2 = ac + (ad + bc)x + bdx^2]$. Multiplying $(a + bi)(c + di)$ in this fashion, we have:

$$
\begin{aligned}
(a + bi)(c + di) &= ac + adi + bic + bidi \\
&= ac + (ad + bc)i + bdi^2 \\
&= ac + (ad + bc)i - bd \\
&= (ac - bd) + (ad + bc)i
\end{aligned}
$$

The process of this manner of solution assumes all kinds of properties including commutativity, and so on. It does, however, generate an answer analogous to that in the material of this section.

Division can be performed in an easy mechanical fashion.

$$
\begin{aligned}
\frac{a + bi}{c + di} &= \frac{a - bi}{c + di}\frac{c - di}{c - di} = \frac{(ac + bd) + (bd - ad)i}{c^2 + d^2} \\
&= \frac{ac + bd}{c^2 + d^2} + \frac{(bd - ad)i}{c^2 + d^2}
\end{aligned}
$$

Again, no proof of the validity of the chain of operations is given. The process does, though, generate the analogous answer.

   a. Rewrite all definitions given in Sections 6.2 and 6.3 in terms of the $a + bi$ form.

   b. Rewrite and redo Exercises 6.2.1 and 6.2.2 in the $a + bi$ form.

6.4 COMPLEX NUMBERS AND EXPONENTS

Exponential notation can be employed as a descriptive symbolism. If n is a positive integer, $(a, b)^n$ means $(a, b)(a, b)^{n-1}$ where $(a, b)^0 = (1, 0)$. The expression $(a, b)^{-n}$ means $1/(a, b)^n$, being interpreted as $(1, 0) \div (a, b)^n$.

Example 6.4.1 $(3, 1)^3 = (3, 1)(3, 1)(3, 1) = (18, 26) = 2(9, 13)$.

Since (a, b) is a point in the plane, it can be written in the form $(r \cos \theta, r \sin \theta)$ where $r = (a^2 + b^2)^{1/2}$ and θ is an appropriate real number (Section 6.1). Then $(a, b) = (r \cos \theta, r \sin \theta) = r(\cos \theta, \sin \theta)$. The latter form is called the **trigonometric form** of (a, b); r is called the **absolute value** of (a, b) and θ is called the **argument.** The absolute value of (a, b) is written $|(a, b)|$.

Example 6.4.2 $(1, 1) = (\sqrt{2} \cos \pi/4, \sqrt{2} \sin \pi/4) = \sqrt{2}(\cos \pi/4, \sin \pi/4)$.
The argument is $\pi/4$ and $|(1, 1)| = \sqrt{2}$.
Let $(a, b) = r(\cos \theta, \sin \theta)$ and $(c, d) = s(\cos \Phi, \sin \Phi)$.   Then,

$$
\begin{aligned}
(a, b)(c, d) &= (r \cos \theta, r \sin \theta)(s \cos \Phi, s \sin \Phi) \\
&= (rs \cos \theta \cos \Phi - rs \sin \theta \sin \Phi, rs \cos \theta \sin \Phi \\
&\qquad\qquad\qquad\qquad\qquad\qquad\qquad + rs \sin \theta \cos \Phi) \\
&= rs(\cos \theta \cos \Phi - \sin \theta \sin \Phi, \cos \theta \sin \Phi + \sin \theta \cos \Phi) \\
&= rs(\cos (\theta + \Phi), \sin (\theta + \Phi))
\end{aligned}
$$

To multiply two pairs, we first write each in trigonometric form. We
then form the product of the respective absolute values with the term
$(\cos (\theta + \Phi), \sin (\theta + \Phi))$. *The argument of the product is the sum of the
two respective arguments.*

Example 6.4.3   Find $(3, 3)(1, 3)$.

$$
\begin{aligned}
(3, 3)(1, \sqrt{3}) &= 3\sqrt{2}\left(\cos \frac{\pi}{4}, \sin \frac{\pi}{4}\right) \cdot 2\left(\cos \frac{\pi}{3}, \sin \frac{\pi}{3}\right) \\
&= 6\sqrt{2}\left(\cos \frac{7\pi}{12}, \sin \frac{7\pi}{12}\right)
\end{aligned}
$$

See Fig. 6.4.1.

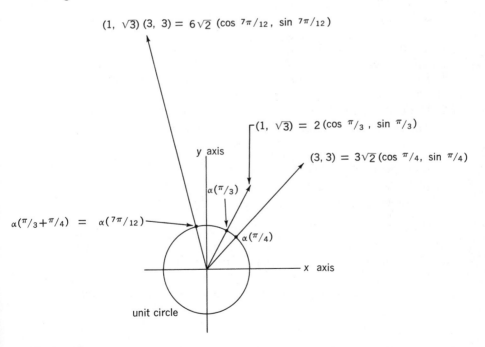

Figure 6.4.1
Products of complex numbers

This result, together with mathematical induction, can be used to prove *De Moivre's Theorem:* if $n \in Z$ and if $(a, b) = r(\cos \theta, \sin \theta)$,

$$(a, b)^n = [r(\cos \theta, \sin \theta)]^n = r^n(\cos n\theta, \sin n\theta)$$

Example 6.4.4 To illustrate De Moivre's Theorem, find $(2, 2)^4$.

$$(2, 2)^4 = \left[2\sqrt{2}\left(\cos \frac{\pi}{4}, \sin \frac{\pi}{4}\right)\right]_4$$
$$= (2\sqrt{2})^4(\cos \pi, \sin \pi) = 64(-1, 0) = (-64, 0)$$

■ **EXERCISES 6.4**

1. a. Give the absolute value of each complex number in Exercises 6.2.1 and 6.2.2, and the absolute value of each answer.
  b. Find the argument of each complex number used in Exercises 6.2.1 and 6.2.2, as well as the argument of each answer.

2. Find each of the following.

   a. $(3, 3)^6$.   **e.** $(0, -2)^4$.
   b. $(2, 0)^4$.   **f.** $(1, \sqrt{3})^6$.
   c. $(-1, 0)^3$.   **g.** $(3\sqrt{3}, 3)^3$.
   d. $(0, 3)^3$.   **h.** $(5, 5)^{-4}$.

3. Show that $[r(\cos a, \sin a)]/[s(\cos b, \sin b)] = r/s(\cos (a - b), \sin (a - b))$, if $s \neq 0$.
4. Use mathematical induction to prove De Moivre's Theorem.
5. The symbol $re^{i\theta}$ is used to represent $r(\cos \theta, \sin \theta)$.

   a. Write each number in Exercises 6.2.1 and 6.2.2 in the form $re^{i\theta}$.
   b. Show that $(re^{i\theta})(se^{i\Phi}) = rse^{i(\theta+\Phi)}$.
   c. Show that $re^{i\theta}/se^{i\Phi} = (r/s)e^{i(\theta-\Phi)}$.
   d. Show that $(re^{i\theta})^n = r^n e^{in\theta}$.

6. The $re^{i\theta}$ form from Exercise 5 is like the polar form in that the r and θ are the determining factors in the makeup of the number. Thus, we call $re^{i\theta}$ the **polar form** of $r(\cos \theta, \sin \theta)$. Write the following in the $a + bi$ form (see Exercise 6.3.8). Then write each in the (a, b) form and locate the point in the plane. (The point $re^{i\theta}$ may be located as $(r, \theta)^*$ in polar form.)

   a. $3e^{\pi i/2}$.   **d.** $6e^{5\pi i/6}$.   **g.** $21e^{-3\pi i/4}$.
   b. $e^{2\pi i}$.   **e.** $2e^{\pi i}$.   **h.** $4e^{-3\pi i/4}$.
   c. $5e^{\pi i/6}$.   **f.** $3e^{3\pi i/2}$.

7. Find in each case at least one suitable complex number for (x, y).

    a. $(x, y)^2 = (0, 1)$.
    b. $(x, y)^2 = (2, 2)$.
    c. $(x, y)^3 = (0, 1)$.
    d. $(x, y)^4 = (8, 8\sqrt{3})$.

6.5 ROOTS OF COMPLEX NUMBERS

As we might expect, exponential notation can be employed in the case of roots. We must define, though, what a root of a complex number ought to be.

Suppose (c, d) is any complex number. If $n \in N$, (a, b) is an **nth root** of (c, d) means (just as we would expect) $(a, b)^n = (c, d)$. We shall also show this same relationship by $(a, b) = (c, d)^{1/n}$. The fractional exponent is used to indicate any nth root. If $(c, d) \neq (0, 0)$, we see that there are precisely n *distinct* nth roots of (c, d). This is quite different from the case for the real numbers (for example, there is only one real cube root of -27).

Let us set about finding the n different nth roots for (c, d). First, write (c, d) in its trigonometric form, say $(c, d) = r(\cos \theta, \sin \theta)$ with $0 \leqslant \theta < 2\pi$. Recall that for any positive integer k,

$$(c, d)^k = r^k(\cos k\theta, \sin k\theta)$$

It would be a natural desire then to have $(a, d)^{1/n} = r^{1/n}(\cos \theta/n, \sin \theta/n)$, where $r^{1/n}$ is the *principal* nth root of r. This result is found by replacing k above by $1/n$. By the definition of $(c, d)^{1/n}$, $((c, d)^{1/n})^n = (c, d)$. Thus, we need to evaluate $(r^{1/n}(\cos \theta/n, \sin \theta/n))^n$. By De Moivre's Theorem,

$$(r^{1/n}(\cos \theta/n, \sin \theta/n))^n = r(\cos \theta, \sin \theta) = (c, d)$$

Necessarily, this analogue of De Moivre's Theorem holds for the case of roots.

Our analysis is not complete, however. Let $k \in Z$ and examine the following.

$$\left[r^{1/n}\left(\cos \frac{\theta + 2k\pi}{n}, \sin \frac{\theta + 2k\pi}{n} \right) \right]^n = r(\cos (\theta + 2k\pi), \sin (\theta + 2k\pi))$$
$$= r(\cos \theta, \sin \theta) = (c, d)$$

Hence, we conclude that each number of the form

$$r^{1/n}\left(\cos \theta + \frac{2k\pi}{n}, \sin \theta + \frac{2k\pi}{n} \right)$$

is an nth root of $(c, d) = r(\cos \theta, \sin \theta)$.

Does there exist an infinite number of nth roots of (c, d)? No. If $k \in \{0, 1, 2, \cdots n - 1\}$, the corresponding values of $r^{1/n}(\cos (\theta + 2\pi k)/n, \sin (\theta + 2\pi k)/n)$ form a set of n distinct roots [if $(c, d) \neq (0, 0)$]. If $k = n$, $r^{1/n}(\cos (\theta + 2n\pi)/n, \sin (\theta + 2n\pi/n) = r^{1/n}(\cos \theta/n, \sin \theta/n)$. The latter is the same root as that given by $k = 0$. Similarly, if $k \in Z - \{0, 1, 2, \cdots n - 1\}$,

$$r^{1/n}\left(\cos \theta + \frac{2k\pi}{n}, \sin \theta + \frac{2k\pi}{n}\right)$$

is one of the roots given when $k \in \{0, 1, 2, \cdots n - 1\}$. (The reader should verify this algebraically.) Consequently, there exist precisely n distinct nth roots of each $(c, d) \neq (0, 0)$.

The following worked example shows an application of the above reasoning.

Example 6.5.1 Find the four fourth roots of $16 = (16, 0) = 16(\cos 0, \sin 0)$. These roots are:

$16^{1/4}(\cos (0 + 0\pi)/4, \sin (0 + 0\pi)/4) = 2(\cos 0, \sin 0) = (2, 0)$
$16^{1/4}(\cos (0 + 2\pi)/4, \sin (0 + 2\pi)/4) = 2(\cos \pi/2, \sin \pi/2) = (0, 2)$
$16^{1/4}(\cos (0 + 4\pi)/4, \sin (0 + 4\pi)/4) = 2(\cos \pi, \sin \pi) = (-2, 0)$
$16^{1/4}(\cos (0 + 6\pi)/4, \sin (0 + 6\pi)/4) = 2(\cos 3\pi/2, \sin 3\pi/2) = (0, -2)$

It is of interest to note that the nth roots of $r(\cos \theta, \sin \theta)$ are all on the circle $x^2 + y^2 = (r^{1/n})^2$ and the difference in the argument of any two "successive" points is $2\pi/n$. The roots from the vertices of a regular n-sided polygon inscribed in the given circle. See Fig. 6.5.1.

The following example together with Fig. 6.5.2 illustrates this point.

Example 6.5.2 Compute the three cube roots of $(4\sqrt{2}, 4\sqrt{2})$.
Now, $(4\sqrt{2}, 4\sqrt{2}) = 8(\cos \pi/4, \sin \pi/4)$ and one cube root is:

$$8^{1/3}\left(\cos \frac{\pi}{12}, \sin \frac{\pi}{12}\right) = 2\left(\cos \frac{\pi}{12}, \sin \frac{\pi}{12}\right)$$

The difference in argument between successive roots is $2\pi/3$. Since $\pi/12 + 2\pi/3 = 3\pi/4$ and $3\pi/4 + 2\pi/3 = 17\pi/12$, we have for the other pair of roots:

$$2(\cos 3\pi/4, \sin 3\pi/4)$$
$$2(\cos 17\pi/12, \sin 17\pi/12)$$

For the geometric interpretation, see Fig. 6.5.2.

The notation $(a, b)^{m/n}$ means $((a, b)^{1/n})^m$, where n is any positive integer and $m \in Z$.

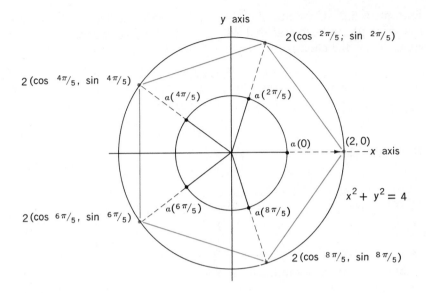

Figure 6.5.1
The five fifth roots of (32, 0)

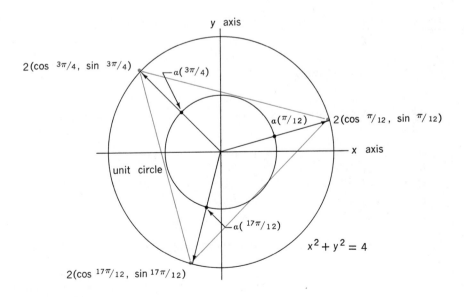

Figure 6.5.2

Example 6.5.3 Compute $(4\sqrt{2}, 4\sqrt{2})^{2/3}$. We know that $(4\sqrt{2}, 4\sqrt{2})^{1/3}$ has three possibilities and that $(4\sqrt{2}, 4\sqrt{2})^{2/3}$ is the square of any cube root. Thus, the choices for $(4\sqrt{2}, 4\sqrt{2})^{2/3}$ are:

$$4(\cos \pi/6, \sin \pi/6)$$
$$4(\cos 3\pi/2, \sin 3\pi/2)$$
$$4(\cos 11\pi/6, \sin 11\pi/6)$$

■ **EXERCISES 6.5**

Find all roots in Exercises 1 to 8.
1. $(16, 0)^{1/2}$.   5. $(50\sqrt{2}, 50\sqrt{2})^{1/2}$.
2. $(8, 0)^{1/3}$.   6. $(1, 0)^{1/10}$.
3. $(0, 27)^{1/3}$.   7. $(1, \sqrt{3})^{1/2}$.
4. $(0, -27)^{1/3}$.   8. $(2\sqrt{3}, 2)^{1/4}$.
Solve the equations in Exercises 9 to 12.
9. $(a, b)^4 = (0, -81)$.
10. $(a, b)^2 = (16, 0)$.
11. $(a, b)^3 = (-125, 0)$.
12. $(a, b)^5 = (-16, 16\sqrt{3})$.
13. Rewrite all the definitions of the section in terms of the $a + bi$ form. Write out the worked examples in the $a + bi$ form and observe the corresponding analysis.
14. Follow the instructions of Exercise 13 for the polar form re^i.

6.6 REVIEW EXERCISES

1. The following are in polar form.   Rewrite them in rectangular coordinates.

a. $(-3, 27\pi/4)^*$.   f. $(-8, -\sin^{-1} \frac{1}{2})^*$.
b. $(2, -21\pi/6)^*$.   g. $(12, \pi - \cos^{-1} (\%_{17}))^*$.
c. $(13, 285\pi)^*$.   h. $(-11, \cos^{-1} (-\frac{22}{13}))^*$.
d. $(21, -23\pi/3)^*$.   i. $(r, \theta)^*$.
e. $(16, \tan^{-1} \frac{3}{4})^*$.

2. The following are in rectangular coordinates. Convert each to polar coordinates.

a. $(25, 60)$.   d. $(29, 29)$.   g. $(-51, 51\sqrt{3})$.
b. $(24, 25)$.   e. $(30, 0)$.   h. $(13, -3/\sqrt{3})$.
c. $(51, 68)$.   f. $(0, -18)$.

3. Describe each of the following curves given in polar form (the set of points of the form $(r, \theta)^*$ satisfying each of the following).

a. $\theta = \pi/4$.   **e.** $r = 6$.
b. $\theta = \pi/6$.   **f.** $r = 17$.
c. $r = 64 \cos \theta$.   **g.** $r = \theta$.
d. $r = 50 \sin \theta$.

4. Give the absolute value and argument of each complex number. Find also the real and imaginary part of each.

a. $3 + 3i$.   **f.** $-4 + 4i$.
b. 6.   **g.** $2 - i$.
c. $2 + 2\sqrt{3}i$.   **h.** $2 + i$.
d. $-6i$.   **i.** $3 - 3i$.
e. -16.

5. Perform the following operations.

a. $(3 - 2i) + (5 + i)$.   **g.** $(3 - i)(4 + 2i)$.
b. $(-2 + i) + (6 - i)$.   **h.** $(3 + 2i)(-3 - 2i)$.
c. $(1 - i) - (-2 + i)$.   **i.** $(a + bi) \div (a + bi)$.
d. $(5 + 3i) - (3 - 3i)$.   **j.** $(4 - 2i) \div (-3 + i)$.
e. $(1 + i)(1 - i)$.   **k.** $(5 + 6i) \div (2 - 3i)$.
f. $i(1 + i)$.   **l.** $(-i)(-3 + i)$.

6. Square each number in Exercise 4.
7. Cube each number in Exercise 4.
8. Find all square roots of each number in Exercise 4.
9. Find all fourth roots of each number in Exercise 4.

6.7 ADVANCED EXERCISES

1. Show that $d((r, \theta)^*, (s, \Phi)^*) = r^2 + s^2 - 2rs \cos(\theta - \Phi)$ where $(r, \theta)^*$ and $(s, \Phi)^*$ are in polar form.
2. Sketch the following curves. They are given in polar coordinates.

a. $r = k(1 + \cos \theta)$.   **d.** $r^2 = k^2 \cos 2\theta$.
b. $r \cos \theta = 2$.   **e.** $r = k/2$.
c. $r^2 = 1 + \sin^2 \theta$.

3. Find the point(s) of intersection of (d) and (e) from Exercise 2.
4. Show that $(a + bi)(c + di) = 0$ implies $a^2 + b^2 = 0$ or $c^2 + d^2 = 0$.
5. Solve the following equations, the solution set being a set of complex numbers.

a. $z^2 + 2z + 1 = 8\sqrt{2} + 8\sqrt{2}$. (Hint: The left side is a perfect square.)

b. $z^2 + 2z = 5 - 2i$.

c. $z^{1/2} = 2 + i$.

d. $z^5 = 32$.

6. Let z and a be complex numbers in the ordered pair form. Describe the set of z where:

a. $|z - a| = 2$.

b. $|z - a| = 20$.

c. $|z - a| < 2$.

d. $|z - a| > 2$.

e. $R(z) = 2$ ($R(z)$ is the real part of z).

f. $I(z) = -2$ ($I(z)$ is the imaginary part of z).

g. $R(z) < 0$.

h. $I(z) > 0$.

i. $I(z)/R(z) = 1$.

j. $R^2(z) + I^2(z) = 4$.

k. $R(z) > 0$ and argument $z = \pi/3$.

Chapter 7
Trigonometry
and
Angles

7.1 RAYS AND ANGLES

This chapter is devoted to a discussion of the trigonometric functions in a geometric setting. Such an approach opens up areas of applications, some of which are usually considered (and rightfully so) to be a natural part of trigonometry. For example, the idea of a vector in the complex plane leads to the notion of an angle. The approach taken in this chapter does not assume a knowledge of any material in Chapter 4, except for Section 4.1.

The trigonometric functions will be approached in two settings, each setting having been traditional for many years. Central in the discussion to be presented is the concept of an angle.

The term "ray" or "half-line" will be used, a definition having been given in Section 2.5. It will be assumed that you have an intuitive feeling for the topic of rays.

Example 7.1.1 Figure 7.1.1 illustrates a ray of slope 1 emanating from the point $(3, 3)$. The ray constitutes part of the line ℓ having equation $y = x$.

Figure 7.1.1
A ray

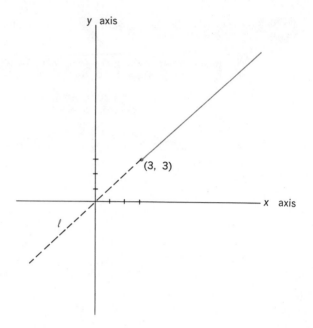

Now, if each of ℓ and ℓ' is a ray emanating from a point P, we say that an **angle** of vertex P is formed. In fact, it will be the case that an infinite number of angles are formed. This should, perhaps, sound surprising. Our text will use a concept of angle that involves more than two rays emanating from a point.

In other words, two rays will not completely characterize a unique angle.

One method we shall employ in order to differentiate between angles is the labeling of the sides of each angle (that is, the rays forming the angle). We shall label one side the **initial side** and the other the **terminal side.**

Given two rays emanating from a point, we may, by our perrogative, label the rays in two different manners. Figure 7.1.2 shows two views of rays ℓ and ℓ' emanating from P. Each view displays one of the posssible ways of titling the rays.

It will be the case that even discerning between initial and terminal sides is not enough to determine a unique angle. To illustrate this condition, consider the two rays ℓ_i and ℓ_t emanating from P (see Fig. 7.1.3). The subscripts i and t denote the initial and terminal sides, respectively. It is of practical value to consider forming the figure by rotating ℓ_i (fixed at P) from its position until it is superimposed over ℓ_t. The amount of rotation incurred is to be called the *measure* of the determined angle.

Figure 7.1.3(b) is a copy of 7.1.3(a) except that an arrow is used to indicate rotation from ℓ_i to ℓ_t.

Figure 7.1.2

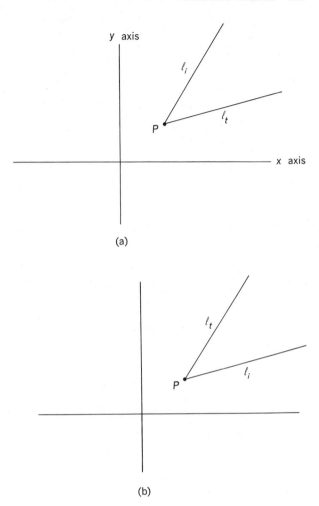

(a)

(b)

Figure 7.1.3(c) shows the same rays but with the rotation's being in the opposite sense (direction).

Figure 7.1.3(d) again displays the same rays. A still different rotation (counterclockwise but in excess of one complete rotation) of ℓ_i is shown.

Thus, to completely determine a unique angle, we must know:

1. The initial and terminal sides.
2. The associated measure.

Now, two angles are said to be **equal** if and only if they have the same measure. Intuitively, we believe that equal angles may be geometrically superimposed. That is the case, although this superimposition is not a sufficient criterion by which we can determine the equality of angles. The

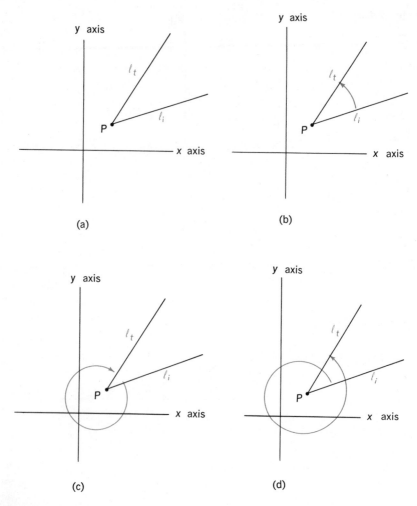

Figure 7.1.3
Different angles

three angles of Figs. 7.1.3(b), 7.1.3(c), and 7.1.3(d) are not equal (it will be seen that, since the rotations of ℓ_i to ℓ_t are different, their measures are not equal). They can, however, be superimposed one over the other. In fact, the angles share common initial and terminal sides. It should seem feasible that the most useful knowledge obtainable (relative to an angle) is the knowledge of its measure.

The reader should take this opportunity to review the above paragraphs in an attempt to find a definition of the term "angle." A description, rather than a definition, has been given; angle will not be defined.

Given an angle, we need to settle on a means of determining a number to be called the measure. The basis for such a determination has been given—the amount of rotation the initial side experiences as it rotates to a position coincident with the terminal side. The number of complete rotations will be the **measure.** The *measure* shall be *positive* if the rotation is counterclockwise and *negative* if it is clockwise. Such a convention agrees with that of Section 4.1 used in attaching a polarity to arc length traversed from one point on a circle to another.

Of particular interest are angles with

1. common vertex $(0, 0)$ and
2. initial side the nonnegative portion of the x axis (right half).

Such angles are said to be in *standard position.* Angles that are in standard position and also share the same terminal side are said to be **coterminal.**

Angles (in standard position) whose terminal sides fall on one of the axes are of particular interest and are called **quadrantal angles.** Such angles have measure a multiple of ¼ rotation. Figures 7.1.4(a) to 7.1.4(e) illustrate a few of the quadrantal angles. Once again, it is important to review the fact that an initial and terminal side can represent an infinite number of angles in the sense that an infinite number of measures may be attached.

Example 7.1.2 Figure 7.1.5 shows an angle with terminal side in the second quadrant. The counterclockwise rotation needed to superimpose the initial side over the terminal side is ⅜ of a rotation. However, an angle (in standard position) of ¹⅜ rotations has this same terminal side. In fact, any number of the form ⅜ + $n(n \in Z)$ represents the rotational measure of an angle coterminal with the two discussed. A similar arrangement exists for any given initial and terminal side.

Two systems of measurement (other than the rotation) are those of the **radian** and the **degree.** These systems are defined in terms of the rotation in the following way:

$$1 \text{ rotation} = 2\pi \text{ radians}$$
$$1 \text{ rotation} = 360° = 360 \text{ degrees}$$

We generalize these statements by multiplying both sides of each "equation" by a.

$$a \text{ rotations} = 2a\pi \text{ radians}$$
$$a \text{ rotations} = a \cdot 360°$$

The statement that 1 rotation = 2π radians is actually incorrect. What we mean is that an angle of 1 rotation has the same radian measure as an angle of measure 2π radians. In other words, an angle in standard position having measure one rotation is equal to an angle in standard position having measure 2π radians. The same analogy holds for the other "equa-

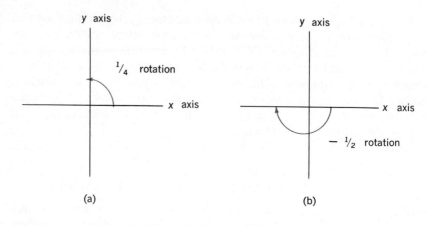

(a)

(b)

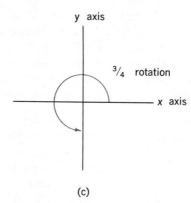

(c)

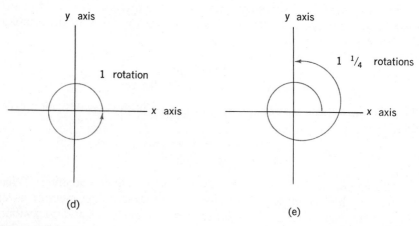

(d)

(e)

Figure 7.1.4
Some quadrantal angles

Figure 7.1.5

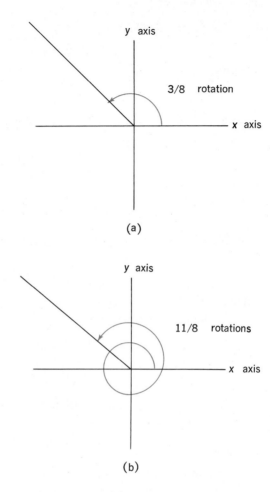

(a)

(b)

tions." Such statements prove useful, however, when converting from one system of measurement to another. The following is an example.

Example 7.1.3   One equality we can deduce is:

$$2\pi \text{ radians} = 360°$$

or alternatively,

$$\pi \text{ radians} = 180°$$

Example 7.1.4   As an example of the correlation between the systems of measures, convert ¾ rotation to radian measure and to degree measure.

$$¾ \text{ rotation} = (¾)(2\pi) \text{ radians} = 3\pi/2 \text{ radians}$$
$$¾ \text{ rotation} = (¾)(360°) = 270°$$

Example 7.1.5 Convert $\pi/4$ radians to degrees.

$$\pi \text{ radians} = 180°$$

Dividing by 4, we have

$$\pi/4 \text{ radians} = 45°$$

Example 7.1.6 As another exercise, convert 5 radians to degrees.

$$\pi \text{ radians} = 180°$$

$$1 \text{ radian} = \left(\frac{180}{\pi}\right)°$$

$$5 \text{ radians} = 5\left(\frac{180}{\pi}\right)° = \left(\frac{900}{\pi}\right)°$$

Now, using 3.14 as an approximation for the irrational number π, we see that

$$5 \text{ radians} = 287° \qquad \text{approximately}$$

Example 7.1.7 To reverse the problem of Example 7.1.6, find the number of radians that corresponds to a measure of 150°.

$$180° = \pi \text{ radians}$$

$$1° = \left(\frac{\pi}{180}\right) \text{ radians}$$

$$150° = 150\left(\frac{\pi}{180}\right) \text{ radians}$$

$$150° = \frac{5\pi}{6} \text{ radians}$$

The degree system of measurement has two subunits of measure. They are the **minute** (′) and **second** (″). They are defined by

$$1° = 60' \qquad \text{and} \qquad 1' = 60''$$

Example 7.1.8 Find the quadrant of termination of an angle in standard position having measure 67 radians.

We can easily find the quadrant of termination for angles of less than one rotation. Thus we determine the number of rotations by

$$\frac{67}{2\pi} = 10.7 \text{ rotations} \qquad \text{approximately}$$

Thus, the angle is coterminal with an angle of approximately .7 rotation and terminates in quadrant III ($.5 < .7 < .75$).

Example 7.1.9 Find the quadrant of termination for an angle in standard position having measure 3976°.

Figure 7.1.6
Arc length

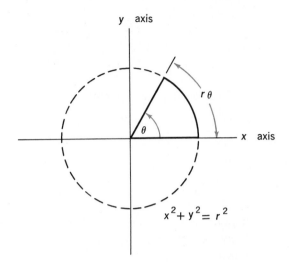

As in Example 7.1.8, we compute the rotational measure:

$$\frac{3976}{360} = 11.02 \qquad \text{approximately}$$

Thus, the angle terminates in quadrant I, coterminal with an angle having approximate measure .02 rotation $(0 < .02 < .5)$.

An interesting relation between the radian measure of an angle and arc length is to be found. Consider a circle $x^2 + y^2 = r^2$ and an angle (in standard position) of radian measure θ. For simplicity, suppose θ satisfies $0 < \theta < 2\pi$. The length of arc subtended by this (central) angle [that is, the length of traversal of the point $(r, 0)$ on the initial side as it is rotated through an angle of measure θ] is given by

$$\text{arc length} = s = r\theta$$

Intuitively, the length of arc s has the same ratio to the circumference $2\pi r$ as the measure θ of the angle θ in radians has to 2π (the radian measure for an angle of one rotation). That is,

$$\frac{s}{2\pi r} = \frac{\theta}{2\pi}$$

or
$$s = r\theta$$

See Fig. 7.1.6. Further discussions on arc length may be found in the advanced exercises of Chapter 4. The above relationship is for radian measure only.

Example 7.1.10 Find the length of the arc on the circle $x^2 + y^2 = 64$ intercepted by an angle (in standard position) of measure $135°$.

$$135° = \frac{3\pi}{4} \text{ radians}$$

$$s = r\theta = 8\left(\frac{3\pi}{4}\right) = 6\pi$$

It is of interest to note in passing that, if the circle is the unit circle,

$$s = r\theta \qquad \text{becomes} \qquad s = \theta$$

That is, the arc length and radian measure of the subtending angle *agree*.

EXERCISES 7.1

All angles are assumed to be in standard position.

1. Draw a diagram of each angle showing initial and terminal side and the rotation where the measure of the angle is:

a. $\frac{1}{8}$ rotation.
b. $\frac{7}{8}$ rotation.
c. $\frac{3}{8}$ rotation.
d. $\frac{5}{8}$ rotation.
e. $\frac{17}{8}$ rotation.
f. $-3\frac{1}{8}$ rotations.
g. $640\frac{3}{8}$ rotations.
h. $-\frac{3}{8}$ rotation.
i. $\frac{1}{12}$ rotation.
j. $\frac{1}{6}$ rotation.
k. $\frac{1}{3}$ rotation.
l. $\frac{5}{12}$ rotation.
m. $\frac{7}{12}$ rotation.
n. $\frac{2}{3}$ rotation.
o. $\frac{5}{6}$ rotation.
p. $\frac{11}{12}$ rotation.

2. Describe in each case in Exercise 1 the complete set of measures of angles coterminal with the given angle.

3. Convert each of the measures of Exercise 1 to radian measure and then rewrite the answers to Exercise 2 in terms of radian measurement.

4. Convert each of the measures of Exercise 1 to degree measure and then rewrite the answers to Exercise 2 in terms of degree measurements.

5. In which quadrant does the terminal side of each angle of the following given measure lie?

a. 1 radian.
b. 2 radians.
c. 3 radians.
d. 4 radians.
e. 5 radians.
f. 6 radians.
g. 7 radians.
h. -25 radians.
i. 105 radians.
j. -87 radians.
k. 237 radians.
l. 1005 radians.

6. Convert each measure in 5 to:

a. Rotations.
b. Degrees.

7. Describe the complete set of measures of angles that are coterminal with each in 5.

8. In which quadrant does the terminal side of each angle of the given measure lie?

a. 70°.	**g.** −971°.
b. 143°.	**h.** 1085°.
c. 206°.	**i.** 23974°.
d. 309°.	**j.** 3500°.
e. −28°.	**k.** −2860°.
f. −388°.	**l.** 19°.

9. Convert each measure in Exercise 8 to:

a. Rotations.   **b.** Radians.

10. Describe the complete set of measures of angles that are coterminal with those in Exercise 8.

11. Find the length of the intercepted arc of the circle $x^2 + y^2 = 4$, where the subtending angle has measure:

a. 1 radian.	**g.** 60°.
b. 3 radians.	**h.** 45°.
c. $\pi/2$ radians.	**i.** 150°.
d. ¾ rotation.	**j.** 225°.
e. ⅛ rotation.	**k.** $5\pi/3$ radians.
f. 30°.	**l.** $5\pi/6$ radians.

12. Redo Exercise 11, where the arc is on the circle.

a. $x^2 + y^2 = 1$.
b. $x^2 + y^2 = 25$.
c. $x^2 + y^2 = 2$.
d. $x^2 + y^2 = 9$.

13. Show that the measure of the length of arc subtended on the circle $x^2 + y^2 = r^2$ by a central angle of 1 radian is r.

14. Let each of the special values given for $\alpha(0, \pi/6, \pi/4, \pi/3, \cdots)$ represent a radian measure. Give, in each case, the equivalent measure in degrees (compare with Exercise 1).

15. Using your information from Exercise 14, write the equation of the line containing the terminal side of θ (in standard position) where the measure of θ is:

a. 0°.	**f.** 120°.
b. 30°.	**g.** 135°.
c. 45°.	**h.** 150°.
d. 60°.	**i.** 180°.
e. 90°.	

16. Suppose an object is traveling at a constant (directed) rate around a circular path (Fig. 7.1.6) of radius r. The linear velocity V_ℓ of the object is given by $V_\ell = \Delta s/t$ where Δs is the arc length traversed during the time interval t. Since the angle (of measure) ω formed also changes with respect to time, it is appropriate to define the object's angular velocity V_ω by $V_\omega = \Delta\omega/t$, $\Delta\omega$ being the change in ω occurring during the time interval t.

a. Find k, where $\Delta s = k\,\Delta\omega$. (There are 3 choices for k depending on the system of measure. Find all 3 values.)
b. Find 3 values for c where $V_\ell = cV_\omega$.

The type of motion described here is called **uniform circular motion.**
17. Find V_ℓ (Exercise 16) where the uniform circular motion is described by:

a. An arc length of 16 ft is traversed each 2 sec.
b. $V_\omega = 3$ radians per sec and the radius of the path is 5 ft.
c. $V_\omega = 16°$ per min and the radius of the path is 20 miles.
d. $V_\omega = \%$ rotations per hr and the radius of the path is 65 yd.

18. Find V_ω (Exercise 16) where the uniform circular motion is given by:

a. An arc length of 18 ft is traversed each second on a path of radius 3 ft.
b. $V_\ell = 19,000$ miles per hour (mph) and the radius of the path is 40 miles.

19. Consider the circle given by $x^2 + y^2 = r^2$. The area of the circle is given by πr^2. What do you believe to be the area of the sector of this circle described by the central angle θ (the sector is seen in Fig. 7.1.7)? For a discussion of the area of a sector, see the advanced exercises of this chapter.

Figure 7.1.7
A sector

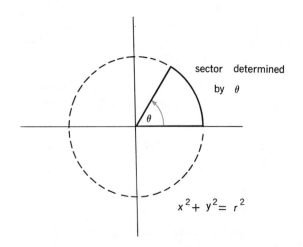

sector   determined
by θ

$x^2 + y^2 = r^2$

7.2 ANGLES AND THE TRIGONOMETRIC FUNCTIONS

The measure of an angle has been considered to be the most important single piece of information we can know.   Thus in the material to follow, no distinction will (in general) be made between an angle and its measure.   For example $\theta = \pi/6$ may be read: θ is an angle whose measure is $\pi/6$ radians. The symbol θ will, by an abuse of the language, be used to represent both an angle and the number that is its measure.   This is actually standard procedure in most texts.   Whenever the measure of an angle is given as a pure number (that is, with no system of measurement indicated), the system of measurement will be understood to be the radian system.

Let θ be an angle in standard position as shown in Fig. 7.2.1.   The point (x, y) is any point (other than the origin) on the terminal side of θ.   Let $r = (x^2 + y^2)^{1/2} = d((0, 0), (x, y))$.   The six trigonometric functions can (will) be defined by the following correspondences:

$$\sin (\theta) = \sin \theta = \frac{y}{r}$$

$$\text{cosine } (\theta) = \cos \theta = \frac{x}{r}$$

$$\text{tangent } (\theta) = \tan \theta = \frac{y}{x} \qquad \text{if } x \neq 0$$

$$\text{cosecant } (\theta) = \csc \theta = \frac{r}{y} \qquad \text{if } y \neq 0$$

$$\text{secant } (\theta) = \sec \theta = \frac{r}{x} \qquad \text{if } x \neq 0$$

$$\text{cotangent } (\theta) = \cot \theta = \frac{x}{y} \qquad \text{if } y \neq 0$$

Because of the notation, we are at liberty to interpret $\sin \theta$ as:

1. the value of the sine of an angle, or
2. the value of the sine of a number (the radian measure of θ).

Furthermore, in order for each correspondence to be representative of a function, it is necessary that for a given θ, $\sin \theta$, $\cos \theta$, $\tan \theta$, and so forth must each be a unique real number.

The value of $\sin \theta$ is apparently tied to the choice of the point (x, y) on the terminal side of the angle.   This is only an appearance, though. Regardless of the choice of point (on the terminal side), the value y/r remains the same.   The truth of this statement will be demonstrated.

If the terminal side of θ is not vertical, the equation of the line containing it is of the form $y = mx$.   (What measures θ will denote angles having

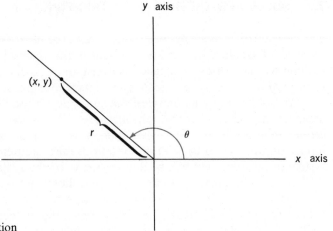

Figure 7.2.1
θ in standard position

a vertical terminal side?) Then $y/x = m =$ the slope of the terminal side. [Note: y/x is $\tan \theta$ and is independent of the point (x, y) chosen.]

Let (x_1, y_1) and (x_2, y_2) be two points on the ray (Fig. 7.2.2). Let $r_1 = (x_1{}^2 + y_1{}^2)^{1/2}$ and $r_2 = (x_2{}^2 + y_2{}^2)^{1/2}$. Remembering that $y_1/x_1 = m = y_2/x_2$, we see:

$$r_1{}^2 = x_1{}^2 + y_1{}^2 = \frac{y_1{}^2}{m^2} + y_1{}^2 = y_1{}^2\left(1 + \frac{1}{m^2}\right)$$

$$r_2{}^2 = x_2{}^2 + y_2{}^2 = \frac{y_2{}^2}{m^2} + y_2{}^2 = y_2{}^2\left(1 + \frac{1}{m^2}\right)$$

$$\frac{y_2{}^2}{r_2{}^2} = \frac{1}{1 + 1/m^2} = \frac{y_1{}^2}{r_1{}^2}$$

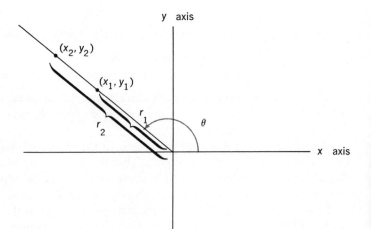

Figure 7.2.2

Since y_1 and y_2 have the same polarity (why?):

$$\frac{y_2}{r_2} = \frac{y_1}{r_1}$$

Necessarily, the choice of the point on the terminal side does not affect the value $\sin\theta$. A similar argument verifies this fact for the other functions. Thus, for each value θ, there is at most one choice for $\sin\theta$, $\cos\theta$, and so on.

Since the choice of (x, y) is arbitrary [as long as it is a point of the terminal side other than $(0, 0)$], we may choose points (x, y) on the unit circle. That is, $r = x^2 + y^2 = 1$. The correspondences then take on the appearance:

$$\sin\theta = \frac{y}{1} = y$$

$$\cos\theta = \frac{x}{1} = x$$

$$\tan\theta = \frac{y}{x} \text{ for } x \neq 0$$

$$\csc\theta = \frac{1}{y} \text{ for } y \neq 0$$

$$\sec\theta = \frac{1}{x} \text{ for } x \neq 0$$

$$\cot\theta = \frac{x}{y} \text{ for } y \neq 0$$

We recall from Chapter 4 that these are the expressions given for the trigonometric functions and that, if θ is a given radian measure, $(x, y) = \alpha(\theta)$. Consequently, we see that our new notion agrees with our former one (from Chapter IV).

The definitions indicate that some of the functions are not well defined at certain values of θ. Tan θ, for instance, is not defined for $x = 0$. The zero value for x is attained only on the y axis. In such a case, if (x, y) is on the terminal side of an angle, that angle must have measure $(2n + 1)\pi/2$ for some integer n [the numbers of the form $(2n + 1)\pi/2$ being the odd multiples of $\pi/2$]. The reader should verify this statement, and also that, when $y = 0$, θ is an angle of radian measure $n\pi$ for some integer n. Knowing these two facts, we are able to list the values θ for which any given trigonometric function fails to be defined.

Several special values of measures are such that we can determine the corresponding values of the trigonometric functions. The quadrantal angles are the easiest.

Figure 7.2.3 illustrates the points $(1, 0)$, $(0, 1)$, $(-1, 0)$, and $(0, -1)$, each being on the terminal side of an angle of measure $2n\pi$, $\pi/2 + 2n\pi$,

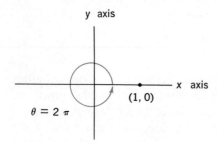

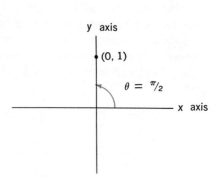

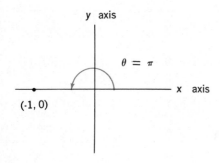

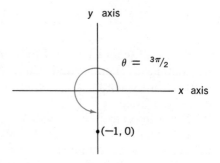

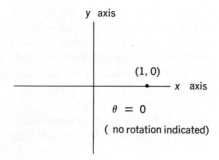

Figure 7.2.3

	0	$\pi/2$	π	$3\pi/2$	2π
sin	0	1	0	-1	0
cos	1	0	-1	0	1
tan	0	✕	0	✕	0
csc	✕	1	✕	-1	✕
sec	1	✕	-1	✕	1
cot	✕	0	✕	0	✕

Figure 7.2.4
Trigonometric functional values for quadrantal angles

$(2n + 1)\pi$, and $3\pi/2 + 2n\pi$, respectively (n an integer). Letting $n = 0$ in each case, we attain the values in the table of Fig. 7.2.4.

We should also be able to visualize the truth that the functional values of coterminal angles are identical. This property is to prove fruitful.

Important also is the concept of the reference angle. In Fig. 7.2.5, an angle θ is shown terminating in the second quadrant ($90° < \theta < 180°$). The angle $180° - \theta$ formed between the terminal side of θ and the left half of the x axis is the **reference angle** for θ. A comparison of θ and $180° - \theta$ (both in standard position) given in Fig. 7.2.6 shows:

$$\sin \theta = \sin (180° - \theta)$$
$$\cos \theta = -\cos (180° - \theta)$$
$$\tan \theta = -\tan (180° - \theta)$$

Similar diagrams of angles and their reference angles are shown in Figs. 7.2.7 and 7.2.8. The reference angles in each case are the *acute angles* (angles between 0 and 90°) made by the x axis and the terminal side of θ.

We see from the diagrams:

$$\sin \theta = -\sin (\theta - 180°)$$
$$\cos \theta = -\cos (\theta - 180°)$$
$$\tan \theta = \tan (\theta - 180°)$$

for θ satisfying $180° < \theta < 270°$. Furthermore,

$$\sin \theta = -\sin (360° - \theta)$$
$$\cos \theta = \cos (360° - \theta)$$
$$\tan \theta = -\tan (360° - \theta)$$

where θ lies between 270° and 360°. For angles θ of measure greater than 360° (or less than 0°), we can determine the reference angle from angles between 0° and 360° coterminal with θ.

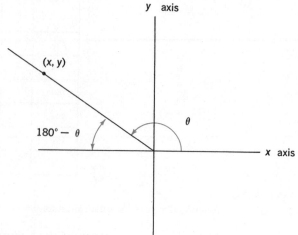

Figure 7.2.5
Reference angle for θ, $90° < \theta < 180°$

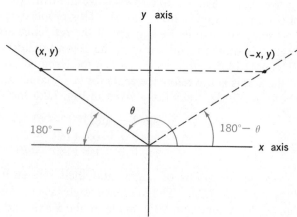

Figure 7.2.6
Another view of the reference angle

Example 7.2.1   Find the sine, cosine, and tangent functional values for θ where

$$(1)\ \theta = 136°, \quad (2)\ \theta = 245°, \quad \text{and} \quad (3)\ \theta = 327°.$$

The solution of 1 follows by observing that the reference angle is $180° - 136° = 44°$ (see Fig. 7.2.9). Thus,

$$\sin 136° = \sin 44° = .6947$$
$$\cos 136° = -\cos 44° = .7193$$
$$\tan 136° = -\tan 44° = -.9657$$

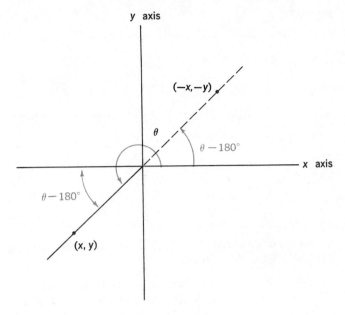

Figure 7.2.7
Reference angle for θ, $180° < \theta < 270°$

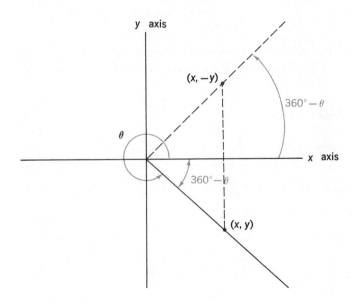

Figure 7.2.8
Reference angle for θ, $270° < \theta < 360°$

Figure 7.2.9

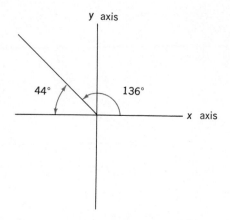

As you can see in Fig. 7.2.10, the reference angle for 245° is 245° − 180° = 65°, whence

$$\sin 245° = -\sin 65° = -.9063$$
$$\cos 245° = -\cos 65° = -.4226$$
$$\tan 245° = \tan 65° = 2.1445$$

The reference angle for 327° is 33° and is pictured in Fig. 7.2.11. Thus,

$$\sin 327° = -\sin 33° = -.5446$$
$$\cos 327° = \cos 33° = .8387$$
$$\tan 327° = -\tan 33° = -.6494$$

The numerical values in the example were taken from the tables of Appendix A. The use of these tables is shown in Appendix D.

Figure 7.2.10

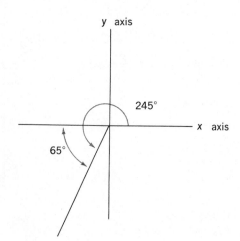

Figure 7.2.11

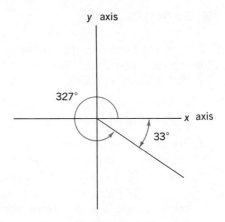

Example 7.2.2 Find the sine, cosine, and tangent functional values for θ, where (1) $\theta = 797°$, and (2) $\theta = -107°$.

The angle θ is coterminal with an angle of measure $77°$ ($797° = 720° + 77°$). Thus,

$$\sin 797° = \sin 77° = .9744$$
$$\cos 797° = \cos 77° = .2250$$
$$\tan 797° = \tan 77° = 4.3315$$

From Fig. 7.2.12, we see that θ terminates in quadrant III and has a reference angle of measure $73°$. Consequently,

$$\sin (-107°) = -\sin 73° = -.9563$$
$$\cos (-107°) = -\cos 73° = -.2924$$
$$\tan (-107°) = \tan 73° = 3.2709$$

Figure 7.2.12

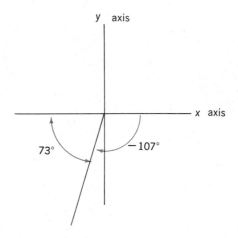

■ **EXERCISES 7.2**

1. List the measures (in radians) for which the tangent is undefined. Cotangent? Cosecant? Secant?

2. Redo Exercise 1 using:

   a. The degree system of measurement.
   b. The rotational system.

3. By an analysis similar to that used in the text, prove that the value of $\cos \theta$ is independent of the point chosen from the terminal side.

4. Let (x, y) be any point (other than $(0, 0)$) in the plane. Let $r = (x^2 + y^2)^{1/2}$ and let θ denote an angle in standard position having as its terminal side the ray from the origin through (x, y). Show that $(x, y) = (r \cos \theta, r \sin \theta)$.

5. Let β and γ be angles in standard position. (See Fig. 7.2.13.) Let $P_1 = (a, b)$ and $P_2 = (c, d)$ be on the terminal sides of β and γ, respectively, both a distance r from $(0, 0)$ (that is, both on $x^2 + y^2 = r^2$). Write (a, b) and (c, d) in the form given in Exercise 4 and then show that:

$$d(P_1 P_2) = r(2 - 2(\cos \beta \cos \gamma + \sin \beta \sin \gamma))^{1/2}$$

6. Let P_1 and P_2 be as in Exercise 5. Then place $(\beta - \gamma)$ in standard position (Fig. 7.2.14). Let $P_3 = (r \cos (\beta - \gamma), r \sin (\beta - \gamma))$ and $P_4 = (r, 0)$. Now P_3 and P_4 are both on $x^2 + y^2 = r^2$ with P_4 on the initial side of $(\beta - \gamma)$ and P_3 on the terminal side. Show

$$d(P_3, P_4) = r(2 - 2 \cos (\beta - \gamma))^{1/2}$$

Figure 7.2.13

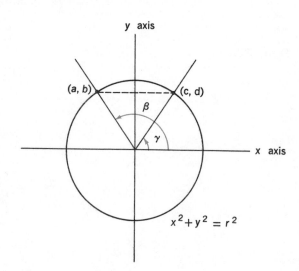

Figure 7.2.14

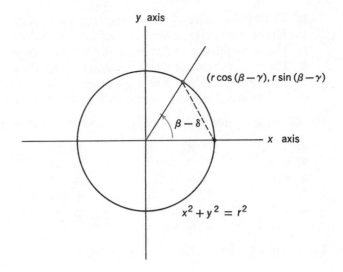

7. Equating $d(P_1, P_2)$ and $d(P_3, P_4)$ from Exercises 5 and 6, show

$$\cos (\beta - \gamma) = \cos \beta \cos \gamma + \sin \beta \sin \gamma$$

8. Using the definitions given in this section, verify the following identities.

a. $\sin \theta \csc \theta = 1.$ **d.** $\sin^2 \theta + \cos^2 \theta = 1.$
b. $\cos \theta \sec \theta = 1.$ **e.** $\tan^2 \theta + 1 = \sec^2 \theta.$
c. $\tan \theta \cot \theta = 1.$ **f.** $\cot^2 \theta + 1 = \csc^2 \theta.$

9. Find the values of each of 6 trigonometric functions for each angle θ (all are taken to be in standard position).

a. $(3, 4)$ is a point on the terminal side of θ.
b. $(5, 12)$ is a point on the terminal side of θ.
c. $(8, 15)$ is a point on the terminal side of θ.
d. $(3, 7)$ is a point on the terminal side of θ.
e. $(-6, 10)$ is a point on the terminal side of θ.
f. $(-5, -12)$ is a point on the terminal side of θ.
g. $(8, -15)$ is a point on the terminal side of θ.
h. The terminal side of θ is horizontal. (Two choices may exist.)
i. The terminal side of θ is vertical. (Two choices may exist.)
j. $(x, 2x)$ is in quadrant 1 and on the terminal side of θ.
k. $(x, 2x)$ is in quadrant 3 and on the terminal side of θ.
l. $(x, -2x)$ is in quadrant 2 and on the terminal side of θ.
m. $(x, -2x)$ is in quadrant 4 and on the terminal side of θ.
n. The terminal side of θ is in quadrant 1 with slope 5.
o. The terminal side of θ is in quadrant 3 with slope 5.

p. The terminal side of θ is in quadrant 4 with slope -2.
q. The terminal side of θ is in quadrant 2 with slope -2.
r. The terminal side of θ is in quadrant 1 with slope ½.
s. The terminal side of θ is in quadrant 2 with slope $-⅔$.
t. The terminal side of θ is in quadrant 3 with slope ¾.
u. The terminal side of θ is in quadrant 4 with slope $-⅝$.

10. Give the degree measure equivalent to each radian measure shown.

a. 0.

b. $\pi/6$.

c. $\pi/4$.

d. $\pi/3$.

e. $\pi/2$.

f. $2\pi/3$.

g. $3\pi/4$.

h. $5\pi/6$.

i. π.

j. $7\pi/6$.

k. $5\pi/4$.

l. $4\pi/3$.

m. $3\pi/2$.

n. $5\pi/3$.

o. $7\pi/4$.

p. $11\pi/6$.

q. 2π.

11. Using Exercise 10, fill in the chart below.

	0°	30°	45°	60°	90°	120°	135°	150°	180°	210°	225°	240°	270°	300°	315°	330°	360°
sin																	
cos																	
tan																	
csc																	
sec																	
cot																	

12. Find the values of the 6 trigonometric functions for an angle of each of the following measures.

a. 1860°.

b. $-240°$.

c. 600°.

d. $-60°$.

e. 2700°.

f. $-585°$.

g. 1575°.

h. $-3015°$.

i. $-405°$.

j. $-1890°$.

k. 7950°.

l. $-1290°$.

m. 930°.

n. 5700°.

o. 7560°.

p. $-90°$.

13. Write each of the points in the form $(r \cos \theta, r \sin \theta)$, $\theta \geqslant 0$.

a. $(8\sqrt{2}, 8\sqrt{2})$.

b. $(15, 15\sqrt{3})$.

c. $(32\sqrt{3}, 32)$.

d. $(0, 21)$.

e. $(20, 0)$.
f. $(-19, 0)$.
g. $(0, -51)$.
h. $(-5, -5)$.
i. $(-10, 10)$.
j. $(19, -19)$.

k. $(-15, 15\sqrt{3})$.
l. $(-81, -81\sqrt{3})$.
m. $(25, -25\sqrt{3})$.
n. $(\sqrt{3}, -1)$.
o. $(-15\sqrt{3}, -15)$.
p. $(-18, 18\sqrt{3})$.

14. Redo Exercise 13 employing for θ an angle of negative measure.

15. Using the chart of Exercise 11 and the definitions of the 6 trigonometric functions given in this section, we may observe relationships involving certain right triangles (a right triangle is one having an angle of measure $90°$). Figure 7.2.15 shows a right triangle in which one acute angle (an angle of measure between $0°$ and $90°$) has measure $45°$. Since the sum total measure of the three angles must be $180°$, the third angle has what measure? After examination of Fig. 7.2.16, determine the value of the unknown quantities from Fig. 7.2.16.

Figure 7.2.15
45° right triangle

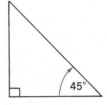

Figure 7.2.16

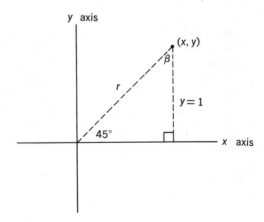

16. Observe Figs. 7.2.17 and 7.2.18. Determine the unknown values from Fig. 7.2.18 (following the instructions in Exercise 15).

Figure 7.2.17
30°-60°-90° right triangle

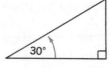

Figure 7.2.18

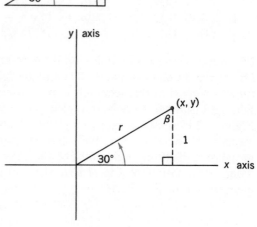

17. Observe Figs. 7.2.19 and 7.2.20. Determine the unknown values from Fig. 7.2.20. Compare these values with those of Exercise 16.

18. In Fig. 7.2.21, the angle (of measure) η is given and the sides of the right triangle are listed with descriptive adjectives. Define the 6 trigonometric functions in terms of these adjectives (see Fig. 7.2.22).

19. Examine Fig. 7.2.23. In all cases below, find the missing values. (In certain instances, reference to Appendix A and the trigonometric tables must be made.)

a. $a = 1, b = 2$.
b. $a = 2, b = 3$.
c. $a = 3, c = 5$.
d. $a = 12, b = 13$.
e. $b = 8, c = 17$.
f. $A = 30°, a = 14$.
g. $A = 45°, b = 12$.
h. $B = 60°, c = 90$.

i. $B = 18°, a = 5$.
j. $B = 70°, b = 45$.
k. $a = 14, A = 28° \ 10'$.
l. $b = 21, B = 42° \ 50'$.
m. $c = 19, B = 71° \ 20'$.
n. $a = 10, B = 13° \ 14'$.
o. $b = 21, A = 29° \ 36'$.

Figure 7.2.19

Figure 7.2.20

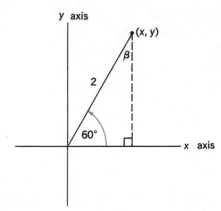

Figure 7.2.21

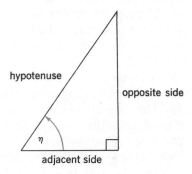

Figure 7.2.22
Meaningful adjectives

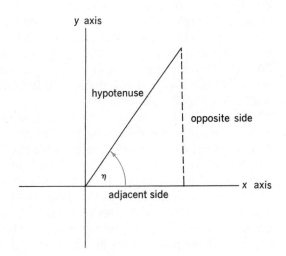

Figure 7.2.23
General right triangle

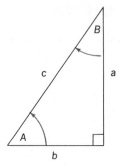

In Exercises 20 to 30 use the information gained from Exercises 15 to 19.
20. A ladder is leaning against the side of a building. The bottom of the ladder is 5 ft from the base of the building and the angle made by the top of the ladder and the wall of the building is 20°. How long is the ladder?
21. The top of a ladder leaning against a building extends 2 ft beyond the top of the building. The ladder makes a 65° angle with the ground at a point 7 ft from the base of the building. How long is the ladder? See Fig. 7.2.24.
22. A rectangle has sides of lengths 25 and 62. Find the length of the diagonals and the angles the diagonals make with the sides.
23. The diagonal of a rectangle makes an angle of 20° with the longer side. If the diagonal is of length 50, find the dimensions of the rectangle.
24. A parallelogram has sides of length 15 and 25 and one angle made by a shorter side and a longer side is 75°. Find the height (perpendicular) between the two longer sides. Between the two shorter sides. (See Fig. 7.2.25.)
25. Three lines intersect in such a way that a triangle is formed. The slopes of the three lines are 1, −5, and −2. Find the measures of the angles of the triangle. (Hint: In general, the slope is the tangent of the angle made by the line as it crosses the *x* axis or any other horizontal line.)
26. A surveyor using a transit notices that the elevation to the top of a tree is 21° as measured at a point 1000 ft from its base. How tall is the tree? (See Fig. 7.2.26.)
27. The angle of elevation from a boat to the light on a lighthouse is 3° 30′. The light is known to be 300 ft above the level of the water. How far is the boat from the lighthouse? (See Fig. 7.2.27.)
28. A surveyor is standing on one bank of a river looking through a transit at a tree on the opposite bank (directly across from him). The surveyor turns his transit 90° clockwise and sights his helper at a measured distance of 1000 ft. His helper marks that point, while the surveyor marks the point immediately below his transit. The surveyor then sets up his transit over the point where he had previously stationed his helper. The

Figure 7.2.24

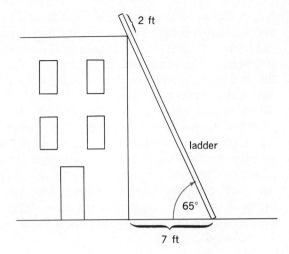

2 ft

ladder

65°

7 ft

Figure 7.2.25
Parallelogram

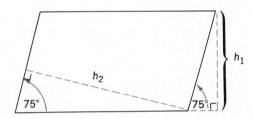

h_1

h_2

75°

75°

Figure 7.2.26

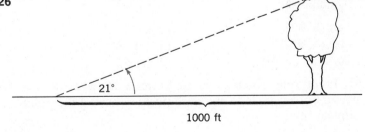

21°

1000 ft

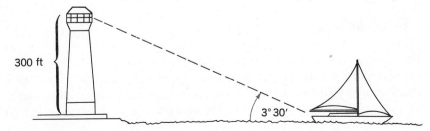

300 ft

3° 30'

Figure 7.2.27

helper moves to the surveyor's original set-up point. Next, the surveyor sights first on his helper at the original transit set-up point and then swings the transit to sight on the same tree as sighted previously. The angle of rotation is measured as 36° in a clockwise direction. How wide is the river?

29. A flagpole is 50 ft tall and the angle of elevation to the sun is 39°. How long a shadow will the pole cast?

30. A certain telephone pole casts a 90-ft shadow when the angle of elevation to the sun is 25°. How tall is the pole?

31. Using the tables of Appendix A and the results of Exercise 4, compute the coordinates of a point on the terminal side of an angle θ (in standard position) where θ is as given in each of (a) to (p) of Exercise 12.

32. Each complex number (Chapter 6) determines a vector and the vector describes an angle in standard position. Compute each complex number where the absolute value and angle described by the vector are, respectively:

a. 7, 30°.   **f.** 2, 61°.
b. 4, 150°.   **g.** 3, 796°.
c. 8, 225°.   **h.** 15, −38°.
d. 9, 270°.   **i.** 20, 183°.
e. 15, 330°.   **j.** 14, 99°.

33. Referring to Exercise 32, determine the measure of the angle formed between two "successive" nth roots of a complex number.

34. Each of the points below is given in polar form. For each point, compute the measure of some angle in standard position having the point on its terminal side.

a. $(6, \pi/4)^*$.   **d.** $(2, \pi)^*$.
b. $(5, \pi/2)^*$.   **e.** $(5, -\pi/6)^*$.
c. $(17, 3\pi/4)^*$.   **f.** $(3, 4\pi/3)^*$.

35. From your observations in Exercise 34, what do you find to be a measure of an angle in standard position having the point $(r, \theta)^*$ on its terminal side?

7.3 LAW OF SINES

The exercises of the last section introduced us to a use of trigonometry relative to (right) triangles. Clearly, we are extremely limited if we cannot attack similar problems when the triangle in question is not a right triangle. Some generalizations and new formulations are then in order. Figure 7.3.1 affords a starting point.

Angle A of the triangle is in standard position and the vertex of angle C

Figure 7.3.1
Nonright triangle

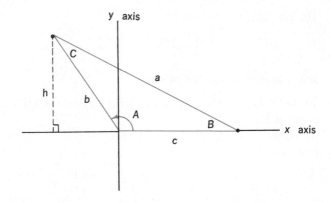

has coordinates ($b \cos A$, $b \sin A$). However, $h/a = \sin B$, whence $h = a \sin B$. Now, h is the value of the second coordinate of the vertex of C. Necessarily, $h = b \sin A$. Thus,

$$b \sin A = a \sin B$$

$$\frac{\sin A}{a} = \frac{\sin B}{b}$$

By placing angle C in standard position (Fig. 7.3.2) we see via a similar procedure: the vertex of B has coordinates ($a \cos C$, $a \sin C$) and $\theta = 180° - A$. Consequently,

$$\sin \theta = \frac{a \sin}{c}$$

$$a \sin C = c \sin \theta = c \sin (180° - A) = c \sin A$$

$$c \sin A = a \sin C$$

$$\frac{\sin A}{a} = \frac{\sin C}{c}$$

Figure 7.3.2

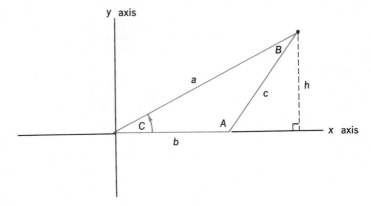

The end result is seen to be

$$\frac{\sin A}{a} = \frac{\sin B}{b} = \frac{\sin C}{c}$$

and is known as the **law of sines.**

Example 7.3.1   Use the law of sines to find the missing values of the triangle of Fig. 7.3.3.

Now, $C = 180° - B - A = 180° - 37° - 28° = 115°$. By the law of sines,

$$\frac{\sin A}{a} = \frac{\sin C}{c}$$

Therefore $a = \dfrac{c \sin A}{\sin C} = \dfrac{c \sin 28°}{\sin 115°} = \dfrac{10(.4695)}{.9063}$

$$= 5.2 \quad \text{approximately}$$

Furthermore, $\dfrac{\sin B}{b} = \dfrac{\sin C}{c}$

and $b = \dfrac{c \sin B}{\sin C} = \dfrac{10(.6018)}{.9063} = 6.7 \quad \text{approximately}$

Employment of the law of sines obviates the knowledge of either (1) two angles and a side opposite one of them, or (2) two sides and an angle opposite one of the sides.

Example 7.3.2   Find the unknown parts of a triangle having the following properties:

$$A = 30°$$
$$a = \tfrac{1}{2}$$
$$b = 1$$

By the law of sines, we know

$$1 = \frac{\tfrac{1}{2}}{\tfrac{1}{2}} = \frac{a}{\sin A} = \frac{b}{\sin B} = \frac{1}{\sin B}$$

Figure 7.3.3

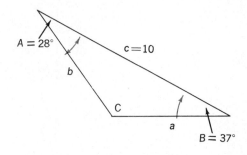

Figure 7.3.4
One triangle formed

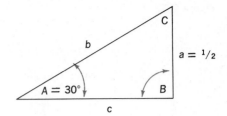

Thus, $B = 90°$ and the triangle is a right triangle as seen in Fig. 7.3.4.

Example 7.3.3   Describe completely any triangle having the following properties: $A = 30°$, $a = ¼$, and $b = 1$.

In Example 7.3.2 above, $a = b \sin A$. Here, though, $a < b \sin A$ and Fig. 7.3.5 shows that *no* triangle is possible.

Figure 7.3.5
No possible triangle

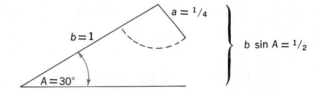

Example 7.3.4   Suppose $A = 30°$, $a = ¾$, and $b = 1$. Find all triangles satisfying these conditions.

Note first that $b \sin A < a < b$. The law of sines reveals that

$$\frac{3}{2} = \frac{¾}{½} = \frac{a}{\sin A} = \frac{b}{\sin B} = \frac{1}{\sin B}$$

Whence $\sin B = ⅔$. However, there are two choices for B satisfying $0° \leqslant B \leqslant 180°$. The tabular values show the possibilities for B to be approximately $41° \, 50'$ and $180° - 41° \, 50' = 138° \, 10'$. (See Fig. 7.3.6.) Thus, $C = 108° \, 10'$ or $11° \, 50'$ depending on the value of B.

Figure 7.3.6
Two possible triangles

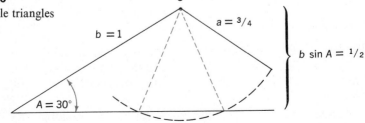

Figure 7.3.7

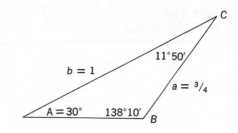

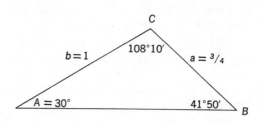

Using the law of sines again, we see that

$$c = \frac{a \sin C}{\sin A} = (\tfrac{3}{4}) \sin C$$

Thus, $c = 1.4253$ if $C = 108°\ 10'$

and $c = .3077$ if $C = 11°\ 50'$

Figure 7.3.7 shows a comparison of the two triangles formed.

Example 7.3.5   Find the unknown parts of any triangle having the following properties:

$$A = 30°$$
$$a = 2$$
$$b = 1$$

Observe that $a > b > b \sin A$. The law of sines shows:

$$2/\tfrac{1}{2} = 1/\sin B$$

Whence $\sin B = \tfrac{1}{4}$

The two choices for B are approximately $14°\ 30'$ and $165°\ 30'$. However, if $B = 165°\ 30', A + B = 195°\ 30'$ and this is impossible. (See Fig. 7.3.8.) Thus, $B = 14°\ 30'$ and $C = 135°\ 30'$. Necessarily, C is approximately 2.8036.

Figure 7.3.9 sums up the results implied in Examples 7.3.3 to 7.3.6.

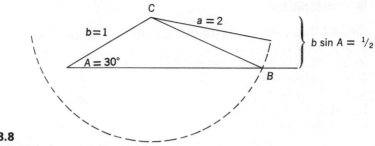

Figure 7.3.8
One possible triangle

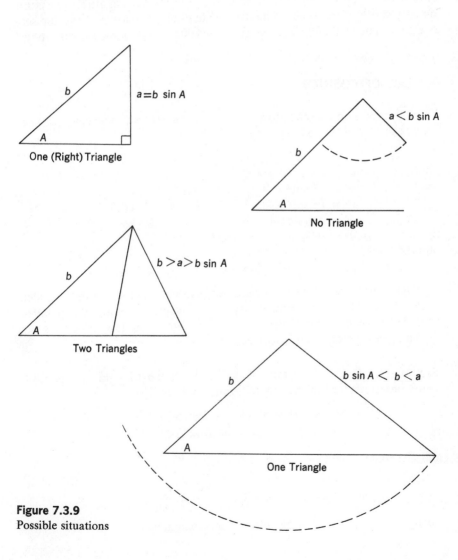

Figure 7.3.9
Possible situations

■ **EXERCISES 7.3**

Find all the missing parts for the triangles in Exercises 1 to 6.
1. $a = 15, b = 21, A = 36°$.
2. $a = 25, A = 25°, B = 98°$.
3. $c = 36, B = 100°, C = 50°$.
4. $b = 30, A = 27°, C = 35°$.
5. $c = 5, b = 6, C = 86°$.
6. $A = 31°, B = 29°, a = 20$.
7. A surveyor observes the angle of elevation to the top of a flag pole. He does so at a point directly north of the pole and then again from a point directly south of the pole. The first angle of elevation is 37° and the second 28°. The two points are separated by 500 ft. How long is the flag pole?

7.4 LAW OF COSINES

Other formulations of the nature of the law of sines exist. The law of cosines is an example. Refer again to Fig. 7.3.1. Since

$$k = -b \cos A$$

and $$a^2 = h^2 + (c + k)^2$$

then $$a^2 = b^2 \sin^2 A + (c - b \cos A)^2$$

$$a^2 = b^2 \sin^2 A + c^2 - 2cb \cos A + b^2 \cos^2 A$$

$$a^2 = b^2 (\sin^2 A + \cos^2 A) + c^2 - 2cb \cos A$$

$$a^2 = b^2 + c^2 - 2cb \cos A$$

Similarly, $$b^2 = a^2 + c^2 - 2ac \cos B$$

$$c^2 = a^2 + b^2 - 2ab \cos C$$

Each statement of the law of cosines involves 3 sides and the included angle. Hence the law of cosines may be used when 2 sides and the included angle are known or when all 3 sides are known. The law of cosines is a "generalized Pythagorean Theorem."

Example 7.4.1 Find the missing values from the triangle in Fig. 7.4.1. Using one statement of the law of cosines, we see

$$a^2 = b^2 + c^2 - 2bc \cos A$$

or $$2bc \cos A = b^2 + c^2 - a^2$$

and alternately $$\cos A = \frac{b^2 + c^2 - a^2}{2bc}$$

Now, $$\cos A = \frac{100 + 225 - 169}{2 \cdot 10 \cdot 15} = \frac{156}{300} = .52$$

Then, $$A = 58° \, 40' \quad \text{approximately}$$

Figure 7.4.1

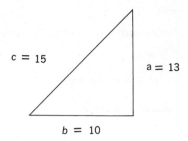

$c = 15$

$a = 13$

$b = 10$

Similarly, $\cos B = \dfrac{a^2 + c^2 - b^2}{2ac} = \dfrac{169 + 225 - 100}{2 \cdot 13 \cdot 15}$

$$= \frac{294}{390} = .7540$$

and, $B = 41° \ 4'$ approximately

This·shows $C = 180° - (58° \ 40' + 41° \ 4') = 80° \ 16'$

The value for B could also have been found by the use of the law of sines.

■ **EXERCISES 7.4**

Compute all the missing parts for the triangles in Exercises 1 to 6.
1. $a = 15, b = 29, C = 13°$.
2. $b = 17, c = 25, A = 32°$.
3. $c = 19, a = 21, B = 50°$.
4. $a = 100, b = 200, C = 10°$.
5. $b = 20, c = 50, A = 97°$.
6. $c = 15, a = 100, B = 112°$.
7. $a = 26, b = 20, c = 30$.
8. $a = 52, b = 66, c = 100$.
9. $a = 60, b = 84, c = 132$.
10. $a = 60, b = 73, B = 28°$.

7.5 LAWS OF TANGENTS

We might expect that there exists a law of tangent. There are two such laws expressed in the material below.
The first law of tangents we shall discuss requires extensive use of identities and algebraic manipulations. From the law of sines, we see

$$\frac{a}{\sin A} = \frac{b}{\sin B}$$

Whence $a \sin B = b \sin A$

and $$2a \sin B = 2b \sin A$$

Necessarily, $$a \sin B - b \sin A = b \sin A - a \sin B$$

Adding $$a \sin A - b \sin B$$

to both sides we have

$$a \sin A - a \sin B + b \sin A - b \sin B$$
$$= a \sin A - b \sin A + a \sin B - b \sin B$$

Factoring, we form the equation

$$(a + b)(\sin A - \sin B) = (a - b)(\sin A + \sin B)$$

which can alternatively be written in the form:

$$\frac{a - b}{a + b} = \frac{\sin A - \sin B}{\sin A + \sin B}$$

From previous identities:

$$\sin A - \sin B = 2 \sin \tfrac{1}{2}(A + B) \cos \tfrac{1}{2}(A + B)$$
$$\sin A + \sin B = 2 \cos \tfrac{1}{2}(A + B) \sin \tfrac{1}{2}(A - B)$$

Reexamining the above equality, we note

$$\frac{a - b}{a + b} = \frac{\sin A - \sin B}{\sin A + \sin B} = \frac{\sin \tfrac{1}{2}(A - B) \cos \tfrac{1}{2}(A + B)}{\cos \tfrac{1}{2}(A - B) \sin \tfrac{1}{2}(A + B)}$$

Alternatively, $$\frac{a - b}{a + b} = \frac{\tan \tfrac{1}{2}(A - B)}{\tan \tfrac{1}{2}(A + B)}$$

We can, in a like manner, deduce two other forms of this law of tangents:

$$\frac{b - c}{b + c} = \frac{\tan \tfrac{1}{2}(B - C)}{\tan \tfrac{1}{2}(B + C)}$$

$$\frac{a - c}{a + c} = \frac{\tan \tfrac{1}{2}(A - C)}{\tan \tfrac{1}{2}(A + C)}$$

This particular result is of value in determining the various parts of a triangle when two sides and the included angle are known.

Example 7.5.1 Find the unknown parts of a triangle, where $a = 12$, $b = 26$, and $C = 76°$.

We know that $A + B = 180° - 76° = 104°$. The law of tangents will be used to calculate $A - B$, from which A and B will be determined.

$$\frac{\tan \frac{1}{2}(A - B)}{\tan \frac{1}{2}(A + B)} = \frac{a - b}{a + b}$$

$$\tan \frac{1}{2}(A - B) = \frac{(a - b) \tan \frac{1}{2}(A + B)}{a + b}$$

$$= \frac{-14 \tan 52°}{38}$$

$$= -.4715 \qquad \text{approximately}$$

Thus, $\frac{1}{2}(A - B) = -25°\ 15'$ (why can $\frac{1}{2}(A - B)$ not be $164°\ 45'$?) and $A - B = -51°\ 30'$. Now, adding the equations

$$A + B = 104°$$
$$A - B = -50°\ 30'$$
we have $\qquad\qquad 2A = 53°\ 30'$
or $\qquad\qquad\quad A = 26°\ 45'$
whence $\qquad\qquad B = 77°\ 15'$

The law of sines shows

$$c = \frac{a \sin C}{\sin A} = \frac{12 \sin 76°}{\sin 26°\ 45'} = 25.86 \qquad \text{approximately}$$

The next law of tangents can be derived by starting with the law of cosines. From

$$a^2 = b^2 + c^2 - 2bc \cos A$$
We see $\qquad\qquad \cos A = \frac{b^2 + c^2 - a^2}{2bc}$

Adding 1 to both sides,

$$1 + \cos A = \frac{b^2 + c^2 - a^2 + 2bc}{2bc}$$

$$= \frac{(b^2 + 2bc + c^2) - a^2}{2bc}$$

$$= \frac{(b + c + a)(b + c - a)}{2bc}$$

Necessarily, $\qquad \dfrac{1 + \cos A}{2} = \dfrac{(b + c + a)(b + c - a)}{4bc}$

Now, $\qquad \cos \dfrac{A}{2} = \left[\dfrac{1 + \cos A}{2}\right]^{1/2} = \dfrac{1}{2}\left[\dfrac{(b + c + a)(b + c - a)}{bc}\right]^{1/2}$

There is no ambiguity in polarity, since it is assumed that A is an angle of a triangle, meaning that $0° < A < 180°$ and $0° < A/2 < 90°$. Starting anew, we write

$$-\cos A = \frac{a^2 - b^2 - c^2}{2bc}$$

$$1 - \cos A = \frac{a^2 - b^2 - c^2 + 2bc}{2bc}$$

$$1 - \cos A = \frac{a^2 - (b^2 - 2bc + c^2)}{2bc}$$

$$= \frac{a^2 - (b - c)^2}{2bc}$$

$$= \frac{(a + b - c)(a - b + c)}{2bc}$$

$$\frac{1 - \cos A}{2} = \frac{(a + b - c)(a - b + c)}{4bc}$$

and $\quad \sin \dfrac{A}{2} = \left[\dfrac{1 - \cos A}{2} \right]^{1/2} = \dfrac{1}{2} \left[\dfrac{(a + b - c)(a - b + c)}{bc} \right]^{1/2}$

One statement of the second law of tangents becomes

$$\tan \frac{A}{2} = \frac{\sin A/2}{\cos A/2} = \frac{\dfrac{1}{2} \left[\dfrac{(a + b - c)(a - b + c)}{bc} \right]^{1/2}}{\dfrac{1}{2} \left[\dfrac{(a + b + c)(b + c - a)}{bc} \right]^{1/2}}$$

$$= \left[\frac{(a + b - c)(a - b + c)}{(a + b + c)(b + c - a)} \right]^{1/2}$$

We often see the *semiperimeter* s of a triangle given as $s = \frac{1}{2}(a + b + c)$. In terms of s, our above result can then be written:

$$\tan \frac{A}{2} = \left[\frac{(s - c)(s - b)}{s(s - a)} \right]^{1/2}$$

Two other similar expressions are:

$$\tan \frac{B}{2} = \left[\frac{(s - a)(s - c)}{s(s - b)} \right]^{1/2}$$

$$\tan \frac{C}{2} = \left[\frac{(s - a)(s - b)}{s(s - c)} \right]^{1/2}$$

The expressions above may be simplified (in appearance) even more by the introduction of

$$\rho = \left[\frac{(s - a)(s - b)(s - c)}{s} \right]^{1/2}$$

The above three expressions become, upon substitution,

$$\tan \frac{A}{2} = \frac{\rho}{s-a}$$

$$\tan \frac{B}{2} = \frac{\rho}{s-b}$$

$$\tan \frac{C}{2} = \frac{\rho}{s-c}$$

forms easily recalled.

This form of a law of tangents is sometimes called a half angle form and is most adapted to use when all 3 sides of a triangle are known. We see that the two laws of tangents duplicate the law of cosines in that they are used under the same circumstances as that law. The laws of tangents are better suited to computation using logarithms. (See Appendix F.)

Example 7.5.2 Compute, using this law of tangents, the angles of a triangle having sides $a = 6$, $b = 7$, and $c = 11$.

First, we compute s, $s - a$, $s - b$, $s - c$, and s.

$$s = \frac{6 + 7 + 11}{2} = 12$$

$$s - a = 6$$
$$s - b = 5$$
$$s - c = 1$$

$$= \left(\frac{(s-a)(s-b)(s-c)}{s} \right)^{1/2}$$

$$= (^{30}\!/_{12})^{1/2}$$

$$= 1.58 \qquad \text{approximately}$$

Thus, $$\tan \frac{A}{2} = \frac{\rho}{s-a} = \frac{1.58}{6} = .2635$$

whence $$\frac{A}{2} = 14° \ 46' \text{ and } A = 29° \ 32'$$

Similarly, $$\tan \frac{B}{2} = \frac{\rho}{s-b} = \frac{1.58}{5} = .316$$

and $$\frac{B}{2} = 17° \ 32' \text{ so that } B = 35° \ 4'$$

Consequently, $$C = 180° - A - B = 115° \ 24'$$

The measures of the parts of any given triangle can be determined if any of the following hold:

1. The measure of 2 angles and the length of any side is known (law of sines).

2. The lengths of 2 sides and the measure of the included angle is known (law of cosines, law of tangents).
3. The lengths of all 3 sides are known (law of cosines, law of tangents).

The aforementioned "ambiguous case" can arise whenever two sides and an angle other than the included angle are known, that is, a unique triangle may not be determined. In such a case, there may exist one triangle, two triangles, or no triangle with the given characteristics.

■ EXERCISES 7.5

Find the missing values in each of Exercises 1 to 13.
1. $a = 15, b = 21, c = 33$.
2. $a = 18, b = 20, c = 35$.
3. $a = 19, b = 50, c = 60$.
4. $a = 15, b = 25, A - B = -15°$.
5. $a = 21, c = 19, A + C = 120°$.
6. $b = 55, c = 80, C - B = 111°$.
7. $a = 1, b = 2, A = 30°$. (Hint: Ambiguous case.)
8. $a = 1, b = 2, A = 60°$. (Hint: Ambiguous case.)
9. $a = 1, b = 2, A = 28°$. (Hint: Ambiguous case.)
10. $a = 1, b = 2, A = 3°$. (Hint: Ambiguous case.)
11. $a = 3, b = 2, A = 25°$. (Hint: Ambiguous case.)
12. $a = 1, b = 2, A = 110°$. (Hint: Ambiguous case.)
13. $a = 3, b = 2, A = 110°$. (Hint: Ambiguous case.)
14. The roof of a house makes an angle of 30° with the horizontal. An antenna on the roof is mounted vertically. The antenna is located 12 ft from the edge of the roof and is 33 ft tall. How far is the top of the antenna from the edge of the roof? See Fig. 7.5.1.
15. A 40-ft flag pole is being erected on a straight slope of 36°. A line is attached to the top of the pole and then threaded through a rung placed in the ground 20 ft up the slope from the base of the pole. The line has markers each foot so that one can measure the length of line from the top of the pole to the rung. How long should the line from the top of the pole to the rung be in order that the pole be vertical?
16. A man sights a ship in the distance and notices that the angle traversed in sighting on one end of the ship and then sighting on the other is 1°. The ship is known to be 300 yd long. Approximately how far away is the ship? (Assume that the man's view is a broadside view of the ship.)
17. The sides of a parallelogram are of lengths 25 and 38, while one diagonal is of length 30. Find the length of the other diagonal and the measures of the angles of the parallelogram.
18. A surveyor adjusts his transit so that when he sights through it, his line

of sight is horizontal. A rodman is holding a rod with graduations marked (starting with 0 at the bottom and increasing in an upward direction). The surveyor sights on the rod and notes that both his cross hairs are on the rod. The rodman "waves" the rod back and forth through a small angle in the direction of the line of sight of the surveyor. How can the surveyor know when the rod passes through a vertical position?

7.6 REVIEW EXERCISES

1. In which quadrant does each of the following angles terminate?

a. $-3\pi/7$.	**f.** 507°.	**k.** 3⅙ rotations.
b. -2.	**g.** $-281°$.	**l.** 5⅔ rotations.
c. 16½.	**h.** $-1076°$.	**m.** 2⅙ rotations.
d. $5\pi/3$.	**i.** 1976°.	**n.** $-13⅓$ rotations.
e. $291\pi/6$.	**j.** 25°.	**o.** $-1½$ rotations.

2. In Exercise 1(a) to 1(e), convert each measure to degree measure. To rotation measure.

3. In Exercise 1(f) to 1(j), convert each measure to radian measure. To rotational measure.

4. In Exercise 1(k) to 1(o), convert each measure to radian measure. To degree measure.

5. Give the polarity for each functional value of each angle in 1.

6. Give each trigonometric functional value for each angle in 1.

7. In which quadrant does θ terminate?

a. $\sin\theta < 0$, $\cos\theta > 0$.	**f.** $\tan\theta < 0$, $\sec\theta < 0$.
b. $\sin\theta < 0$, $\tan\theta < 0$.	**g.** $\cot\theta > 0$, $\csc\theta > 0$.
c. $\sin\theta > 0$, $\tan\theta > 0$.	**h.** $\csc\theta > 0$, $\sec\theta < 0$.
d. $\sin\theta < 0$, $\sec\theta < 0$.	**i.** $\csc\theta < 0$, $\cot\theta < 0$.
e. $\tan\theta < 0$, $\sec\theta < 0$.	

8. In each case, what is the slope of the ray forming the terminal side of the angle θ (in standard position)?

a. $\theta = \pi$.	**g.** $\theta = 1$.	**m.** $\theta = 2⅓$ rotations.
b. $\theta = \pi/4$.	**h.** $\theta = 5$.	**n.** $\theta = 16½$ rotations.
c. $\theta = 2/\pi$.	**i.** $\theta = 28°$.	**o.** $\theta = ⅔$ rotation.
d. $\theta = 5\pi/4$.	**j.** $\theta = 16°$.	**p.** $\theta = 3⅞$ rotations.
e. $\theta = 30°$.	**k.** $\theta = 419°$.	**q.** $\theta = 50$ rotations.
f. $\theta = 240°$.	**l.** $\theta = 672°$.	**r.** $\theta = -2½$ rotations.

9. In each case, find the length of arc subtended by the central angle.

a. $\theta = 3\pi/4, r = 6$.
b. $\theta = 7\pi/6, r = 8$.
c. $\theta = 25\pi/2, r = 10$.
d. $\theta = 3, r = \frac{1}{2}$.
e. $\theta = \frac{1}{2}$ rotation, $r - 5$.
f. $\theta = \frac{2}{3}$ rotation, $r = 8$.

g. $\theta = \frac{3}{4}$ rotation, $r = 4$.
h. $\theta = 1\frac{1}{3}$ rotations, $r = 1$.
i. $\theta = 72°, r = 18$.
j. $\theta = 36°, r = 12$.
k. $\theta = 15°, r = 8$.
l. $\theta = 20°, r = 4$.

10. Describe the set of angles coterminal with each given below.

a. $213°$.
b. $-64°$.
c. $2951°$.
d. $2\pi/3$.
e. $\pi/6$.

f. $-\pi/3$.
g. $\pi/4$.
h. $-\pi/6$.
i. $\pi/2$.

11. Write each of the following points in the $(r \cos \theta, r \sin \theta)$ form.

a. $(1, 2)$.
b. $(3, 5)$.
c. $(6, 2)$.
d. $(-4, 3)$.
e. $(-2, 1)$.

f. $(-1, -1)$.
g. $(-3, 1)$.
h. $(1, -3)$.
i. $(3, 0)$.

12. Find the missing parts of each triangle.

a. $a = 33, b = 44, c = 55$.
b. $a = 16, b = 30, c = 34$.
c. $a = 11, b = 23, c = 32$.
d. $A = 30°, C = 90°, b = 6$.
e. $A = 30°, c = 13, b = 8$.
f. $A = 33°, B = 81°, c = 19$.
g. $A = 19°, C = 64°, a = 21$.
h. $A = 45°, a = 2, b = 2$.
i. $A = 32°, c = 12, a = 8$.
j. $A = 29°, c = 10, a = 14$.
k. $A = 110°, c = 12, a = 16$.
l. $A = 115°, c = 15, a = 11$.

13. A grandfather clock has a pendulum of length 3 ft. While keeping time, the pendulum moves through an arc of 4 in. Through what angle does the clock's pendulum swing?

14. A sprocket of 4-in. diameter is rigidly attached to the axis (center) of a bicycle tire of diameter 26 in. The sprocket is driven at such a rate as to make 60 revolutions per minute (rpm). How many feet per minute is the bicycle propelled under these conditions?

15. The sprocket of Exercise 13 is chain-driven by a sprocket of 9-in.

diameter. Pedals are connected to the larger sprocket. A cyclist is pedal-
ing at a speed to cause the pedals to describe 20 rpm. How fast is the
bicycle moving?

16. A tall building casts a shadow of 1500 ft when the angle of inclination
of the sun is 23°. How tall is the building?

17. A sheer cliff overlooks the sea. The angle of depression from the cliff
to a boat is 12°. A man drops a rock from the cliff and notes that the
sound of the rocks hitting the water returns 6 sec after the rock was
released. If the equation for the distance s a body falls in t seconds is
$s = 16t^2$ and the speed of sound is 1100 ft per second, how far from this
vantage point is the boat? How far is the boat from the bottom of the cliff?

18. A circular band with radius of 6 in. is cut and an additional 1 ft added
to make a larger circular band. Another circular band of radius 2000 miles
has its circumference lengthened by the insertion of a 1-ft piece. Which of
the bands experienced the greatest increase in radius?

19. A certain hill has a side of slope 2. At a point 1000 ft from the base
of this slope, an observer notes that the angle of inclination to the top is
46°. How high is the peak of the hill?

20. In which of the following cases may the law of sines be used?
   a. Only the 3 sides of the triangle are known.
   b. Only 2 sides and an angle opposite one of these sides are known.
   c. Only 2 sides and the included angle are known.
   d. Only 2 angles and a side are known.

21. In which of the cases in Exercise 20 above may the law of cosines be
used?

22. In which of the cases in Exercise 20 may a law of tangents be used?
In each case, specify which law is the one used.

23. A ship travels due east at a rate of 18 mph, while another travels 20°
west of north at a rate of 21 mph. If the two started from the same bay
at 8:00 in the morning, how far apart will they be at 4:00 in the afternoon?

24. Two ships are anchored at ports P_1 and P_2. The port P_2 is 1000 miles
due east from P_1. Both ships leave their respective ports at the same time.
The ship leaving P_1 travels 138 east of north at a rate of 118 mph, while
the other travels 25° west of north at a speed of 20 mph. Are they on a
collision course?

25. Show that the area of the triangle in Fig. 7.6.1 is $\frac{1}{2}bc \sin A =
\frac{1}{2}ac \sin B = \frac{1}{2}ab \sin C$. (Hint: The area of a triangle is $\frac{1}{2}bh$, where b is
the length of any side and h is the altitude from the vertex of the angle
opposite.)

26. Use Exercise 25 together with the law of sines to show that the area
of the triangle in Fig. 7.6.1 is given by

$$\frac{a^2 \sin C \sin B}{2 \sin A} = \frac{b^2 \sin C \sin A}{2 \sin B} = \frac{c^2 \sin A \sin B}{2 \sin A}$$

Figure 7.6.1

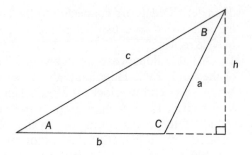

27. Using the identity $\sin^2 A + \cos^2 A = 1$, the law of cosines, and Exercise 26, show that the area T of the triangle in Fig. 7.6.1 satisfies:

$$T^2 = s(s - a)(s - b)(s - c)$$
$$= (s\rho)^2$$

or $$T = s\rho$$

where s and ρ are as defined in Section 7.5.

28. See Fig. 7.6.2. The area of the sector shown is $\frac{1}{2}\theta r^2$, where θ is measured in radians. The area of the segment (that part of the sector outside the triangle formed) can be determined by subtracting the area of the triangle from the area of the sector. Show that the area of the triangle is $\frac{1}{2}r^2 \sin \theta$ and the area of the segment is $\frac{1}{2}r^2(\theta - \sin \theta)$.

Figure 7.6.2
A segment of a circle

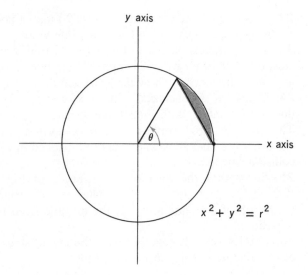

7.7 ADVANCED EXERCISES

1. Show in Fig. 7.7.1 that the point $(1, a)$ is such that $a = \tan \theta$.
2. Such that the point $(b, 1)$ in Fig. 7.7.1 is the point $(\cot \theta, 1)$.
3. Show that $(c, 0)$ in Fig. 7.7.2 is $(\sec \theta + \tan \theta, 0)$.
4. Show that $(0, d)$ in Fig. 7.7.2 is $(0, \csc \theta + \cot \theta)$.

Figure 7.7.1

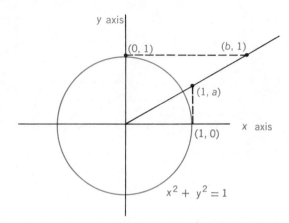

Figure 7.7.2

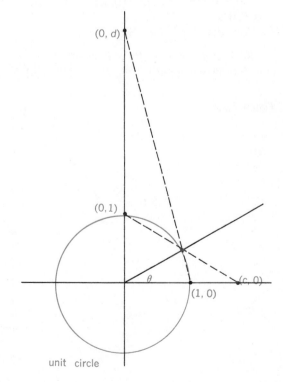

Figure 7.7.3

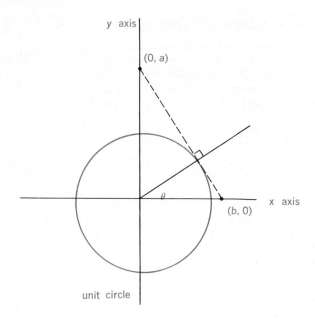

5. Show that the point $(0, a)$ in Fig. 7.7.3 is $(0, \csc \theta)$ and $(b, 0)$ is $(\sec \theta, 0)$.

6. Consider Fig. 7.7.4. Show that:

    a. If $d(P_0, P_1) = d(P_1, P_2)$ and $P_0 = ((\sqrt{5} - 1)/2, 0)$, then the triangles formed by P_0, P_1, and P_2 and $(0, 0)$, P_1, and P_2 are similar.

Figure 7.7.4

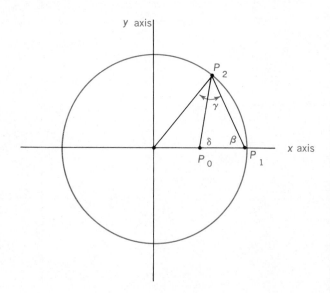

b. $\gamma = \beta = \delta$. (The triangles in (a) are isosceles.)
c. $\beta = 2\theta$. (The triangle determined by $(0,0)$, P_0, and P_2 is isosceles.)

Given that $\beta = \frac{1}{2}(\pi - \theta)$, show that:

d. $\theta = \frac{1}{4}(\pi - \theta)$.
e. $\theta = \pi/5$.

Then find:

f. $\sin \pi/5$.   **i.** $\csc \pi/5$.
g. $\cos \pi/5$.   **j.** $\sec \pi/5$.
h. $\tan \pi/5$.   **k.** $\cot \pi/5$.

7. From Fig. 7.7.5, deduce the following (the area of a sector subtended by θ is given as $A = \theta r^2/2$). The following hold for $0 < \theta < \pi/2$.

a. $\sin \theta < \theta < \tan \theta$.
b. $1 < (\theta/\sin \theta) < \sec \theta$.
c. $\cos \theta < (\sin \theta/\theta) < 1$.
d. $\cos \theta < (\theta/\tan \theta) < 1$.
e. $\theta(1 - \theta^2)^{1/2} < \sin \theta < \theta$.
f. $\theta < \tan \theta < (1 - \theta^2)^{1/2}$.

Figure 7.7.5

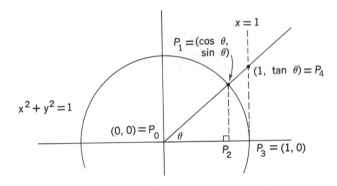

(Hint: The area of the triangle determined by P_0, P_1, and P_2 is less than the area of the sector described by θ, which in turn is less than the area of the triangle with vertices P_1, P_3, and P_4.)

8. Show that the sum total of the measures of the angles of any triangle is $180°$.

Chapter 8
Vector
Spaces

8.1 INTRODUCTION

We have already mentioned, in an intuitive and geometric sense, the concept of a vector. The complex numbers, for example, formed a "space" of vectors, so to speak. A natural desire then is to extend this geometric and intuitive notion to a more general (and abstract) one that is algebraic in nature. This is the concept of a *vector space*. We will then return to a geometric setting.

We are guided in our formulation by the behavior of the complex numbers (vectors) relative to addition and scalar multiplication. We wish also to include the notion of length (this notion is not really a part of the structure attributed to vector spaces). Since this is the case, we are not to be surprised to discover (when we finish) that the complex numbers form a vector space.

To define a space of vectors, we need vectors. Thus we shall have need of a set V, each element of which is a vector. The scalar multiplication discussed for complex numbers was multiplication of a complex number by a real number. The real numbers are called *scalars,* while R is called the *scalar field.* Hence, we define:

The statement that **V** is a vector space over the (scalar) field K means that the following are satisfied:

1. $\mathbf{v}_1 + \mathbf{v}_2 \in \mathbf{V}$.
2. $\mathbf{v}_1 + \mathbf{v}_2 = \mathbf{v}_2 + \mathbf{v}_1$.
3. $(\mathbf{v}_1 + \mathbf{v}_2) + \mathbf{v}_3 = \mathbf{v}_1 + (\mathbf{v}_2 + \mathbf{v}_3)$.
4. There is a zero vector **0** such that $\mathbf{v} + \mathbf{0} = \mathbf{v}$ for all **v** in **V**.
5. For each $\mathbf{v} \in \mathbf{V}$ there is a vector $-\mathbf{v}$ such that $\mathbf{v} + (-\mathbf{v}) = \mathbf{0}$.
6. $r\mathbf{v} \in \mathbf{V}$.
7. $0 \cdot \mathbf{v} = \mathbf{0}$, $1 \cdot \mathbf{v} = \mathbf{v}$, and $-1 \cdot \mathbf{v} = -\mathbf{v}$. (0 and 1 are, respectively, the additive and multiplicative identities for K.)
8. $r(\mathbf{v}_1 + \mathbf{v}_2) = r\mathbf{v}_1 + r\mathbf{v}_2$.
9. $(r_1 + r_2)\mathbf{v} = r_1\mathbf{v} + r_2\mathbf{v}$.
10. $(r_1 r_2)\mathbf{v} = r_1(r_2\mathbf{v})$.

One geometric interpretation has been given the term vector (the complex vector in the plane). The addition of vectors was described as "head to tail" addition as pictured in Fig. 6.2.5. If we consider the (geometric) projection of a vector (point) **v** onto the x axis or y axis, we may consider the result a vector. The projection onto the x axis will be called $\mathbf{v}_x$, the x **component** of **v** while $\mathbf{v}_y$ is the y **component** or projection (see Fig. 8.1.1). We find that for any vector **v**, $\mathbf{v} = \mathbf{v}_x + \mathbf{v}_y$.

Example 8.1.1   If **v** is the vector $(5, -4)$, $\mathbf{v}_x = (5, 0)$, while $\mathbf{v}_y = (0, -4)$. (See Fig. 8.1.2.)

Figure 8.1.1

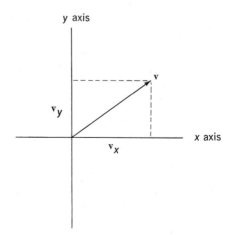

Figure 8.1.2

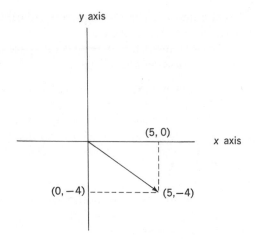

■ **EXERCISES 8.1**

1. Prove each of (a) to (j), the ordered pairs being taken to be complex numbers. Proof of all of these shows that the complex numbers form a vector space over the real numbers.

a. $(a, b) + (c, d)$ is a unique complex number.
b. $(a, b) + (c, d) = (c, d) + (a, b)$.
c. $[(a, b) + (c, d)] + (e, f) = (a, b) + [(c, d) + (e, f)]$.
d. $(a, b) + (0, 0) = (a, b)$.
e. $(a, b) + (-a, -b) = (0, 0)$.
f. $r(a, b)$ is a unique complex number.
g. $0(a, b) = (0, 0)$, $1 \cdot (a, b) = (a, b)$, $(-1)(a, b) = (-a, -b)$.
h. $r((a, b) + (c, d)) = r(a, b) = r(c, d)$.
i. $(r_1 + r_2)(a, b) = r_1(a, b) + r_2(a, b)$.
j. $(r_1 r_2)(a, b) = r_1(r_2(a, b))$.

2. Show that the complex numbers form a vector space over themselves (that is, the complex field is a vector space where the field is the field of complex numbers and scalar multiplication is multiplication of the complex numbers). [Hint: Prove Exercise 1 with complex numbers replacing r, r_1, and r_2 and with $(0, 0)$, $(1, 0)$, and $(-1, 0)$ replacing 0, 1, and -1, respectively.]

3. Show that the real numbers form a vector space over the field of rational numbers.

4. Find v_x and v_y for each of the following vectors in the plane.

a. $(4, 1)$.   **c.** $(\frac{1}{2}, -\frac{2}{3})$.
b. $(-2, \sqrt{3})$.   **d.** $(7, 14)$.

e. $(0, 1)$.   **g.** $(0, 0)$.
f. $(1, 0)$.   **h.** $(5, 5)$.

5. Perform the following operations in the plane.

a. $(2, 3) + (-1, 7)$.   **d.** $26(1, 3)$.
b. $(4, 5) - (7, 3)$.   **e.** $11((3, 1) + (-2, -3))$.
c. $(15, 5) + (-(3, 4))$.   **f.** $(5 + 7)(1, 2)$.

6. Draw a sketch of each vector in Exercise 5 and draw a picture of the answer to each part of Exercise 5.

7. Let $\mathbf{i} = (1, 0)$ and $\mathbf{j} = (0, 1)$. Describe in terms of $\mathbf{i}$ and $\mathbf{j}$ the vector $(1, 1)$. Then describe each vector (a, b) of the plane in terms of $\mathbf{i}$ and $\mathbf{j}$. If $\mathbf{v}$ is a vector in the plane, how do $\mathbf{v}_x$ and $\mathbf{v}_y$ compare with $\mathbf{i}$ and $\mathbf{j}$?

8. Write each of the following vectors in terms of $\mathbf{i}$ and $\mathbf{j}$ as given in Exercise 7.

a. $(1, 2)$.   **e.** $(\sqrt{2}, 1)$.
b. $(-1, 0)$.   **f.** $(-1, \sqrt{3}/2)$.
c. $(2, -3)$.   **g.** $(0, 0)$.
d. $(-\frac{1}{2}, -\frac{1}{3})$.   **h.** (a, b).

9. Describe the addition of two vectors in the plane in terms of $\mathbf{i}$ and $\mathbf{j}$. (See Exercise 7.)

10. Let $\mathbf{r}$ and $\mathbf{s}$ be the vectors $(1/\sqrt{2}, 1/\sqrt{2})$ and $(-1/\sqrt{2}, 1/\sqrt{2})$, respectively. Show that each vector (a, b) can be written in the form $k\mathbf{r} + c\mathbf{s}$ (we say that $\mathbf{r}$ and $\mathbf{s}$ **span** the space). Give another pair of vectors spanning the plane.

11. If (a, b) and (c, d) lie on the same line through the origin, show that they do not span the plane. (See Exercise 10.)

12. Show that (a, b) and $(0, 0)$ cannot span the plane. (See Exercises 10 and 11.)

8.2 INNER PRODUCT

Our discussions of vector spaces have, thus far, involved only the concepts of addition of vectors and scalar multiplication. At this time we wish to introduce a *multiplication of vectors*. Contrary to what might be expected, the definition renders the product of two vectors a *scalar*, not a vector.

The **inner product** of two vectors from a vector space **V** is an operation satisfying:

1. the inner product $\mathbf{v}_1 \cdot \mathbf{v}_2$ is a real number;
2. $\mathbf{v}_1 \cdot \mathbf{v}_1 \geqslant 0$, $\mathbf{v}_1 \cdot \mathbf{v}_1 = 0$ if and only if $\mathbf{v}_1 = \mathbf{0}$;
3. $\mathbf{v}_1 \cdot \mathbf{v}_2 = \mathbf{v}_2 \cdot \mathbf{v}_1$;

4. $(k\mathbf{v}_1) \cdot \mathbf{v}_2 = k(\mathbf{v}_1 \cdot \mathbf{v}_2)$; and
5. $(\mathbf{v}_1 + \mathbf{v}_2) \cdot \mathbf{v}_3 = \mathbf{v}_1 \cdot \mathbf{v}_3 + \mathbf{v}_2 \cdot \mathbf{v}_3$.

Although the list of requirements seems long and at first glance, strange, the list is well motivated. The requirements (3), (4), and (5) simply ensure that the inner product is commutative and that it behaves "properly" with respect to vector addition and scalar multiplication.

A good deal of the motivation for this collection of restrictions comes from the case of the Euclidean plane. The inner product for the plane is given by $(a, b) \cdot (c, d) = ac + bd$. Consequently $(a, b) \cdot (a, b) = a^2 + b^2$ and restrictions (1) and (2) are seen to hold. Conditions (3) to (5) also follow easily.

Since $(a, b) \cdot (a, b) = a^2 + b^2 = d^2((0, 0), (a, b))$, $(a, b) \cdot (a, b)$ is seen to be the *square of the length of the vector*. Furthermore, $(a, b) \cdot (c, d) = ac + bd = |(a, b)| \, |(c, d)| \cos \theta$, where θ is the angle between (a, b) and (c, d) as shown in Fig. 8.2.1 and $|(p, q)| = (p^2 + q^2)^{1/2}$. This observation follows from the law of cosines.

Refer to Fig. 8.2.1 during the following discussion.

$$g^2 = e^2 + f^2 - ef \cos \theta$$
$$= (c - a)^2 + (d - b)^2$$

Thus,

$$c^2 - 2ac + a^2 + d^2 - 2db + b^2$$
$$= a^2 + b^2 + c^2 + d^2 - 2|(a, b)| \, |(c, d)| \cos \theta$$

or alternatively

$$-2ac - 2db = -2|(a, b)| \, |(c, d)| \cos \theta$$

Figure 8.2.1

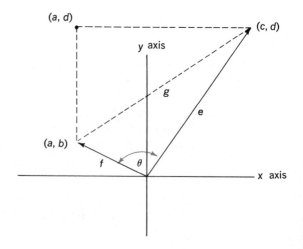

from which we easily obtain

$$(a, b) \cdot (c, d) = ac + bd = |(a, b)| \, |(c, d)| \cos \theta$$

Often, in more general vector spaces over the real field, we wish to define an angle θ, $0° \leqslant \theta < 180°$, between two vectors v_1 and v_2 by

$$\cos \theta = \frac{v_1 \cdot v_2}{|v_1| \, |v_2|}$$

In particular, it seems appropriate to define: v_1 and v_2 are **orthogonal** (perpendicular) if $v_1 \cdot v_2 = 0$ (that is, if $\cos \theta = 0$). The definition is consistent with both our discussion here and with the trigonometry we have developed [that is, $\cos (2n + 1)\pi/2 = 0$].

Example 8.2.1 If $(2, 4)$ and $(3, 1)$ are vectors in the Euclidean plane, $(2, 4) \cdot (3, 1) = 6 + 4 = 10$, $(2, 4) \cdot (2, 4) = 4 + 16 = 20$, and $(3, 1) \cdot (3, 1) = 9 + 1 = 10$. Thus, $|(2, 4)| = \sqrt{20} = 2\sqrt{5}$ and $|(3, 1)| = \sqrt{10}$. Furthermore, if θ is the angle between v_1 and v_2,

$$\cos \theta = \frac{10}{2\sqrt{5}\sqrt{10}} = \frac{1}{\sqrt{2}}$$

whence $\theta = 45°$

Example 8.2.2 Find a **unit vector** (vector of length one) orthogonal to $(1, 2)$.

If (a, b) is orthogonal to $(1, 2)$,

$$0 = \frac{(a, b) \cdot (1, 2)}{|(a, b)| \, |(1, 2)|}$$

Thus, $0 = (a, b) \cdot (1, 2) = a + 2b$. However, $a^2 + b^2 = 1$ whenever the vector is a unit vector so that we have the two equations

$$a + 2b = 0$$
$$a^2 + b^2 = 1$$

The solutions to the pair of equations are seen to be $(-2/\sqrt{5}, 1/\sqrt{5})$ and $(2/\sqrt{5}, -1/\sqrt{5})$. Each is a vector orthogonal to $(1, 2)$. See Fig. 8.2.2.

■ EXERCISES 8.2

1. Compute each of the following:

a. $(0, 2) \cdot (1, 1)$.

b. $(1, 3) \cdot (2, 4)$.

c. $(\frac{1}{2}, -1) \cdot (\frac{1}{2}, 1)$.

d. $(-1, -1) \cdot (2, 2)$.

e. $(\frac{1}{2}, \frac{1}{2}) \cdot (\frac{1}{2}, -\frac{1}{2})$.

f. $|(1, 2)|$.

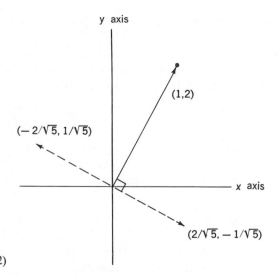

Figure 8.2.2
Unit vectors orthogonal to (1, 2)

g. $|(-1, -2)|$.
h. $|(\frac{1}{2}, \frac{2}{3})|$.
i. $|\mathbf{v}|$, where $\mathbf{v}$ is a unit vector.

j. $|(4, 2)|$.
k. $|k(2, 3)|$.

2. Compute the angle between each of the following pairs of vectors:

a. $(1, 0), (0, 1)$.
b. $(1, 1), (-2, -2)$.
c. $(1, \sqrt{3}), (\sqrt{3}, 1)$.
d. $(1, 1), (-\sqrt{3}, 1)$.
e. $(3, 4), (8, 11)$.

f. $(-2, 3), (1, -6)$.
g. $(\frac{1}{2}, \frac{1}{2}), (-1, -3)$.
h. $(0, 1), (1, 2)$.
i. $(-1, 0), (0, -1)$.

j. $\mathbf{v}_1$ and $\mathbf{v}_2$, where the two vectors are orthogonal.

3. Compute for each vector below 2 orthogonal unit vectors (of length 1).

a. $(1, 0)$.
b. $(-1, 0)$.
c. $(1, 1)$.
d. $(1, 3)$.

e. $(-2, -2\sqrt{3})$.
f. $(4, 5)$.
g. $(-1, -7)$.
h. $(1, 2)$.

4. Write each of the following in terms of the vectors **i** and **j** (the points are given in polar coordinates). See Exercise 8.1.7.

a. $(6, \pi/4)^*$.
b. $(-1, \pi/3)^*$.
c. $(2, -\pi/6)^*$.
d. $(5, 3\pi/4)^*$.
e. $(-1, 7\pi/6)^*$.
f. $(\frac{2}{3}, 2731\pi/3)^*$.

g. $(2, \sin^{-1} \frac{3}{5})^*$.
h. $(5, \cos^{-1}(-\frac{15}{17}))^*$.
i. $(-2, \tan^{-1}(-\frac{5}{12}))^*$.

5. Compute $\mathbf{v} \cdot \mathbf{v}$, where $\mathbf{v}$ is as in Exercise 4.

6. Let θ be the angle between $(a, b) \neq (0, 0)$ and $(1, 0)$. Compute θ for each vector in Exercise 4. θ is called the **direction angle** for (a, b).

7. Show that $\mathbf{v}/|\mathbf{v}|$ is a vector of unit length having the same direction angle as $\mathbf{v}$, where $\mathbf{v}$ is a nonzero vector in the plane.

8. The **work** done in moving a body from the origin along the x axis a distance d is given by fd, where f is a constant force acting on the body along the x axis (see Fig. 8.2.3). If $\mathbf{f}$ is a vector force acting as in Fig. 8.2.4, $\mathbf{f}_x$ is the vector force acting along the x axis. Show that in Fig. 8.2.4 if we treat $\mathbf{d}$ as a vector, the work done in moving the object to the indicated point is $-\mathbf{f} \cdot \mathbf{d}$.

Figure 8.2.3

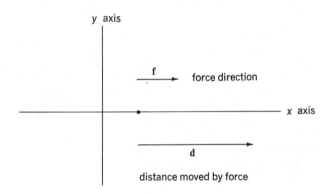

Figure 8.2.4

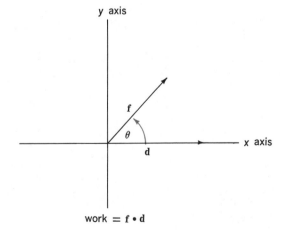

8.3 PROJECTIONS OF VECTORS

We have previously discussed the concept of projections of vectors onto lines. The vectors were points in the plane and the projections were onto the x and y axes. We now wish to examine the idea of projecting such vectors onto any line in the plane. Since we are interested mainly in the projection of one vector onto another, we will restrict ourselves to projections onto rays passing through the origin.

See Fig. 8.3.1. For the plane the *projection of* **v** *in the direction* θ is a vector $\mathbf{v}_\theta$ given by $\mathbf{v}_\theta = (\mathbf{v} \cdot \mathbf{u}_\theta)\mathbf{u}_\theta$, where $\mathbf{u}_\theta$ is a unit vector with direction angle θ. (See Fig. 8.3.2.)

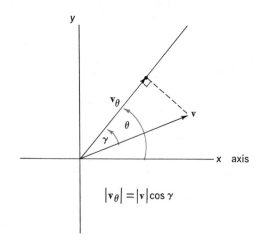

$$|\mathbf{v}_\theta| = |\mathbf{v}|\cos\gamma$$

Figure 8.3.1
Projection of **v** in the direction θ

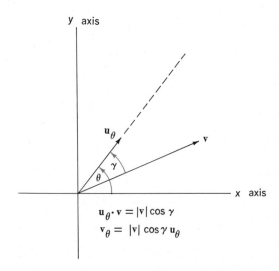

$$\mathbf{u}_\theta \cdot \mathbf{v} = |\mathbf{v}|\cos\gamma$$
$$\mathbf{v}_\theta = |\mathbf{v}|\cos\gamma\, \mathbf{u}_\theta$$

Figure 8.3.2
Unit vector in the direction θ

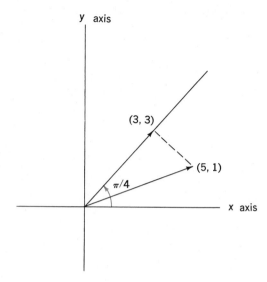

Figure 8.3.3
Projection in the direction $\pi/4$

Example 8.3.1 Let $\mathbf{v} = (5, 1)$ with $\theta = \pi/4$ radians. Compute $\mathbf{v}_\theta$.
The unit vector $\mathbf{u}_\theta$ is $(1/\sqrt{2}, 1/\sqrt{2})$ so that

$$\begin{aligned}
\mathbf{v}_\theta &= (\mathbf{v} \cdot \mathbf{u}_\theta)\mathbf{u}_\theta \\
&= ((5, 1) \cdot (1/\sqrt{2}, 1/\sqrt{2}))(1/\sqrt{2}, 1/\sqrt{2}) \\
&= (5/\sqrt{2} + 1/\sqrt{2})(1/\sqrt{2}, 1/\sqrt{2}) \\
&= (3, 3)
\end{aligned}$$

(See Fig. 8.3.3.)
A more general example follows.

Example 8.3.2 Find $\mathbf{v}_\theta$ where $\mathbf{v} = (a, b)$.

Since $\quad \mathbf{u}_\theta = (\cos\theta, \sin\theta)$
$$\begin{aligned}
\mathbf{v}_\theta &= ((a, b) \cdot (\cos\theta, \sin\theta))(\cos\theta, \sin\theta) \\
&= (a\cos\theta + b\sin\theta)(\cos\theta, \sin\theta) \\
&= (a\cos^2\theta + b\cos\theta\sin\theta, a\cos\theta\sin\theta + b\sin^2\theta)
\end{aligned}$$

(See Fig. 8.3.4.)

Example 8.3.3 Compute the projection of $\mathbf{v} = (3, 4)$ in the direction of
the ray containing the vector $\mathbf{v}_1 = (8, 15)$.
The unit vector $\mathbf{u}_1$ in the same direction as $\mathbf{v}_1$ is

$$\mathbf{u}_1 = \frac{\mathbf{v}_1}{|\mathbf{v}_1|} = \frac{\mathbf{v}_1}{(\mathbf{v}_1 \cdot \mathbf{v}_1)^{1/2}} = \frac{\mathbf{v}_1}{17} = (\tfrac{8}{17}, \tfrac{15}{17})$$

Figure 8.3.4

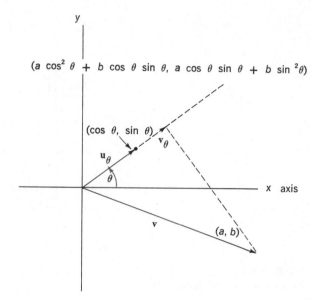

$(a \cos^2 \theta + b \cos \theta \sin \theta, a \cos \theta \sin \theta + b \sin^2 \theta)$

$(\cos \theta, \sin \theta)$

$\mathbf{v}_\theta$

$\mathbf{u}_\theta$

θ

x axis

$\mathbf{v}$

(a, b)

y

Figure 8.3.5

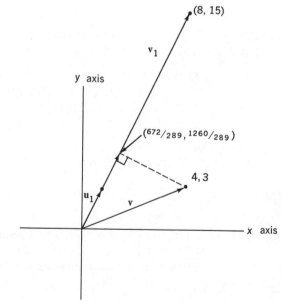

$(8, 15)$

$\mathbf{v}_1$

y axis

$(^{672}/_{289}, ^{1260}/_{289})$

$4, 3$

$\mathbf{u}_1$

$\mathbf{v}$

x axis

Necessarily, the projection of $\mathbf{v}$ in this direction is

$$(\mathbf{v} \cdot \mathbf{u}_1)\mathbf{u}_1 = ((3, 4) \cdot (^8/_{17}, ^{15}/_{17}))(^8/_{17}, ^{15}/_{17})$$
$$= {}^{84}/_{17}(^8/_{17}, ^{15}/_{17}) = (^{672}/_{289}, ^{1260}/_{289})$$

(See Fig. 8.3.5.)

Figure 8.3.6

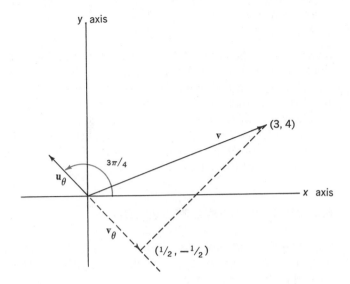

Example 8.3.4 Find v_θ where $v = (3, 4)$ and $\theta = 3\pi/4$.
The unit vector $u_\theta = (-1/\sqrt{2}, 1/\sqrt{2})$ so that

$$v_\theta = ((3, 4)(-1/\sqrt{2}, 1/\sqrt{2}))(-1/\sqrt{2}, 1/\sqrt{2})$$
$$= (\tfrac{1}{2}, -\tfrac{1}{2})$$

Observe Fig. 8.3.6 and note that v_θ lies in the fourth quadrant while u_θ lies in the second. We might say (in a fashion similar to occurrences in polar form) that v actually projects onto the "negative extension" of u_θ.

You should observe that the line connecting v and v_θ is in each case perpendicular to the line containing v.

■ **EXERCISES 8.3**

1. Let v be given as in each case below and compute the unit vector in the same direction as v.

a. (6, 0).   e. (1, −3).
b. (0, 4).   f. (7, 24).
c. (−5, 0).   g. (a, b).
d. (−3, −4).

2. Let v_1 be as given in each case below and compute the projection of v_1 in the direction of v, where v is as given in Exercise 1.

a. (4, 0).   c. (1, 2).
b. (0, 5).   d. (−2, 3).

e. $(-1, -4)$.   **g.** $(3, -2)$.
f. $(-1, -1)$.   **h.** $(\frac{1}{2}, \frac{2}{3})$.

3. In some of the cases of Exercise 2 the projections were $(0, 0)$. When does this occur? Justify your remarks.
4. What is the projection of a vector onto itself?
5. Give an estimate of $|v_\theta|$ in terms of $|v|$.
6. Show that in each of Examples 8.1.1, 8.1.3, and 8.1.4 the line containing the vector and its projection is perpendicular to the line containing the original vector. (Hint: Compare slopes.) Prove that this holds in general (that is, this holds for Example 8.3.2).

8.4 MORE ON GEOMETRIC VECTORS

Our geometric concept of a vector has thus far been limited to directed line segments emanating from the origin. We wish now to extend the idea to other directed line segments.

If P_1 and P_2 are points of the plane, the vector from P_1 to P_2 is as pictured in Fig. 8.4.1. The angle made with the horizontal is the directed angle for the vector and the length of the vector is $d(P_1, P_2)$.

Example 8.4.1   Construct the vector from $(1, 3)$ to $(3, 5)$, determining its direction angle and length.

The construction is shown in Fig. 8.4.2. Since $d((1, 3), (3, 5)) = 2\sqrt{2}$, the vector has this length. The value $\tan \theta$ as seen in Fig. 8.4.2 is one implying that $\theta = 45°$ (θ is a positive angle of less than $90°$).

Two vectors (in this latter sense) are **equivalent** if they have the same direction angle and the same length. (See Fig. 8.4.3.) Two vectors are

Figure 8.4.1
Vector from P_1 to P_2

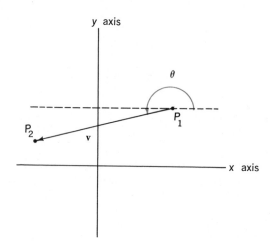

Figure 8.4.2

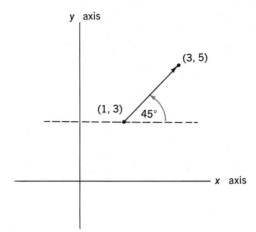

Figure 8.4.3
Equivalent vectors

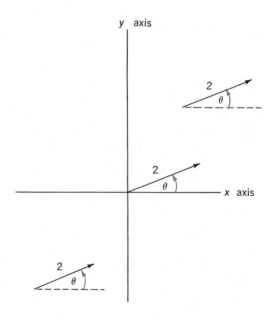

parallel if they have the same direction angle and **antiparallel** if their direction angles differ by 180°. Two vectors are **collinear** if they are parallel or antiparallel. Figures 8.4.4 and 8.4.5 depict these conditions.

Example 8.4.2   Construct a vector emanating from (0, 0) and equivalent to the vector in Example 8.4.1.

The direction angle is 45° so that the vector lies on the line $y = x$. Since the point (2, 2) is a distance $2\sqrt{2}$ from the origin, the vector as shown in Fig. 8.4.6 is the correct vector.

Figure 8.4.4
Vectors parallel and collinear

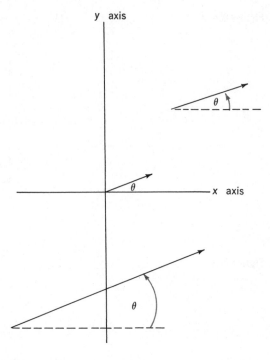

Figure 8.4.5
Vectors antiparallel and collinear

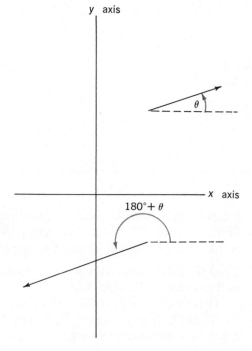

Figure 8.4.6

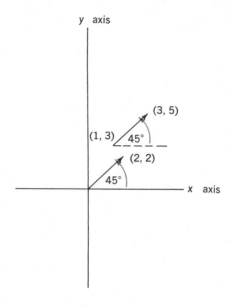

Example 8.4.3 Construct two vectors emanating from the point $(3, 6)$: one parallel to the vector of Example 8.4.1 with length $3\sqrt{2}$ and one antiparallel to the vector of Example 8.4.1 and having length $3\sqrt{2}$.

The direction angle for the vector in Example 8.4.1 is $45°$ so that the slope of the line containing the desired vectors is $\tan 45° = 1$. Since the line passes through $(3, 6)$, its equation is $y - 6 = x - 3$ or $y = x + 3$. The terminal points (x, y) of the two desired vectors must satisfy both

$$y = x + 3$$
and $$3\sqrt{2} = [(x - 3)^2 + (y - 6)]^{1/2}$$

The solution set for the two equations is found to be $\{(6, 9),\ (0, 3)\}$. Thus, the parallel and antiparallel vectors v_1 and v_2, respectively, are as in Fig. 8.4.7.

We keep the same notion of horizontal and vertical components as before. Figure 8.4.8 illustrates the concepts.

A **zero vector** is a vector having zero length (the direction angle is of no consequence).

We still wish to add vectors in a head to tail fashion. Figure 8.4.9 relates the head to tail concept to our new concept of vectors. Scalar multiplication has the same significance as previously (that is, the scalar multiplication affects the length of the vector and, if the scalar involved is negative, the direction angle is changed by $180°$). Figure 8.4.10 shows various scalar multiples of a vector. In particular, the negative of a vector is one having the same length and initial point but one that is antiparallel to the original vector.

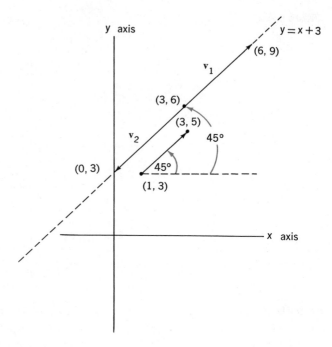

Figure 8.4.7

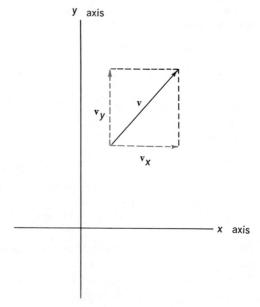

Figure 8.4.8
Horizontal and vertical components

Figure 8.4.9

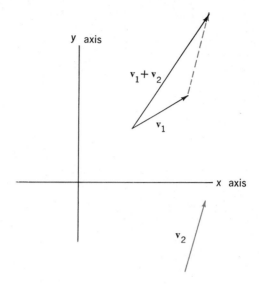

Figure 8.4.10
Vector addition

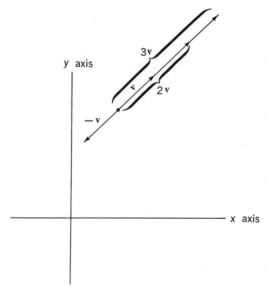

The concept of subtraction of one vector from another is shown in Fig. 8.4.11.

In the above discussion, you should sense that the most important features of a geometric vector are its length and direction angle. Thus, for all practical purposes, all equivalent vectors may be treated as one. Given any vector, we may "translate" it to an equivalent vector at the origin. In

Figure 8.4.11

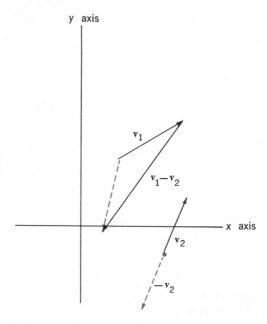

most applications (see the following section) we shall do precisely this. We can, in effect, forget about the coordinate system and treat only the geometric objects and considerations.

■ EXERCISES 8.4

1. Sketch the construction of the following vectors.

a. v_1 is the vector from $(1, 3)$ to $(2, 4)$.
b. v_2 is the vector from $(0, 5)$ to $(7, 11)$.
c. v_3 is the vector from $(-1, 2)$ to $(-3, -4)$.

2. Describe the scalar product kv where v is as in Exercise 1 and k is given by

a. $k = 3$.   **c.** $k = -4$.
b. $k = -1$.   **d.** $k = 0$.

3. Add all possible pairs of vectors (pictorially) from Exercise 1.
4. Draw a sketch of and describe:

a. $v_1 - v_2$.   **e.** $3v_1 - 4v_2$.
b. $v_1 - v_3$.   **f.** $-v_2 - 3v_3$.
c. $v_2 - v_3$.   **g.** $-4v_3 + v_1$.
d. $v_3 - v_2$.

5. Sketch vectors equivalent to v_1, v_2, and v_3 from Exercise 1 emanating from

   a. $(0, 0)$.   **c.** $(3, 5)$.
   b. $(1, 1)$.   **d.** $(-2, -1)$.

6. Sketch vectors antiparallel to v_1, v_2, and v_3 from Exercise 1 emanating from $(0, 0)$ and having unit length.

7. Describe all vectors emanating from $(1, 2)$ collinear with

   a. v_1 from Exercise 1.
   b. v_2 from Exercise 1.
   c. v_3 from Exercise 1.

and having unit length.

8. Describe the horizontal and vertical components of the vectors from Exercise 1.

9. Give the terminal point of a vector emanating from P with direction angle θ and length d where:

   a. $P = (0, 0)$, $\theta = 81°$, $d = 2$.
   b. $P = (-1, 2)$, $\theta = 60°$, $d = 5$.
   c. $P = (-3, -1)$, $\theta = -45°$, $d = \sqrt{2}$.
   d. $P = (2, -1)$, $\theta = 150°$, $d = \sqrt{3}$.
   e. $P = (-1, -2)$, $\theta = 56°$, $d = 2\sqrt{3}$.

8.5 APPLICATIONS USING VECTORS

In the physical world, many entities may be viewed as vectors. It is our purpose here to examine a few of these ideas.

Speed is interpreted as a rate of change of position with respect to time (distance traveled per unit time), while velocity is interpreted as a "directed speed," that is, speed is a scalar while velocity is a vector. The length of a velocity vector is the speed associated with that velocity.

Example 8.5.1 A plane is flying horizontally with a heading of 45° west of north and is flying at maximum speed (600 miles per hour with no wind). A wind is blowing across the path of the plane toward a direction 15° north of west and a speed of 80 miles per hour. Represent the two velocities as vectors and find the actual velocity of the airplane (this velocity is the resultant of the two vectors given).

Figure 8.5.1 shows the two vectors (both emanating from the origin) and the resultant. The terminal point of v_1 is found as $(600 \cos 135°, 600 \sin 135°)$, while the terminal point of v_2 is $(80 \cos 165°, 80 \sin 165°)$. The resultant is found by addition.

Figure 8.5.1

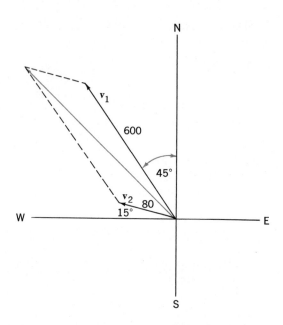

We might say that we found the resultant vector by adding the respective horizontal and vertical components of the two vectors.

A force acting on an object is pictured as a vector.

Example 8.5.2 A force of 100 lb is pulling a steel ball downward through a magnetic field. The field exerts a horizontal force of 20 lb on the ball. Determine the ball's velocity through the field.

Consult Fig. 8.5.2. The two forces are shown as vectors (again emanating from the origin) and the resultant is computed easily, since the vectors are perpendicular.

Figure 8.5.2

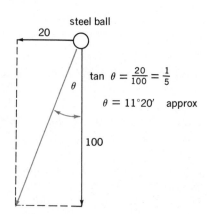

It is sometimes of value to note that the resultant of two vectors can be described by

$$|v_1 + v_2|^2 = |v_1|^2 + |v_2|^2 + |v_1| |v_2| \cos \theta$$

(where θ is the angle between v_1 and v_2) and

$$\frac{|v_1|}{\sin (\alpha - \theta)} = \frac{|v_1 + v_2|}{\sin (180° - \theta)} = \frac{|v_2|}{\sin (2\theta - \alpha)}$$

See Fig. 8.5.3 and Exercise 8.5.1.

We may define the work W done by a force f in displacing an object a distance d (treated as a vector) in the direction of the force's action as being given by $W = f \cdot d$, the inner product of the two vectors. Figure 8.5.4 shows how this expression arises alternatively as the product of the length of the component of force acting in the direction of movement with the distance the force moves the object.

Example 8.5.3   A force of 65 lb acting as in Fig. 8.5.5 moves an object constrained to move along the x axis.   Find the work done by the force in moving the object 15 feet.

The distance vector d is given by $(15, 0)$, while the force vector f is seen to be $(39, 52)$.   Thus $W = f \cdot d = (39, 52)(15, 0) = 39 \cdot 15 = 585$ ft lb.

Figure 8.5.3

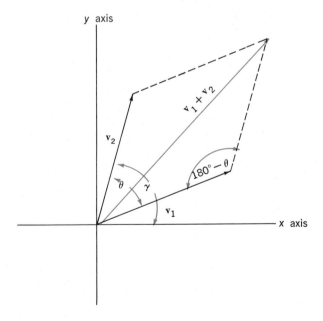

Figure 8.5.4

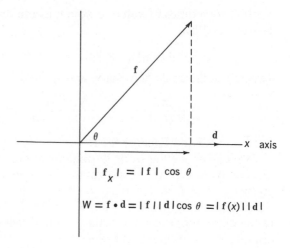

$$|f_x| = |f| \cos \theta$$

$$W = f \cdot d = |f||d|\cos \theta = |f(x)||d|$$

Figure 8.5.5

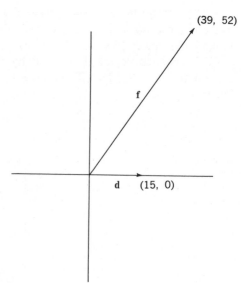

(39, 52)

f

d   (15, 0)

■ **EXERCISES 8.5**

1. Prove that the resultant of the two vectors shown in Fig. 8.5.3 is as specified in the section. (Use the law of cosines and law of sines.)

2. A plane has its nose pointing in a direction 30° west of south and is flying with an air speed (speed in calm air) of 500 mph, while a wind is blowing due north at 50 mph. The drift angle of the plane is the angle between the plane's heading (direction it is pointing) and its course (direction it is flying). Find the velocity of the plane and its drift angle.

3. In Exercise 2, what heading must the plane have if its course is to be 30° west of south?

4. A boat is heading directly across a river that is 5 miles wide. The current has a speed of 35 mph, while the boat is capable of traveling 12 mph in water with no current. If the boat travels full speed ahead, what is its velocity on the river? How far down stream does it dock on the opposite side?

5. In Exercise 4, describe the heading of the boat that will cause it to dock directly opposite its debarkation point.

6. A steel ball is falling through a magnetic field. The gravitational force tending to make the ball drop is 5 lb, while a horizontal force of 3 lb is exerted by the field. Find the resultant force acting on the ball. If the field is 10 ft deep (that is, the ball drops 10 ft in traveling through the field), find the work done (by the resultant force) in moving the ball vertically. Find the work done in moving the ball along its course as it passes through the field.

7. A force shown in Fig. 8.5.6 moves an object constrained to move along the lines $y = x$. Find the work done by the force in moving the object 8 ft.

8. A balloon released in still air rises at a rate of 38 ft per sec. If a horizontal wind blowing due east has a speed of 60 mph (88 ft per sec), find the velocity of the ascending balloon.

9. A pitcher can throw only fast balls that do not drop perceptibly from his point of aim. He throws the ball at a speed of approximately 100 mph.

Figure 8.5.6

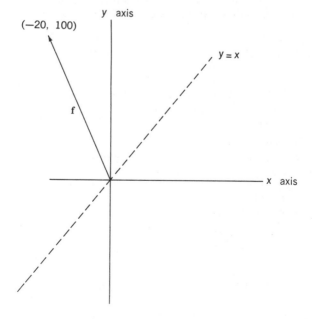

(−20, 100)

y axis

y = *x*

f

x axis

A wind is blowing from the direction of first base toward a point halfway between third base and home plate and has a speed of 25 mph. The batter is standing 1 ft to the (pitcher's) right of the center of the plate. Will the ball hit the batter if the pitcher aims at the center of the plate? Where should the pitcher aim so that the ball travels over the center of the plate?
10. A swimmer can swim 3 mph if no current is present and can swim for no more than 3 hr. He is attempting to swim across the mouth of a river opening into the sea. If the current flows to sea at a rate of 1½ mph and the mouth is 3 miles across, what heading should the swimmer take in order to reach the other side without being washed out to sea? He might feel that he can play safe and aim his swim farther upstream than this calculation. What is the farthest point upstream that he can aim for and not be washed out to sea?

8.6 REVIEW EXERCISES

1. Give the horizontal and vertical components of each of the following vectors in the plane.

a. $(0, 0)$.   **e.** $(2, -1) + (3, -4)$.
b. $(1, 0)$.   **f.** $6(-1, -5)$.
c. $(0, 1)$.   **g.** $(6 + 2)(2, 3)$.
d. $(3, -4)$.   **h.** $8((2, 1) + (-1, 3))$.

2. Find the projection of each vector in 1 onto the ray forming the terminal side of an angle (in standard position) having measure

a. $0°$.   **d.** $-60°$.
b. $\pi/4$ radians.   **e.** $270°$.
c. $135°$.   **f.** $-150°$.

3. Describe the horizontal and vertical components of a complex number z in terms of $R(z)$ and $I(z)$.
4. Let $v_1 = (1, 1)$ and $v_2 = (1, -3)$. Write each vector from 1 in the form $av_1 + bv_2$.
5. Compute the following inner products.

a. $(0, 0) \cdot (a, b)$.
b. $(1, 0) \cdot (a, b)$.
c. $(0, 1) \cdot (a, b)$.
d. $(1, 1) \cdot (2, -3)$.
e. $(a, b) \cdot (a, b)$.
f. $(a, b) \cdot (c, d)$, where the two are orthogonal.

6. Find the length of each vector in 1.

7. Find unit vectors orthogonal to each vector [except $(0, 0)$] from 1.

8. Compute the angle between each pair of vectors.

   a. $(1, 0)$ and $(3, 4)$.
   b. $(2, 5)$ and $(-1, 6)$.
   c. $(4, 7)$ and $(4, -3)$.
   d. $(-2, -2)$ and $(-1, -\sqrt{3})$.
   e. $(1, -3)$ and $(5, 6)$.
   f. $(2, -9)$ and $(5, -1)$.

9. In each case below, write an equivalent vector emanating from the origin and a vector having the same length but antiparallel to each.

   a. The vector from $(8, 4)$ to $(0, -2)$.
   b. The vector from $(3, -1)$ to $(1, 3)$.
   c. The vector from $(-2, 3)$ to $(-2, 4)$.
   d. The vector from $(-5, -6)$ to $(-3, -1)$.
   e. The vector from $(3, \frac{1}{2})$ to $(2, 1)$.
   f. The vector from $(-2, 1)$ to $(1, -7)$.

10. Describe the horizontal and vertical components of each vector in 9.

11. An object is located at the origin $(0, 0)$. A force of magnitude 10 acts along a line making a $60°$ angle with the positive x axis. Another force of magnitude 8 acts at an angle of $135°$. Represent these forces as vectors. The resultant of these vectors is the resulting or equivalent force. Find the magnitude, direction, and x and y components of the resulting force.

12. An airplane has a ground speed of 500 mph and the pilot's instruments indicate a heading of $30°$ east of north with no wind blowing. A wind starts to blow due east with a speed of 35 mph. Represent each speed as a vector and compute the resultant to find the actual speed and course the plane will then follow.

13. A man decides to row to a point directly across a stream. His rate of rowing is 2 mph and the current has a rate of 1 mph. What course should the man attempt to row in order to reach his destination in the least possible time?

14. A weather crew releases a balloon which ascends vertically at a rate of 25 ft per sec. A wind is blowing horizontally with a speed of 10 ft per sec. What is the angle of ascent of the balloon in the wind?

8.7 ADVANCED EXERCISES

1. Let Ω be the set $\{(m, n) : m, n \in Z, n \neq 0\}$. That is to say, Ω is the set of ordered pairs of integers for which the second element is different from zero. Define equality, addition, subtraction, multiplication, and division as below:

$(a, b) = (c, d)$ if and only if $ad = bc$

$(a, b) + (c, d) = (ad + bc, bd)$

$(a, b) - (c, d)$ is any pair $(e, f) \in \Omega$ such that $(c, d) + (e, f) = (a, b)$

$(a, b)(c, d) = (ac, bd)$

$(a, b) \div (c, d)$ is any pair $(e, f) \in \Omega$ so that $(c, d)(e, f) = (a, b)$, provided that if (g, h) is also a solution, $(e, f) = (g, h)$

Show the following.

a. If $x \in Z$, $x \neq 0$, $(a, b) = (ax, bx)$.
b. "$=$" is an equivalence relation.
c. Ω is closed relative to addition.
d. Ω is closed for multiplication.
e. $(a, b) + (c, b) = (a + c, b)$.
f. $(a, b) + (0, d) = (a, b)$.
g. $(a, b)(1, 1) = (a, b)$.
h. $(a, b) \div (c, d) = (a, b)(d, c)$ if $c \neq 0$.
i. $(a, b) \div (0, d)$ is not defined.
j. $(a, b) - (c, d) = (a, b) + (-c, d)$.
k. Ω is closed for subtraction.
l. $(a, b) + (c, d) = (c, d) + (a, b)$.
m. $(a, b)(c, d) = (c, d)(a, b)$.
n. $(a, b) + [(c, d) + (e, f)] = [(a, b) + (c, d)] + (e, f)$.
o. $(a, b)[(c, d)(e, f)] = [(a, b)(c, d)](e, f)$.
p. $(a, b)[(c, d) + (e, f)] = [(a, b)(c, d)] + [(a, b)(e, f)]$.

Furthermore, define scalar multiplication by:

$$r(a, b) = (ra, b) \qquad \text{for } r \in Z$$

Now, prove that the definitions make Ω "like a vector space over the integers."

2. Show that each Euclidean n-space is a vector space over the reals, where

$$(a_1, a_2, a_3, \cdots, a_n) + (b_1, b_2, \cdots, b_n)$$
$$= (a_1 + b_1, a_2 + b_2, a_3 + b_3, \cdots, a_n + b_n)$$

and $\qquad k \cdot (a_1, a_2, \cdots, a_n) = (ka_1, ka_2, ka_3, \cdots, ka_n)$

3. Referring to Exercise 2, we define

$$(a_1, a_2, a_3, \cdots, a_n) \cdot (b_1, b_2, b_3, \cdots, b_n) = \sum_{i=1}^{n} (a_i b_i)$$

What is the relation between $(a_1, a_2, a_3, \cdots, a_n) \cdot (a_1, a_2, \cdots, a_n)$ and the distance concept? What is the angle between $(a_1, a_2, \cdots, a_n)$ and $(b_1, b_2, \cdots, b_n)$?

4. Consider the plane as a vector space over the reals and show that if $(a, b) \cdot (c, d) = 0$ with neither (a, b) nor (c, d) the zero vector, then (a, b) and (c, d) span the space.

5. If (a, b) and (c, d) are nonzero and not collinear, they span the space. Prove this assertion.

6. Show that, if (a, b) and (c, d) are collinear vectors of the plane, then $(a, b) = k(c, d)$ for some real number k.

7. Show that if (a, b), (c, d), and (e, f) are vectors in the plane, then there are real numbers $x, y,$ and z (not all zero) so that $x(a, b) + y(c, d) + z(e, f) = (0, 0)$. We say that the set $\{(a, b), (c, d), (e, f)\}$ is **linearly dependent.**

8. Vectors $v_1, v_2, v_3, \cdots, v_n$ are **linearly independent** if, whenever $a_1 v_1 + a_2 v_2 + a_3 v_3 + \cdots + a_n v_n = 0$, all the a_i are zero. For each space R^n, pick out a set of n linearly independent vectors having unit length.

9. A vector space is said to have dimension $n \in N$ if there exists a basis (a set of vectors spanning the space) of n linearly independent vectors. Show that every basis of R^3 has at least 3 elements and that a linearly independent basis (see Exercises 7 and 8) has exactly 3 elements, that is, R^3 has dimension 3.

Appendix A: Tables

Values of Trigonometric Functions*

Degrees	Radians	Sine	Tangent	Cotangent	Cosine		
0° 00′	.0000	.0000	.0000		1.0000	1.5708	90° 00′
10′	.0029	.0029	.0029	343.77	1.0000	1.5679	50′
20′	.0058	.0058	.0058	171.89	1.0000	1.5650	40′
30′	.0087	.0087	.0087	114.59	1.0000	1.5621	30′
40′	.0116	.0116	.0116	85.940	.9999	1.5592	20′
50′	.0145	.0145	.0145	68.750	.9999	1.5563	10′
1° 00′	.0175	.0175	.0175	57.290	.9998	1.5533	89° 00′
10′	.0204	.0204	.0204	49.104	.9998	1.5504	50′
20′	.0233	.0233	.0233	42.964	.9997	1.5475	40′
30′	.0262	.0262	.0262	38.188	.9997	1.5446	30′
40′	.0291	.0291	.0291	34.368	.9996	1.5417	20′
50′	.0320	.0320	.0320	31.242	.9995	1.5388	10′
2° 00′	.0349	.0349	.0349	28.636	.9994	1.5359	88° 00′
10′	.0378	.0378	.0378	26.432	.9993	1.5330	50′
20′	.0407	.0407	.0407	24.542	.9992	1.5301	40′
30′	.0436	.0436	.0437	22.904	.9990	1.5272	30′
40′	.0465	.0465	.0466	21.470	.9989	1.5243	20′
50′	.0495	.0494	.0495	20.206	.9988	1.5213	10′
3° 00′	.0524	.0523	.0524	19.081	.9986	1.5184	87° 00′
10′	.0553	.0552	.0553	18.075	.9985	1.5155	50′
20′	.0582	.0581	.0582	17.169	.9983	1.5126	40′
30′	.0611	.0610	.0612	16.350	.9981	1.5097	30′
40′	.0640	.0640	.0641	15.605	.9980	1.5068	20′
50′	.0669	.0669	.0670	14.924	.9978	1.5039	10′
4° 00′	.0698	.0698	.0699	14.301	.9976	1.5010	86° 00′
10′	.0727	.0727	.0729	13.727	.9974	1.4981	50′
20′	.0756	.0756	.0758	13.197	.9971	1.4952	40′
30′	.0785	.0785	.0787	12.706	.9969	1.4923	30′
40′	.0814	.0814	.0816	12.251	.9967	1.4893	20′
50′	.0844	.0843	.0846	11.826	.9964	1.4864	10′
5° 00′	.0873	.0872	.0875	11.430	.9962	1.4835	85° 00′
10′	.0902	.0901	.0904	11.059	.9959	1.4806	50′
20′	.0931	.0929	.0934	10.712	.9957	1.4777	40′
30′	.0960	.0958	.0963	10.385	.9954	1.4748	30′
40′	.0989	.0987	.0992	10.078	.9951	1.4719	20′
50′	.1018	.1016	.1022	9.7882	.9948	1.4690	10′
6° 00′	.1047	.1045	.1051	9.5144	.9945	1.4661	84° 00′
10′	.1076	.1074	.1080	9.2553	.9942	1.4632	50′
20′	.1105	.1103	.1110	9.0098	.9939	1.4603	40′
30′	.1134	.1132	.1139	8.7769	.9936	1.4573	30′
40′	.1164	.1161	.1169	8.5555	.9932	1.4544	20′
50′	.1193	.1190	.1198	8.3450	.9929	1.4515	10′
7° 00′	.1222	.1219	.1228	8.1443	.9925	1.4486	83° 00′
10′	.1251	.1248	.1257	7.9530	.9922	1.4457	50′
20′	.1280	.1276	.1287	7.7704	.9918	1.4428	40′
30′	.1309	.1305	.1317	7.5958	.9914	1.4399	30′
40′	.1338	.1334	.1346	7.4287	.9911	1.4370	20′
50′	.1367	.1363	.1376	7.2687	.9907	1.4341	10′
8° 00′	.1396	.1392	.1405	7.1154	.9903	1.4312	82° 00′
10′	.1425	.1421	.1435	6.9682	.9899	1.4283	50′
20′	.1454	.1449	.1465	6.8269	.9894	1.4254	40′
30′	.1484	.1478	.1495	6.6912	.9890	1.4224	30′
40′	.1513	.1507	.1524	6.5606	.9886	1.4195	20′
50′	.1542	.1536	.1554	6.4348	.9881	1.4166	10′
9° 00′	.1571	.1564	.1584	6.3138	.9877	1.4137	81° 00′
		Cosine	Cotangent	Tangent	Sine	Radians	Degrees

*From *Introduction to Mathematical Analysis* by R. E. Johnson, N. H. McCoy, and A. F. O'Neill. Copyright 1962 by Holt, Rinehart and Winston, Inc. Reprinted by permission.

Values of Trigonometric Functions (cont.)

Degrees	Radians	Sine	Tangent	Cotangent	Cosine		
9° 00'	.1571	.1564	.1584	6.3138	.9877	1.4137	81° 00'
10'	.1600	.1593	.1614	6.1970	.9872	1.4108	50'
20'	.1629	.1622	.1644	6.0844	.9868	1.4079	40'
30'	.1658	.1650	.1673	5.9758	.9863	1.4050	30'
40'	.1687	.1679	.1703	5.8708	.9858	1.4021	20'
50'	.1716	.1708	.1733	5.7694	.9853	1.3992	10'
10° 00'	.1745	.1736	.1763	5.6713	.9848	1.3963	80° 00'
10'	.1774	.1765	.1793	5.5764	.9843	1.3934	50'
20'	.1804	.1794	.1823	5.4845	.9838	1.3904	40'
30'	.1833	.1822	.1853	5.3955	.9833	1.3875	30'
40'	.1862	.1851	.1883	5.3093	.9827	1.3846	20'
50'	.1891	.1880	.1914	5.2257	.9822	1.3817	10'
11° 00'	.1920	.1908	.1944	5.1446	.9816	1.3788	79° 00'
10'	.1949	.1937	.1974	5.0658	.9811	1.3759	50'
20'	.1978	.1965	.2004	4.9894	.9805	1.3730	40'
30'	.2007	.1994	.2035	4.9152	.9799	1.3701	30'
40'	.2036	.2022	.2065	4.8430	.9793	1.3672	20'
50'	.2065	.2051	.2095	4.7729	.9787	1.3643	10'
12° 00'	.2094	.2079	.2126	4.7046	.9781	1.3614	78° 00'
10'	.2123	.2108	.2156	4.6382	.9775	1.3584	50'
20'	.2153	.2136	.2186	4.5736	.9769	1.3555	40'
30'	.2182	.2164	.2217	4.5107	.9763	1.3526	30'
40'	.2211	.2193	.2247	4.4494	.9757	1.3497	20'
50'	.2240	.2221	.2278	4.3897	.9750	1.3468	10'
13° 00'	.2269	.2250	.2309	4.3315	.9744	1.3439	77° 00'
10'	.2298	.2278	.2339	4.2747	.9737	1.3410	50'
20'	.2327	.2306	.2370	4.2193	.9730	1.3381	40'
30'	.2356	.2334	.2401	4.1653	.9724	1.3352	30'
40'	.2385	.2363	.2432	4.1126	.9717	1.3323	20'
50'	.2414	.2391	.2462	4.0611	.9710	1.3294	10'
14° 00'	.2443	.2419	.2493	4.0108	.9703	1.3265	76° 00'
10'	.2473	.2447	.2524	3.9617	.9696	1.3235	50'
20'	.2502	.2476	.2555	3.9136	.9689	1.3206	40'
30'	.2531	.2504	.2586	3.8667	.9681	1.3177	30'
40'	.2560	.2532	.2617	3.8208	.9674	1.3148	20'
50'	.2589	.2560	.2648	3.7760	.9667	1.3119	10'
15° 00'	.2618	.2588	.2679	3.7321	.9659	1.3090	75° 00'
10'	.2647	.2616	.2711	3.6891	.9652	1.3061	50'
20'	.2676	.2644	.2742	3.6470	.9644	1.3032	40'
30'	.2705	.2672	.2773	3.6059	.9636	1.3003	30'
40'	.2734	.2700	.2805	3.5656	.9628	1.2974	20'
50'	.2763	.2728	.2836	3.5261	.9621	1.2945	10'
16° 00'	.2793	.2756	.2867	3.4874	.9613	1.2915	74° 00'
10'	.2822	.2784	.2899	3.4495	.9605	1.2886	50'
20'	.2851	.2812	.2931	3.4124	.9596	1.2857	40'
30'	.2880	.2840	.2962	3.3759	.9588	1.2828	30'
40'	.2909	.2868	.2994	3.3402	.9580	1.2799	20'
50'	.2938	.2896	.3026	3.3052	.9572	1.2770	10'
17° 00'	.2967	.2924	.3057	3.2709	.9563	1.2741	73° 00'
10'	.2996	.2952	.3089	3.2371	.9555	1.2712	50'
20'	.3025	.2979	.3121	3.2041	.9546	1.2683	40'
30'	.3054	.3007	.3153	3.1716	.9537	1.2654	30'
40'	.3083	.3035	.3185	3.1397	.9528	1.2625	20'
50'	.3113	.3062	.3217	3.1084	.9520	1.2595	10'
18° 00'	.3142	.3090	.3249	3.0777	.9511	1.2566	72° 00'
		Cosine	Cotangent	Tangent	Sine	Radians	Degrees

Values of Trigonometric Functions (cont.)

Degrees	Radians	Sine	Tangent	Cotangent	Cosine		
18° 00'	.3142	.3090	.3249	3.0777	.9511	1.2566	72° 00'
10'	.3171	.3118	.3281	3.0475	.9502	1.2537	50'
20'	.3200	.3145	.3314	3.0178	.9492	1.2508	40'
30'	.3229	.3173	.3346	2.9887	.9483	1.2479	30'
40'	.3258	.3201	.3378	2.9600	.9474	1.2450	20'
50'	.3287	.3228	.3411	2.9319	.9465	1.2421	10'
19° 00'	.3316	.3256	.3443	2.9042	.9455	1.2392	71° 00'
10'	.3345	.3283	.3476	2.8770	.9446	1.2363	50'
20'	.3374	.3311	.3508	2.8502	.9436	1.2334	40'
30'	.3403	.3338	.3541	2.8239	.9426	1.2305	30'
40'	.3432	.3365	.3574	2.7980	.9417	1.2275	20'
50'	.3462	.3393	.3607	2.7725	.9407	1.2246	10'
20° 00'	.3491	.3420	.3640	2.7475	.9397	1.2217	70° 00'
10'	.3520	.3448	.3673	2.7228	.9387	1.2188	50'
20'	.3549	.3475	.3706	2.6985	.9377	1.2159	40'
30'	.3578	.3502	.3739	2.6746	.9367	1.2130	30'
40'	.3607	.3529	.3772	2.6511	.9356	1.2101	20'
50'	.3636	.3557	.3805	2.6279	.9346	1.2072	10'
21° 00'	.3665	.3584	.3839	2.6051	.9336	1.2043	69° 00'
10'	.3694	.3611	.3872	2.5826	.9325	1.2014	50'
20'	.3723	.3638	.3906	2.5605	.9315	1.1985	40'
30'	.3752	.3665	.3939	2.5386	.9304	1.1956	30'
40'	.3782	.3692	.3973	2.5172	.9293	1.1926	20'
50'	.3811	.3719	.4006	2.4960	.9283	1.1897	10'
22° 00'	.3840	.3746	.4040	2.4751	.9272	1.1868	68° 00'
10'	.3869	.3773	.4074	2.4545	.9261	1.1839	50'
20'	.3898	.3800	.4108	2.4342	.9250	1.1810	40'
30'	.3927	.3827	.4142	2.4142	.9239	1.1781	30'
40'	.3956	.3854	.4176	2.3945	.9228	1.1752	20'
50'	.3985	.3881	.4210	2.3750	.9216	1.1723	10'
23° 00'	.4014	.3907	.4245	2.3559	.9205	1.1694	67° 00'
10'	.4043	.3934	.4279	2.3369	.9194	1.1665	50'
20'	.4072	.3961	.4314	2.3183	.9182	1.1636	40'
30'	.4102	.3987	.4348	2.2998	.9171	1.1606	30'
40'	.4131	.4014	.4383	2.2817	.9159	1.1577	20'
50'	.4160	.4041	.4417	2.2637	.9147	1.1548	10'
24° 00'	.4189	.4067	.4452	2.2460	.9135	1.1519	66° 00'
10'	.4218	.4094	.4487	2.2286	.9124	1.1490	50'
20'	.4247	.4120	.4522	2.2113	.9112	1.1461	40'
30'	.4276	.4147	.4557	2.1943	.9100	1.1432	30'
40'	.4305	.4173	.4592	2.1775	.9088	1.1403	20'
50'	.4334	.4200	.4628	2.1609	.9075	1.1374	10'
25° 00'	.4363	.4226	.4663	2.1445	.9063	1.1345	65° 00'
10'	.4392	.4253	.4699	2.1283	.9051	1.1316	50'
20'	.4422	.4279	.4734	2.1123	.9038	1.1286	40'
30'	.4451	.4305	.4770	2.0965	.9026	1.1257	30'
40'	.4480	.4331	.4806	2.0809	.9013	1.1228	20'
50'	.4509	.4358	.4841	2.0655	.9001	1.1199	10'
26° 00'	.4538	.4384	.4877	2.0503	.8988	1.1170	64° 00'
10'	.4567	.4410	.4913	2.0353	.8975	1.1141	50'
20'	.4596	.4436	.4950	2.0204	.8962	1.1112	40'
30'	4625	.4462	.4986	2.0057	.8949	1.1083	30'
40'	.4654	.4488	.5022	1.9912	.8936	1.1054	20'
50'	.4683	.4514	.5059	1.9768	.8923	1.1025	10'
27° 00'	.4712	.4540	.5095	1.9626	.8910	1.0996	63° 00'
		Cosine	Cotangent	Tangent	Sine	Radians	Degrees

Values of Trigonometric Functions (cont.)

Degrees	Radians	Sine	Tangent	Cotangent	Cosine		
27° 00′	.4712	.4540	.5095	1.9626	.8910	1.0996	63° 00′
10′	.4741	.4566	.5132	1.9486	.8897	1.0966	50′
20′	.4771	.4592	.5169	1.9347	.8884	1.0937	40′
30′	.4800	.4617	.5206	1.9210	.8870	1.0908	30′
40′	.4829	.4643	.5243	1.9074	.8857	1.0879	20′
50′	.4858	.4669	.5280	1.8940	.8843	1.0850	10′
28° 00′	.4887	.4695	.5317	1.8807	.8829	1.0821	62° 00′
10′	.4916	.4720	.5354	1.8676	.8816	1.0792	50′
20′	.4945	.4746	.5392	1.8546	.8802	1.0763	40′
30′	.4974	.4772	.5430	1.8418	.8788	1.0734	30′
40′	.5003	.4797	.5467	1.8291	.8774	1.0705	20′
50′	.5032	.4823	.5505	1.8165	.8760	1.0676	10′
29° 00′	.5061	.4848	.5543	1.8040	.8746	1.0647	61° 00′
10′	.5091	.4874	.5581	1.7917	.8732	1.0617	50′
20′	.5120	.4899	.5619	1.7796	.8718	1.0588	40′
30′	.5149	.4924	.5658	1.7675	.8704	1.0559	30′
40′	.5178	.4950	.5696	1.7556	.8689	1.0530	20′
50′	.5207	.4975	.5735	1.7437	.8675	1.0501	10′
30° 00′	.5236	.5000	.5774	1.7321	.8660	1.0472	60° 00′
10′	.5265	.5025	.5812	1.7205	.8646	1.0443	50′
20′	.5294	.5050	.5851	1.7090	.8631	1.0414	40′
30′	.5323	.5075	.5890	1.6977	.8616	1.0385	30′
40′	.5352	.5100	.5930	1.6864	.8601	1.0356	20′
50′	.5381	.5125	.5969	1.6753	.8587	1.0327	10′
31° 00′	.5411	.5150	.6009	1.6643	.8572	1.0297	59° 00′
10′	.5440	.5175	.6048	1.6534	.8557	1.0268	50′
20′	.5469	.5200	.6088	1.6426	.8542	1.0239	40′
30′	.5498	.5225	.6128	1.6319	.8526	1.0210	30′
40′	.5527	.5250	.6168	1.6212	.8511	1.0181	20′
50′	.5556	.5275	.6208	1.6107	.8496	1.0152	10′
32° 00′	.5585	.5299	.6249	1.6003	.8480	1.0123	58° 00′
10′	.5614	.5324	.6289	1.5900	.8465	1.0094	50′
20′	.5643	.5348	.6330	1.5798	.8450	1.0065	40′
30′	.5672	.5373	.6371	1.5697	.8434	1.0036	30′
40′	.5701	.5398	.6412	1.5597	.8418	1.0007	20′
50′	.5730	.5422	.6453	1.5497	.8403	.9977	10′
33° 00′	.5760	.5446	.6494	1.5399	.8387	.9948	57° 00′
10′	.5789	.5471	.6536	1.5301	.8371	.9919	50′
20′	.5818	.5495	.6577	1.5204	.8355	.9890	40′
30′	.5847	.5519	.6619	1.5108	.8339	.9861	30′
40′	.5876	.5544	.6661	1.5013	.8323	.9832	20′
50′	.5905	.5568	.6703	1.4919	.8307	.9803	10′
34° 00′	.5934	.5592	.6745	1.4826	.8290	.9774	56° 00′
10′	.5963	.5616	.6787	1.4733	.8274	.9745	50′
20′	.5992	.5640	.6830	1.4641	.8258	.9716	40′
30′	.6021	.5664	.6873	1.4550	.8241	.9687	30′
40′	.6050	.5688	.6916	1.4460	.8225	.9657	20′
50′	.6080	.5712	.6959	1.4370	.8208	.9628	10′
35° 00′	.6109	.5736	.7002	1.4281	.8192	.9599	55° 00′
10′	.6138	.5760	.7046	1.4193	.8175	.9570	50′
20′	.6167	.5783	.7089	1.4106	.8158	.9541	40′
30′	.6196	.5807	.7133	1.4019	.8141	.9512	30′
40′	.6225	.5831	.7177	1.3934	.8124	.9483	20′
50′	.6254	.5854	.7221	1.3848	.8107	.9454	10′
36° 00′	.6283	.5878	.7265	1.3764	.8090	.9425	54° 00′
		Cosine	Cotangent	Tangent	Sine	Radians	Degrees

Values of Trigonometric Functions (cont.)

Degrees	Radians	Sine	Tangent	Cotangent	Cosine		
36° 00′	.6283	.5878	.7265	1.3764	.8090	.9425	54° 00′
10′	.6312	.5901	.7310	1.3680	.8073	.9396	50′
20′	.6341	.5925	.7355	1.3597	.8056	.9367	40′
30′	.6370	.5948	.7400	1.3514	.8039	.9338	30′
40′	.6400	.5972	.7445	1.3432	.8021	.9308	20′
50′	.6429	.5995	.7490	1.3351	.8004	.9279	10′
37° 00′	.6458	.6018	.7536	1.3270	.7986	.9250	53° 00′
10′	.6487	.6041	.7581	1.3190	.7969	.9221	50′
20′	.6516	.6065	.7627	1.3111	.7951	.9192	40′
30′	.6545	.6088	.7673	1.3032	.7934	.9163	30′
40′	.6574	.6111	.7720	1.2954	.7916	.9134	20′
50′	.6603	.6134	.7766	1.2876	.7898	.9105	10′
38° 00′	.6632	.6157	.7813	1.2799	.7880	.9076	52° 00′
10′	.6661	.6180	.7860	1.2723	.7862	.9047	50′
20′	.6690	.6202	.7907	1.2647	.7844	.9018	40′
30′	.6720	.6225	.7954	1.2572	.7826	.8988	30′
40′	.6749	.6248	.8002	1.2497	.7808	.8959	20′
50′	.6778	.6271	.8050	1.2423	.7790	.8930	10′
39° 00′	.6807	.6293	.8098	1.2349	.7771	.8901	51° 00′
10′	.6836	.6316	.8146	1.2276	.7753	.8872	50′
20′	.6865	.6338	.8195	1.2203	.7735	.8843	40′
30′	.6894	.6361	.8243	1.2131	.7716	.8814	30′
40′	.6923	.6383	.8292	1.2059	.7698	.8785	20′
50′	.6952	.6406	.8342	1.1988	.7679	.8756	10′
40° 00′	.6981	.6428	.8391	1.1918	.7660	.8727	50° 00′
10′	.7010	.6450	.8441	1.1847	.7642	.8698	50′
20′	.7039	.6472	.8491	1.1778	.7623	.8668	40′
30′	.7069	.6494	.8541	1.1708	.7604	.8639	30′
40′	.7098	.6517	.8591	1.1640	.7585	.8610	20′
50′	.7127	.6539	.8642	1.1571	.7566	.8581	10′
• 41° 00′	.7156	.6561	.8693	1.1504	.7547	.8552	49° 00′
10′	.7185	.6583	.8744	1.1436	.7528	.8523	50′
20′	.7214	.6604	.8796	1.1369	.7509	.8494	40′
30′	.7243	.6626	.8847	1.1303	.7490	.8465	30′
40′	.7272	.6648	.8899	1.1237	.7470	.8436	20′
50′	.7301	.6670	.8952	1.1171	.7451	.8407	10′
42° 00′	.7330	.6691	.9004	1.1106	.7431	.8378	48° 00′
10′	.7359	.6713	.9057	1.1041	.7412	.8348	50′
20′	.7389	.6734	.9110	1.0977	.7392	.8319	40′
30′	.7418	.6756	.9163	1.0913	.7373	.8290	30′
40′	.7447	.6777	.9217	1.0850	.7353	.8261	20′
50′	.7476	.6799	.9271	1.0786	.7333	.8232	10′
43° 00′	.7505	.6820	.9325	1.0724	.7314	.8203	47° 00′
10′	.7534	.6841	.9380	1.0661	.7294	.8174	50′
20′	.7563	.6862	.9435	1.0599	.7274	.8145	40′
30′	.7592	.6884	.9490	1.0538	.7254	.8116	30′
40′	.7621	.6905	.9545	1.0477	.7234	.8087	20′
50′	.7650	.6926	.9601	1.0416	.7214	.8058	10′
44° 00′	.7679	.6947	.9657	1.0355	.7193	.8029	46° 00′
10′	.7709	.6967	.9713	1.0295	.7173	.7999	50′
20′	.7738	.6988	.9770	1.0235	.7153	.7970	40′
30′	.7767	.7009	.9827	1.0176	.7133	.7941	30′
40′	.7796	.7030	.9884	1.0117	.7112	.7912	20′
50′	.7825	.7050	.9942	1.0058	.7092	.7883	10′
45° 00′	.7854	.7071	1.0000	1.0000	.7071	.7854	45° 00′
		Cosine	Cotangent	Tangent	Sine	Radians	Degrees

Appendix B: Tables

Logarithms of Numbers*

N	0	1	2	3	4	5	6	7	8	9
1.0	.0000	.0043	.0086	.0128	.0170	.0212	.0253	.0294	.0334	.0374
1.1	.0414	.0453	.0492	.0531	.0569	.0607	.0645	.0682	.0719	.0755
1.2	.0792	.0828	.0864	.0899	.0934	.0969	.1004	.1038	.1072	.1106
1.3	.1139	.1173	.1206	.1239	.1271	.1303	.1335	.1367	.1399	.1430
1.4	.1461	.1492	.1523	.1553	.1584	.1614	.1644	.1673	.1703	.1732
1.5	.1761	.1790	.1818	.1847	.1875	.1903	.1931	.1959	.1987	.2014
1.6	.2041	.2068	.2095	.2122	.2148	.2175	.2201	.2227	.2253	.2279
1.7	.2304	.2330	.2355	.2380	.2405	.2430	.2455	.2480	.2504	.2529
1.8	.2553	.2577	.2601	.2625	.2648	.2672	.2695	.2718	.2742	.2765
1.9	.2788	.2810	.2833	.2856	.2878	.2900	.2923	.2945	.2967	.2989
2.0	.3010	.3032	.3054	.3075	.3096	.3118	.3139	.3160	.3181	.3201
2.1	.3222	.3243	.3263	.3284	.3304	.3324	.3345	.3365	.3385	.3404
2.2	.3424	.3444	.3464	.3483	.3502	.3522	.3541	.3560	.3579	.3598
2.3	.3617	.3636	.3655	.3674	.3692	.3711	.3729	.3747	.3766	.3784
2.4	.3802	.3820	.3838	.3856	.3874	.3892	.3909	.3927	.3945	.3962
2.5	.3979	.3997	.4014	.4031	.4048	.4065	.4082	.4099	.4116	.4133
2.6	.4150	.4166	.4183	.4200	.4216	.4232	.4249	.4265	.4281	.4298
2.7	.4314	.4330	.4346	.4362	.4378	.4393	.4409	.4425	.4440	.4456
2.8	.4472	.4487	.4502	.4518	.4533	.4548	.4564	.4579	.4594	.4609
2.9	.4624	.4639	.4654	.4669	.4683	.4698	.4713	.4728	.4742	.4757
3.0	.4771	.4786	.4800	.4814	.4829	.4843	.4857	.4871	.4886	.4900
3.1	.4914	.4928	.4942	.4955	.4969	.4983	.4997	.5011	.5024	.5038
3.2	.5051	.5065	.5079	.5092	.5105	.5119	.5132	.5145	.5159	.5172
3.3	.5185	.5198	.5211	.5224	.5237	.5250	.5263	.5276	.5289	.5302
3.4	.5315	.5328	.5340	.5353	.5366	.5378	.5391	.5403	.5416	.5428
3.5	.5441	.5453	.5465	.5478	.5490	.5502	.5514	.5527	.5539	.5551
3.6	.5563	.5575	.5587	.5599	.5611	.5623	.5635	.5647	.5658	.5670
3.7	.5682	.5694	.5705	.5717	.5729	.5740	.5752	.5763	.5775	.5786
3.8	.5798	.5809	.5821	.5832	.5843	.5855	.5866	.5877	.5888	.5899
3.9	.5911	.5922	.5933	.5944	.5955	.5966	.5977	.5988	.5999	.6010
4.0	.6021	.6031	.6042	.6053	.6064	.6075	.6085	.6096	.6107	.6117
4.1	.6128	.6138	.6149	.6160	.6170	.6180	.6191	.6201	.6212	.6222
4.2	.6232	.6243	.6253	.6263	.6274	.6284	.6294	.6304	.6314	.6325
4.3	.6335	.6345	.6355	.6365	.6375	.6385	.6395	.6405	.6415	.6425
4.4	.6435	.6444	.6454	.6464	.6474	.6484	.6493	.6503	.6513	.6522
4.5	.6532	.6542	.6551	.6561	.6571	.6580	.6590	.6599	.6609	.6618
4.6	.6628	.6637	.6646	.6656	.6665	.6675	.6684	.6693	.6702	.6712
4.7	.6721	.6730	.6739	.6749	.6758	.6767	.6776	.6785	.6794	.6803
4.8	.6812	.6821	.6830	.6839	.6848	.6857	.6866	.6875	.6884	.6893
4.9	.6902	.6911	.6920	.6928	.6937	.6946	.6955	.6964	.6972	.6981
5.0	.6990	.6998	.7007	.7016	.7024	.7033	.7042	.7050	.7059	.7067
5.1	.7076	.7084	.7093	.7101	.7110	.7118	.7126	.7135	.7143	.7152
5.2	.7160	.7168	.7177	.7185	.7193	.7202	.7210	.7218	.7226	.7235
5.3	.7243	.7251	.7259	.7267	.7275	.7284	.7292	.7300	.7308	.7316
5.4	.7324	.7332	.7340	.7348	.7356	.7364	.7372	.7380	.7388	.7396
N	0	1	2	3	4	5	6	7	8	9

*From *Introduction to Mathematical Analysis* by R. E. Johnson, N. H. McCoy, and A. F. O'Neill. Copyright 1962 by Holt, Rinehart and Winston, Inc. Reprinted by permission.

Logarithms of Numbers (cont.)

N	0	1	2	3	4	5	6	7	8	9
5.5	.7404	.7412	.7419	.7427	.7435	.7443	.7451	.7459	.7466	.7474
5.6	.7482	.7490	.7497	.7505	.7513	.7520	.7528	.7536	.7543	.7551
5.7	.7559	.7566	.7574	.7582	.7589	.7597	.7604	.7612	.7619	.7627
5.8	.7634	.7642	.7649	.7657	.7664	.7672	.7679	.7686	.7694	.7701
5.9	.7709	.7716	.7723	.7731	.7738	.7745	.7752	.7760	.7767	.7774
6.0	.7782	.7789	.7796	.7803	.7810	.7818	.7825	.7832	.7839	.7846
6.1	.7853	.7860	.7868	.7875	.7882	.7889	.7896	.7903	.7910	.7917
6.2	.7924	.7931	.7938	.7945	.7952	.7959	.7966	.7973	.7980	.7987
6.3	.7993	.8000	.8007	.8014	.8021	.8028	.8035	.8041	.8048	.8055
6.4	.8062	.8069	.8075	.8082	.8089	.8096	.8102	.8109	.8116	.8122
6.5	.8129	.8136	.8142	.8149	.8156	.8162	.8169	.8176	.8182	.8189
6.6	.8195	.8202	.8209	.8215	.8222	.8228	.8235	.8241	.8248	.8254
6.7	.8261	.8267	.8274	.8280	.8287	.8293	.8299	.8306	.8312	.8319
6.8	.8325	.8331	.8338	.8344	.8351	.8357	.8363	.8370	.8376	.8382
6.9	.8388	.8395	.8401	.8407	.8414	.8420	.8426	.8432	.8439	.8445
7.0	.8451	.8457	.8463	.8470	.8476	.8482	.8488	.8494	.8500	.8506
7.1	.8513	.8519	.8525	.8531	.8537	.8543	.8549	.8555	.8561	.8567
7.2	.8573	.8579	.8585	.8591	.8597	.8603	.8609	.8615	.8621	.8627
7.3	.8633	.8639	.8645	.8651	.8657	.8663	.8669	.8675	.8681	.8686
7.4	.8692	.8698	.8704	.8710	.8716	.8722	.8727	.8733	.8739	.8745
7.5	.8751	.8756	.8762	.8768	.8774	.8779	.8785	.8791	.8797	.8802
7.6	.8808	.8814	.8820	.8825	.8831	.8837	.8842	.8848	.8854	.8859
7.7	.8865	.8871	.8876	.8882	.8887	.8893	.8899	.8904	.8910	.8915
7.8	.8921	.8927	.8932	.8938	.8943	.8949	.8954	.8960	.8965	.8971
7.9	.8976	.8982	.8987	.8993	.8998	.9004	.9009	.9015	.9020	.9025
8.0	.9031	.9036	.9042	.9047	.9053	.9058	.9063	.9069	.9074	.9079
8.1	.9085	.9090	.9096	.9101	.9106	.9112	.9117	.9122	.9128	.9133
8.2	.9138	.9143	.9149	.9154	.9159	.9165	.9170	.9175	.9180	.9186
8.3	.9191	.9196	.9201	.9206	.9212	.9217	.9222	.9227	.9232	.9238
8.4	.9243	.9248	.9253	.9258	.9263	.9269	.9274	.9279	.9284	.9289
8.5	.9294	.9299	.9304	.9309	.9315	.9320	.9325	.9330	.9335	.9340
8.6	.9345	.9350	.9355	.9360	.9365	.9370	.9375	.9380	.9385	.9390
8.7	.9395	.9400	.9405	.9410	.9415	.9420	.9425	.9430	.9435	.9440
8.8	.9445	.9450	.9455	.9460	.9465	.9469	.9474	.9479	.9484	.9489
8.9	.9494	.9499	.9504	.9509	.9513	.9518	.9523	.9528	.9533	.9538
9.0	.9542	.9547	.9552	.9557	.9562	.9566	.9571	.9576	.9581	.9586
9.1	.9590	.9595	.9600	.9605	.9609	.9614	.9619	.9624	.9628	.9633
9.2	.9638	.9643	.9647	.9652	.9657	.9661	.9666	.9671	.9675	.9680
9.3	.9685	.9689	.9694	.9699	.9703	.9708	.9713	.9717	.9722	.9727
9.4	.9731	.9736	.9741	.9745	.9750	.9754	.9759	.9763	.9768	.9773
9.5	.9777	.9782	.9786	.9791	.9795	.9800	.9805	.9809	.9814	.9818
9.6	.9823	.9827	.9832	.9836	.9841	.9845	.9850	.9854	.9859	.9863
9.7	.9868	.9872	.9877	.9881	.9886	.9890	.9894	.9899	.9903	.9908
9.8	.9912	.9917	.9921	.9926	.9930	.9934	.9939	.9943	.9948	.9952
9.9	.9956	.9961	.9965	.9969	.9974	.9978	.9983	.9987	.9991	.9996
N	0	1	2	3	4	5	6	7	8	9

Appendix C: Tables

Logarithms of Trigonometric Functions*

Degrees	Log_{10} Sine	Log_{10} Tangent	Log_{10} Cotangent	Log_{10} Cosine	
0° 00′					90° 00′
10′	.4637 − 3	.4637 − 3	2.5363	.0000	50′
20′	.7648 − 3	.7648 − 3	2.2352	.0000	40′
30′	9408 − 3	.9409 − 3	2.0591	.0000	30′
40′	.0658 − 2	.0658 − 2	1.9342	.0000	20′
50′	.1627 − 2	.1627 − 2	1.8373	.0000	10′
1° 00′	.2419 − 2	.2419 − 2	1.7581	.9999 − 1	89° 00′
10′	.3088 − 2	.3089 − 2	1.6911	.9999 − 1	50′
20′	.3668 − 2	.3669 − 2	1.6331	.9999 − 1	40′
30′	.4179 − 2	.4181 − 2	1.5819	.9999 − 1	30′
40′	.4637 − 2	.4638 − 2	1.5362	.9998 − 1	20′
50′	.5050 − 2	.5053 − 2	1.4947	.9998 − 1	10′
2° 00′	.5428 − 2	.5431 − 2	1.4569	.9997 − 1	88° 00′
10′	.5776 − 2	.5779 − 2	1.4221	.9997 − 1	50′
20′	.6097 − 2	.6101 − 2	1.3899	.9996 − 1	40′
30′	.6397 − 2	.6401 − 2	1.3599	.9996 − 1	30′
40′	.6677 − 2	.6682 − 2	1.3318	.9995 − 1	20′
50′	.6940 − 2	.6945 − 2	1.3055	.9995 − 1	10′
3° 00′	.7188 − 2	.7194 − 2	1.2806	.9994 − 1	87° 00′
10′	.7423 − 2	.7429 − 2	1.2571	.9993 − 1	50′
20′	.7645 − 2	.7652 − 2	1.2348	.9993 − 1	40′
30′	.7857 − 2	.7865 − 2	1.2135	.9992 − 1	30′
40′	.8059 − 2	.8067 − 2	1.1933	.9991 − 1	20′
50′	.8251 − 2	.8261 − 2	1.1739	.9990 − 1	10′
4° 00′	.8436 − 2	.8446 − 2	1.1554	.9989 − 1	86° 00′
10′	.8613 − 2	.8624 − 2	1.1376	.9989 − 1	50′
20′	.8783 − 2	.8795 − 2	1.1205	.9988 − 1	40′
30′	.8946 − 2	.8960 − 2	1.1040	.9987 − 1	30′
40′	.9104 − 2	.9118 − 2	1.0882	.9986 − 1	20′
50′	.9256 − 2	.9272 − 2	1.0728	.9985 − 1	10′
5° 00′	.9403 − 2	.9420 − 2	1.0580	.9983 − 1	85° 00′
10′	.9545 − 2	.9563 − 2	1.0437	.9982 − 1	50′
20′	.9682 − 2	.9701 − 2	1.0299	.9981 − 1	40′
30′	.9816 − 2	.9836 − 2	1.0164	.9980 − 1	30′
40′	.9945 − 2	.9966 − 2	1.0034	.9979 − 1	20′
50′	.0070 − 1	.0093 − 1	.9907	.9977 − 1	10′
6° 00′	.0192 − 1	.0216 − 1	.9784	.9976 − 1	84° 00′
10′	.0311 − 1	.0336 − 1	.9664	.9975 − 1	50′
20′	.0426 − 1	.0453 − 1`	.9547	.9973 − 1	40′
30′	.0539 − 1	.0567 − 1	.9433	.9972 − 1	30′
40′	.0648 − 1	.0678 − 1	.9322	.9971 − 1	20′
50′	.0755 − 1	.0786 − 1	.9214	.9969 − 1	10′
7° 00′	.0859 − 1	.0891 − 1	.9109	.9968 − 1	83° 00′
10′	.0961 − 1	.0995 − 1	.9005	.9966 − 1	50′
20′	.1060 − 1	.1096 − 1	.8904	.9964 − 1	40′
30′	.1157 − 1	.1194 − 1	.8806	.9963 − 1	30′
40′	.1252 − 1	.1291 − 1	.8709	.9961 − 1	20′
50′	.1345 − 1	.1385 − 1	.8615	.9959 − 1	10′
8° 00′	.1436 − 1	.1478 − 1	.8522	.9958 − 1	82° 00′
10′	.1525 − 1	.1569 − 1	.8431	.9956 − 1	50′
20′	.1612 − 1	.1658 − 1	.8342	.9954 − 1	40′
30′	.1697 − 1	.1745 − 1	.8255	.9952 − 1	30′
40′	.1781 − 1	.1831 − 1	.8169	.9950 − 1	20′
50′	.1863 − 1	.1915 − 1	.8085	.9948 − 1	10′
9° 00′	.1943 − 1	.1997 − 1	.8003	.9946 − 1	81° 00′
	Log_{10} Cosine	Log_{10} Cotangent	Log_{10} Tangent	Log_{10} Sine	Degrees

*From *Introduction to Mathematical Analysis* by R. E. Johnson, N. H. McCoy, and A. F. O'Neill. Copyright 1962 by Holt, Rinehart and Winston, Inc. Reprinted by permission.

Logarithms of Trigonometric Functions (cont.)

Degrees	Log₁₀ Sine	Log₁₀ Tangent	Log₁₀ Cotangent	Log₁₀ Cosine	
9° 00′	.1943 − 1	.1997 − 1	.8003	.9946 − 1	81° 00′
10′	.2022 − 1	.2078 − 1	.7922	.9944 − 1	50′
20′	.2100 − 1	.2158 − 1	.7842	.9942 − 1	40′
30′	.2176 − 1	.2236 − 1	.7764	.9940 − 1	30′
40′	.2251 − 1	.2313 − 1	.7687	.9938 − 1	20′
50′	.2324 − 1	.2389 − 1	.7611	.9936 − 1	10′
10° 00′	.2397 − 1	.2463 − 1	.7537	.9934 − 1	80° 00′
10′	.2468 − 1	.2536 − 1	.7464	.9931 − 1	50′
20′	.2538 − 1	.2609 − 1	.7391	.9929 − 1	40′
30′	.2606 − 1	.2680 − 1	.7320	.9927 − 1	30′
40′	.2674 − 1	.2750 − 1	.7250	.9924 − 1	20′
50′	.2740 − 1	.2819 − 1	.7181	.9922 − 1	10′
11° 00′	.2806 − 1	.2887 − 1	.7113	.9919 − 1	79° 00′
10′	.2870 − 1	.2953 − 1	.7047	.9917 − 1	50′
20′	.2934 − 1	.3020 − 1	.6980	.9914 − 1	40′
30′	.2997 − 1	.3085 − 1	.6915	.9912 − 1	30′
40′	.3058 − 1	.3149 − 1	.6851	.9909 − 1	20′
50′	.3119 − 1	.3212 − 1	.6788	.9907 − 1	10′
12° 00′	.3179 − 1	.3275 − 1	.6725	.9904 − 1	78° 00′
10′	.3238 − 1	.3336 − 1	.6664	.9901 − 1	50′
20′	.3296 − 1	.3397 − 1	.6603	.9899 − 1	40′
30′	.3353 − 1	.3458 − 1	.6542	.9896 − 1	30′
40′	.3410 − 1	.3517 − 1	.6483	.9893 − 1	20′
50′	.3466 − 1	.3576 − 1	.6424	.9890 − 1	10′
13° 00′	.3521 − 1	.3634 − 1	.6366	.9887 − 1	77° 00′
10′	.3575 − 1	.3691 − 1	.6309	.9884 − 1	50′
20′	.3629 − 1	.3748 − 1	.6252	.9881 − 1	40′
30′	.3682 − 1	.3804 − 1	.6196	.9878 − 1	30′
40′	.3734 − 1	.3859 − 1	.6141	.9875 − 1	20′
50′	.3786 − 1	.3914 − 1	.6086	.9872 − 1	10′
14° 00′	.3837 − 1	.3968 − 1	.6032	.9869 − 1	76° 00′
10′	.3887 − 1	.4021 − 1	.5979	.9866 − 1	50′
20′	.3937 − 1	.4074 − 1	.5926	.9863 − 1	40′
30′	.3986 − 1	.4127 − 1	.5873	.9859 − 1	30′
40′	.4035 − 1	.4178 − 1	.5822	.9856 − 1	20′
50′	.4083 − 1	.4230 − 1	.5770	.9853 − 1	10′
15° 00′	.4130 − 1	.4281 − 1	.5719	.9849 − 1	75° 00′
10′	.4177 − 1	.4331 − 1	.5669	.9846 − 1	50′
20′	.4223 − 1	.4381 − 1	.5619	.9843 − 1	40′
30′	.4269 − 1	.4430 − 1	.5570	.9839 − 1	30′
40′	.4314 − 1	.4479 − 1	.5521	.9836 − 1	20′
50′	.4359 − 1	.4527 − 1	.5473	.9832 − 1	10′
16° 00′	.4403 − 1	.4575 − 1	.5425	.9828 − 1	74° 00′
10′	.4447 − 1	.4622 − 1	.5378	.9825 − 1	50′
20′	.4491 − 1	.4669 − 1	.5331	.9821 − 1	40′
30′	.4533 − 1	.4716 − 1	.5284	.9817 − 1	30′
40′	.4576 − 1	.4762 − 1	.5238	.9814 − 1	20′
50′	.4618 − 1	.4808 − 1	.5192	.9810 − 1	10′
17° 00′	.4659 − 1	.4853 − 1	.5147	.9806 − 1	73° 00′
10′	.4700 − 1	.4898 − 1	.5102	.9802 − 1	50′
20′	.4741 − 1	.4943 − 1	.5057	.9798 − 1	40′
30′	.4781 − 1	.4987 − 1	.5013	.9794 − 1	30′
40′	.4821 − 1	.5031 − 1	.4969	.9790 − 1	20′
50′	.4861 − 1	.5075 − 1	.4925	.9786 − 1	10′
18° 00′	.4900 − 1	.5118 − 1	.4882	.9782 − 1	72° 00′
	Log₁₀ Cosine	Log₁₀ Cotangent	Log₁₀ Tangent	Log₁₀ Sine	Degrees

Logarithms of Trigonometric Functions (cont.)

Degrees	Log₁₀ Sine	Log₁₀ Tangent	Log₁₀ Cotangent	Log₁₀ Cosine	
18° 00′	.4900 −1	.5118 −1	.4882	.9782 −1	72° 00′
10′	.4939 −1	.5161 −1	.4839	.9778 −1	50′
20′	.4977 −1	.5203 −1	.4797	.9774 −1	40′
30′	.5015 −1	.5245 −1	.4755	.9770 −1	30′
40′	.5052 −1	.5287 −1	.4713	.9765 −1	20′
50′	.5090 −1	.5329 −1	.4671	.9761 −1	10′
19° 00′	.5126 −1	.5370 −1	.4630	.9757 −1	71° 00′
10′	.5163 −1	.5411 −1	.4589	.9752 −1	50′
20′	.5199 −1	.5451 −1	.4549	.9748 −1	40′
30′	.5235 −1	.5491 −1	.4509	.9743 −1	30′
40′	.5270 −1	.5531 −1	.4469	.9739 −1	20′
50′	.5306 −1	.5571 −1	.4429	.9734 −1	10′
20° 00′	.5341 −1	.5611 −1	.4389	.9730 −1	70° 00′
10′	.5375 −1	.5650 −1	.4350	.9725 −1	50′
20′	.5409 −1	.5689 −1	.4311	.9721 −1	40′
30′	.5443 −1	.5727 −1	.4273	.9716 −1	30′
40′	.5477 −1	.5766 −1	.4234	.9711 −1	20′
50′	.5510 −1	.5804 −1	.4196	.9706 −1	10′
21° 00′	.5543 −1	.5842 −1	.4158	.9702 −1	69° 00′
10′	.5576 −1	.5879 −1	.4121	.9697 −1	50′
20′	.5609 −1	.5917 −1	.4083	.9692 −1	40′
30′	.5641 −1	.5954 −1	.4046	.9687 −1	30′
40′	.5673 −1	.5991 −1	.4009	.9682 −1	20′
50′	.5704 −1	.6028 −1	.3972	.9677 −1	10′
22° 00′	.5736 −1	.6064 −1	.3936	.9672 −1	68° 00′
10′	.5767 −1	.6100 −1	.3900	.9667 −1	50′
20′	.5798 −1	.6136 −1	.3864	.9661 −1	40′
30′	.5828 −1	.6172 −1	.3828	.9656 −1	30′
40′	.5859 −1	.6208 −1	.3792	.9651 −1	20′
50′	.5889 −1	.6243 −1	.3757	.9646 −1	10′
23° 00′	.5919 −1	.6279 −1	.3721	.9640 −1	67° 00′
10′	.5948 −1	.6314 −1	.3686	.9635 −1	50′
20′	.5978 −1	.6348 −1	.3652	.9629 −1	40′
30′	.6007 −1	.6383 −1	.3617	.9624 −1	30′
40′	.6036 −1	.6417 −1	.3583	.9618 −1	20′
50′	.6065 −1	.6452 −1	.3548	.9613 −1	10′
24° 00′	.6093 −1	.6486 −1	.3514	.9607 −1	66° 00′
10′	.6121 −1	.6520 −1	.3480	.9602 −1	50′
20′	.6149 −1	.6553 −1	.3447	.9596 −1	40′
30′	.6177 −1	.6587 −1	.3413	.9590 −1	30′
40′	.6205 −1	.6620 −1	.3380	.9584 −1	20′
50′	.6232 −1	.6654 −1	.3346	.9579 −1	10′
25° 00′	.6259 −1	.6687 −1	.3313	.9573 −1	65° 00′
10′	.6286 −1	.6720 −1	.3280	.9567 −1	50′
20′	.6313 −1	.6752 −1	.3248	.9561 −1	40′
30′	.6340 −1	.6785 −1	.3215	.9555 −1	30′
40′	.6366 −1	.6817 −1	.3183	.9549 −1	20′
50′	.6392 −1	.6850 −1	.3150	.9543 −1	10′
26° 00′	.6418 −1	.6882 −1	.3118	.9537 −1	64° 00′
10′	.6444 −1	.6914 −1	.3086	.9530 −1	50′
20′	.6470 −1	.6946 −1	.3054	.9524 −1	40′
30′	.6495 −1	.6977 −1	.3023	.9518 −1	30′
40′	.6521 −1	.7009 −1	.2991	.9512 −1	20′
50′	.6546 −1	.7040 −1	.2960	.9505 −1	10′
27° 00′	.6570 −1	.7072 −1	.2928	.9499 −1	63° 00′
	Log₁₀ Cosine	Log₁₀ Cotangent	Log₁₀ Tangent	Log₁₀ Sine	Degrees

Logarithms of Trigonometric Functions (cont.)

Degrees	Log₁₀ Sine	Log₁₀ Tangent	Log₁₀ Cotangent	Log₁₀ Cosine	
27° 00′	.6570 −1	.7072 −1	.2928	.9499 −1	63° 00′
10′	.6595 −1	.7103 −1	.2897	.9492 −1	50′
20′	.6620 −1	.7134 −1	.2866	.9486 −1	40′
30′	.6644 −1	.7165 −1	.2835	.9479 −1	30′
40′	.6668 −1	.7196 −1	.2804	.9473 −1	20′
50′	.6692 −1	.7226 −1	.2774	.9466 −1	10′
28° 00′	.6716 −1	.7257 −1	.2743	.9459 −1	62° 00′
10′	.6740 −1	.7287 −1	.2713	.9453 −1	50′
20′	.6763 −1	.7317 −1	.2683	.9446 −1	40′
30′	.6787 −1	.7348 −1	.2652	.9439 −1	30′
40′	.6810 −1	.7378 −1	.2622	.9432 −1	20′
50′	.6833 −1	.7408 −1	.2592	.9425 −1	10′
29° 00′	.6856 −1	.7438 −1	.2562	.9418 −1	61° 00′
10′	.6878 −1	.7467 −1	.2533	.9411 −1	50′
20′	.6901 −1	.7497 −1	.2503	.9404 −1	40′
30′	.6923 −1	.7526 −1	.2474	.9397 −1	30′
40′	.6946 −1	.7556 −1	.2444	.9390 −1	20′
50′	.6968 −1	.7585 −1	.2415	.9383 −1	10′
30° 00′	.6990 −1	.7614 −1	.2386	.9375 −1	60° 00′
10′	.7012 −1	.7644 −1	.2356	.9368 −1	50′
20′	.7033 −1	.7673 −1	.2327	.9361 −1	40′
30′	.7055 −1	.7701 −1	.2299	.9353 −1	30′
40′	.7076 −1	.7730 −1	.2270	.9346 −1	20′
50′	.7097 −1	.7759 −1	.2241	.9338 −1	10′
31° 00′	.7118 −1	.7788 −1	.2212	.9331 −1	59° 00′
10′	.7139 −1	.7816 −1	.2184	.9323 −1	50′
20′	.7160 −1	.7845 −1	.2155	.9315 −1	40′
30′	.7181 −1	.7873 −1	.2127	.9308 −1	30′
40′	.7201 −1	.7902 −1	.2098	.9300 −1	20′
50′	.7222 −1	.7930 −1	.2070	.9292 −1	10′
32° 00′	.7242 −1	.7958 −1	.2042	.9284 −1	58° 00′
10′	.7262 −1	.7986 −1	.2014	.9276 −1	50′
20′	.7282 −1	.8014 −1	.1986	.9268 −1	40′
30′	.7302 −1	.8042 −1	.1958	.9260 −1	30′
40′	.7322 −1	.8070 −1	.1930	.9252 −1	20′
50′	.7342 −1	.8097 −1	.1903	.9244 −1	10′
33° 00′	.7361 −1	.8125 −1	.1875	.9236 −1	57° 00′
10′	.7380 −1	.8153 −1	.1847	.9228 −1	50′
20′	.7400 −1	.8180 −1	.1820	.9219 −1	40′
30′	.7419 −1	.8208 −1	.1792	.9211 −1	30′
40′	.7438 −1	.8235 −1	.1765	.9203 −1	20′
50′	.7457 −1	.8263 −1	.1737	.9194 −1	10′
34° 00′	.7476 −1	.8290 −1	.1710	.9186 −1	56° 00′
10′	.7494 −1	.8317 −1	.1683	.9177 −1	50′
20′	.7513 −1	.8344 −1	.1656	.9169 −1	40′
30′	.7531 −1	.8371 −1	.1629	.9160 −1	30′
40′	.7550 −1	.8398 −1	.1602	.9151 −1	20′
50′	.7568 −1	.8425 −1	.1575	.9142 −1	10′
35° 00′	.7586 −1	.8452 −1	.1548	.9134 −1	55° 00′
10′	.7604 −1	.8479 −1	.1521	.9125 −1	50′
20′	.7622 −1	.8506 −1	.1494	.9116 −1	40′
30′	.7640 −1	.8533 −1	.1467	.9107 −1	30′
40′	.7657 −1	.8559 −1	.1441	.9098 −1	20′
50′	.7675 −1	.8586 −1	.1414	.9089 −1	10′
36° 00′	.7692 −1	.8613 −1	.1387	.9080 −1	54° 00′
	Log₁₀ Cosine	Log₁₀ Cotangent	Log₁₀ Tangent	Log₁₀ Sine	Degrees

Logarithms of Trigonometric Functions (cont.)

Degrees	Log_{10} Sine	Log_{10} Tangent	Log_{10} Cotangent	Log_{10} Cosine	
36° 00′	.7692 −1	.8613 −1	.1387	.9080 −1	54° 00′
10′	.7710 −1	.8639 −1	.1361	.9070 −1	50′
20′	.7727 −1	.8666 −1	.1334	.9061 −1	40′
30′	.7744 −1	.8692 −1	.1308	.9052 −1	30′
40′	.7761 −1	.8718 −1	.1282	.9042 −1	20′
50′	.7778 −1	.8745 −1	.1255	.9033 −1	10′
37° 00′	.7795 −1	.8771 −1	.1229	.9023 −1	53° 00′
10′	.7811 −1	.8797 −1	.1203	.9014 −1	50′
20′	.7828 −1	.8824 −1	.1176	.9004 −1	40′
30′	.7844 −1	.8850 −1	.1150	.8995 −1	30′
40′	.7861 −1	.8876 −1	.1124	.8985 −1	20′
50′	.7877 −1	.8902 −1	.1098	.8975 −1	10′
38° 00′	.7893 −1	.8928 −1	.1072	.8965 −1	52° 00′
10′	.7910 −1	.8954 −1	.1046	.8955 −1	50′
20′	.7926 −1	.8980 −1	.1020	.8945 −1	40′
30′	.7941 −1	.9006 −1	.0994	.8935 −1	30′
40′	.7957 −1	.9032 −1	.0968	.8925 −1	20′
50′	.7973 −1	.9058 −1	.0942	.8915 −1	10′
39° 00′	.7989 −1	.9084 −1	.0916	.8905 −1	51° 00′
10′	.8004 −1	.9110 −1	.0890	.8895 −1	50′
20′	.8020 −1	.9135 −1	.0865	.8884 −1	40′
30′	.8035 −1	.9161 −1	.0839	.8874 −1	30′
40′	.8050 −1	.9187 −1	.0813	.8864 −1	20′
50′	.8066 −1	.9212 −1	.0788	.8853 −1	10′
40° 00′	.8081 −1	.9238 −1	.0762	.8843 −1	50° 00′
10′	.8096 −1	.9264 −1	.0736	.8832 −1	50′
20′	.8111 −1	.9289 −1	.0711	.8821 −1	40′
30′	.8125 −1	.9315 −1	.0685	.8810 −1	30′
40′	.8140 −1	.9341 −1	.0659	.8800 −1	20′
50′	.8155 −1	.9366 −1	.0634	.8789 −1	10′
41° 00′	.8169 −1	.9392 −1	.0608	.8778 −1	49° 00′
10′	.8184 −1	.9417 −1	.0583	.8767 −1	50′
20′	.8198 −1	.9443 −1	.0557	.8756 −1	40′
30′	.8213 −1	.9468 −1	.0532	.8745 −1	30′
40′	.8227 −1	.9494 −1	.0506	.8733 −1	20′
50′	.8241 −1	.9519 −1	.0481	.8722 −1	10′
42° 00′	.8255 −1	.9544 −1	.0456	.8711 −1	48° 00′
10′	.8269 −1	.9570 −1	.0430	.8699 −1	50′
20′	.8283 −1	.9595 −1	.0405	.8688 −1	40′
30′	.8297 −1	.9621 −1	.0379	.8676 −1	30′
40′	8311 −1	.9646 −1	.0354	.8665 −1	20′
50′	.8324 −1	.9671 −1	.0329	.8653 −1	10′
43° 00′	.8338 −1	.9697 −1	.0303	.8641 −1	47° 00′
10′	.8351 −1	.9722 −1	.0278	.8629 −1	50′
20′	.8365 −1	.9747 −1	.0253	.8618 −1	40′
30′	.8378 −1	.9772 −1	.0228	.8606 −1	30′
40′	.8391 −1	.9798 −1	.0202	.8594 −1	20′
50′	.8405 −1	.9823 −1	.0177	.8582 −1	10′
44° 00′	.8418 −1	.9848 −1	.0152	.8569 −1	46° 00′
10′	.8431 −1	.9874 −1	.0126	.8557 −1	50′
20′	.8444 −1	.9899 −1	.0101	.8545 −1	40′
30′	.8457 −1	.9924 −1	.0076	.8532 −1	30′
40′	.8469 −1	.9949 −1	.0051	.8520 −1	20′
50′	.8482 −1	.9975 −1	.0025	.8507 −1	10′
45° 00′	.8495 −1	.0000	.0000	.8495 −1	45° 00′
	Log_{10} Cosine	Log_{10} Cotangent	Log_{10} Tangent	Log_{10} Sine	Degrees

Appendix D
Using Tables
of Trigonometric
Functions

The tables of Appendix A allow us to make "reasonable" estimates of functional values for the six trigonometric functions.

The tables have (column) headings of two types. Those at the top of the tables complement those at the foot. Note, however, that the column topped by *Sine* has at its foot, *Cosine*. Each column headed by a trigonometric function has at its foot the corresponding cofunction.

The left column is headed **Degrees** and the next is headed **Radians.** Neither of these columns has a title at its foot. However, the foot of the two columns farthest to the right repeats degree, and radian headings. The left column ranges from 0° to 45°, while the right goes from 45° to 90°. The corresponding ranges for the radian columns are 0 to .7854, a decimal approximation for $\pi/4$, and $\pi/4$ to $\pi/2$ (approximated by 1.5708) for the second from left and second from right columns, respectively.

The pairs of columns have complementary entries in any given row. For example, the row containing 3° on the left contains $90° - 3° = 87°$ on the right, while the corresponding radian columns record .0524 and $\pi/2 - .0524 = 1.5184$, approximately.

To read the tables, we use the appropriate column determined as that column in which the argument of the function is found.

Example D.1 Find sin .4189.

Since $0 < .4189 < \pi/4$, the entry .4189 is found in the second from the left column (never use the degree columns unless the degree symbol is used to denote that type of measure). When we find .4189, we go to the *right,* to the column headed *sine* (*radian* heads the column in which we find .4189). The value found there is .4067. Thus, we write sin .4189 = .4067.

Example D.2 Find cos 1.0123.

The value 1.0123 lies between $\pi/4$ and $\pi/2$, whence 1.0123 will be found in the next to the right column. Since *Radian* appears at the *foot* of the column, we proceed to the *left* to the column with cosine at its foot. Here we read cos 1.0123 = .5299.

Example D.3 Find tan 9.9484.

Using the approximation 3.1416 for π, we may reduce the number 9.9484 to one less than $\pi/2$ (the maximum value of our tables); tan $A =$ tan $(A - n\pi)$, since π is a period for the tangent. Now, 3π is approximately 9.4248, whence tan 9.9484 = tan (9.9484 − 9.4248) = tan .5236. The fact that $0 < .5236 < .7854$ dictates the use of the column headed **Radian.** Searching out .5236 and going to the *right* to the column headed **tangent,** we see tan .5236 = .5774. Thus, tan 9.9484 = .5774.

We see that the tables are easily interpreted whenever the values are found therein or are reducible to values found there. Suppose, though, we are asked to find sin .6. The values sin .5992 and sin .6021 are found in the tables, but sin .6000 is not to be found. The process in the next example is called **(linear) interpolation** and is one method of ending our dilemma.

Example D.4 Find sin .6.

The chart below illustrates the example.

Our value sin .6 is .5640 $+ b =$ sin .5992 $+ b$. The value b is found by solving

$$\frac{a}{A} = \frac{b}{B}$$

Whence $b = aB/A$. Since $a = 8$ (actually .0008, but the decimal can be ignored throughout), $B = 24$, and $A = 29$,

$$b = 8 \cdot \frac{24}{29}$$

$$= 7 \text{ approximately}$$

Consequently, $\sin .6 = .5640 + .0007 = .5647$. Interpolation in a reverse sense is shown next.

Example D.5   Find θ where $0 < \theta < \pi/2$ and $\cos \theta = .5010$.

Upon examining the column headed **cosine**, we find that the values range from 1.000 down to .7071. Thus, our θ must be found in the column having cosine at its *foot*.

We find in the latter column the values .5000 and .5025. Our schematic diagram (just like that of Example 4) is:

The radian values are found on the *right*, since we are working with the column with cosine at the foot.

We need only solve $a/A = b/B$. The solution shows $a = bA/B = 15 \cdot 29/25 = 17$, approximately. We then conclude that $\theta = 1.0443 + .0017 = 1.0460$.

Appendix E
Using Tables
of Logarithms

The logarithmic functions are both interesting and valuable. We propose in this appendix to view one such function. Since we are primarily interested in the computational aspect of the function, we shall concentrate here rather than on its function-theoretic properties.

If $x > 0$ is any real number, $\log_{10} x$ (read the **logarithm of x with respect to the base ten**) is the unique number y where $10^y = x$. We could write the relationship:

$$10^{\log_{10} x} = x$$

The logarithm is seen to be an exponent. Since

$$xy = 10^{\log_{10} x} \, 10^{\log_{10} y} = 10^{(\log_{10} x + \log_{10} y)}$$

Thus,
$$\log_{10} xy = \log_{10} x + \log_{10} y$$

For example
$$\log_{10} 6 = \log_{10} (3 \cdot 2) = \log_{10} 3 + \log_{10} 2$$

Furthermore,
$$\frac{x}{y} = \frac{10^{\log_{10} x}}{10^{\log_{10} y}} = 10^{(\log_{10} x - \log_{10} y)}$$

so that
$$\log_{10} \frac{x}{y} = \log_{10} x - \log_{10} y$$

In addition,

$$x^a = (10^{\log_{10} x})^a = 10^{a \log_{10} x}$$

whence

$$\log_{10} x^a = a \log_{10} x$$

As examples of the last two properties,

$$\log_{10} \tfrac{3}{2} = \log_{10} 3 - \log_{10} 2$$

and

$$\log_{10} 9 = \log_{10} 3^2 = 2 \log_{10} 3$$

Example E.1   If $\log_{10} 2 = .3010$ and $\log_{10} 3 = .4771$, find $\log_{10} 6$, $\log_{10} \tfrac{3}{2}$, and $\log_{10} 9$.

From above,

$$\log_{10} 6 = \log_{10} 3 + \log_{10} 2 = .3010 + .4771 = .7781$$
$$\log_{10} \tfrac{3}{2} = \log_{10} 3 - \log_{10} 2 = .0361$$
$$\log_{10} 9 = 2 \log_{10} 3 = 2(.4771) = .9542$$

The tables of Appendix B give approximate values for logarithms of numbers between 1 and 10 only. We have, then, three remaining purposes for this appendix:

1. to show how the tabular values are interpreted,
2. to show how to use the tables to find logarithms for numbers other than those between 1 and 10, and
3. to illustrate a computational use of logarithms.

We might partially attain objective (2) by observing that each positive number can be written in the form $x(10^n)$, where x is between 1 and 10 and n is an integer. This form is known as *scientific notation*.

As examples of the above:

$$674 = 6.74(10^2)$$
$$.042 = 4.2(10^{-2})$$

Thus, $\log_{10} 674 = \log_{10} 6.74(10^2) = \log_{10} 6.74 + \log_{10} 10^2 = 2 + \log_{10} 6.74$. (This comes about by the identity $\log_{10} 10^a = a$.)   Furthermore, $\log_{10} .042 = \log_{10} 4.2 + \log_{10} 10^{-2} = -2 + \log_{10} 4.2$.

Example E.2   Find $\log_{10} 300$ and $\log_{10} .002$ using the values in Example 1.

$$\log_{10} 300 = \log_{10} 3(10^2) = 2 + \log_{10} 3 = 2.4771$$
$$\log_{10} .002 = \log_{10} 2(10^{-3}) = -3 + \log_{10} 2 = -3 + .3010$$

Observe that $\log_{10} .002$ was left in the form $-3 + .3010$ and was not written -2.6990. The reason for this is that the decimal portion as found in our tables *must be positive*. The integer added to the decimal is called the **characteristic;** the *positive* decimal is called the **mantissa.**

In the tables themselves, we find the column headed **N** to consist of two-digit integers while each of the other columns is headed by a single digit. If we wish to find a logarithm for a number involving the digit combination 1-3-5, we would go down the column **N** to 1.3 and then move horizontally to the column headed 5. There we find the decimal .1303. This tabular value is $\log_{10} 1.35$.

Example E.3   As an alternative example, $\log_{10} 64.7$ is found as $1 + \log_{10} 6.47$ and since $\log_{10} 6.47 = .8109$ (down column **N** to 6.4 and across to column 7), $\log 64.7 = 1.8109$.

Suppose we are asked to find $\log_{10} 581.6$. Our tables only supply values for 581 and 582. A four-digit number is not to be found. However, interpreting 581 as 5810 and 582 as 5820, we find that 5816 lies between. The $\log_{10} 5.810$ is given as .7642 and $\log_{10} 5.820$ is given as .7649. Our mantissa must lie between.

The following diagram is helpful.

The expressions a, A, and B are the differences between the bracketed values; b is the difference between 7642 and the value associated with $\log_{10} 5.816$.

We **interpolate linearly** when we solve the expression

$$\frac{a}{A} = \frac{b}{B}$$

In the example above, we have

$$\frac{6}{10} = \frac{b}{7}$$

or $b = 4$ (to the nearest integer). Consequently, we write $\log_{10} 5.816 = .7642 + .0004 = .7646$. Thus, $\log_{10} 581.6$ has an interpolated value 2.7646.

On the other hand, we might be asked to find a value for N where $\log_{10} N = -1.5960$.

The tables show only *positive* mantissas so that we ought to (must) write $\log_{10} N = -2 + .4040$. We search the table for a mantissa .4040, only to find none.

We do, however, find mantissas .4031 and .4048. Setting up a chart as before, we see

In this case A, b, and B are known. The equality

$$\frac{a}{A} = \frac{b}{B}$$

becomes

$$\frac{a}{10} = \frac{9}{17}$$

whence a is approximately 5. Thus $\log 2.535 = .4040$ and $N = 2.535(10^{-2}) = .02535$, since the characteristic for N is -2.

Example E.4 Use the logarithmic tables to find a value for

$$N = \frac{636(.0712)}{(156)^{1/2}}$$

where four-digit accuracy is desired.

The properties of logarithms show

$$\begin{aligned}
\log_{10} N &= \log_{10} 636 + \log_{10} .0712 - \log_{10} (156)^{1/2} \\
&= \log_{10} 636 + \log_{10} .0712 - \tfrac{1}{2} \log_{10} 156 \\
&= 2.8035 + (-2) + .8525 - \tfrac{1}{2}(2.1931) \\
&= .5594
\end{aligned}$$

A check of Appendix B confronts us with the chart

Our computations show

$$\frac{a}{10} = \frac{7}{12}$$

Whence a is chosen as 6 and $N = 3.626$. (The decimal is between the first two digits, since the characteristic is *zero*.)

The tables of Appendix C are treated much as are the tables of Appendix B and interpolation is the same. The values given are logarithms (base ten) of trigonometric functional values.

If α is $41°$, for instance, $\log_{10} \sin \alpha = \log_{10} \sin 41°$ is found in Appendix C by finding $41° \ 00'$ in the left column and moving to the column headed "$\log_{10}$ sine." We see $\log_{10} \sin 41° = .8169 - 1$.

The tables are "double-ended," in that the angles complementary to those in the left column are found in the right column and the heading at the foot of each column is the cofunction for that at the top (just like the tables of Appendix A).

Example E.5 If $\log_{10} \cos \alpha = .9037 - 1$, find α where $0 < \alpha < 90°$. The chart

$$
10 -
\begin{cases}
a \begin{cases} 36°\,40' \\ \alpha \end{cases} \\
\;\;36°\,50'
\end{cases}
\qquad
\begin{cases}
.9042 - 1 \\ .9037 - 1 \\ .9033 - 1
\end{cases}
\;5
\;\Big\}\;9
$$

shows $a = 6$ (approximately), whence $\alpha = 36°\ 46'$.

The following appendix makes use of these tables in various problems.

Appendix F
Solving Problems Using Appendixes A, B, and C

In this appendix we shall work sample problems of the nature encountered in Chapter 7, showing the use of logarithms and the tables of Appendixes A, B, and C for computational short cuts.

Example F.1   Refer to Fig. F.1.   Compute the value of c in the figure.

$$\cos \alpha = \cos 31° = \frac{c}{19}$$

$$c = 19 \cos 31°$$
$$\log_{10} c = \log_{10} 19 + \log_{10} \cos 31°$$
$$= 1.2788 + (.9331 - 1)$$
$$= 1.2119$$

Figure F.1

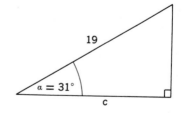

$$\alpha = 31°$$
19
c

From Appendix B we see $c = 16.29$.   The interpolation is seen in the following chart.

Example F.2   Find angle β of Fig. F.2.
    We know that

$$\sin \beta = \frac{16}{59}$$

$$\log_{10} \sin \beta = \log_{10} 16 - \log_{10} 59$$
$$= 1.2041 - 1.7709$$

Since $1.7709 > 1.2041$ and we wish to have a *positive* mantissa, we write 1.2041 as $2.2041 - 1$, whence

$$\log_{10} \sin \beta = 2.2041 - 1 - 1.7709$$

$$= .4332 - 1$$

Appendix C gives the following chart.

Necessarily $a/10 = {}^{18}\!/_{45}$ and $a = 4$.   Thus, $\beta = 15° \, 44'$.

Figure F.2

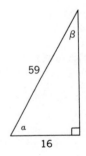

Example F.3   Find c in Fig. F.3.

By the law of sines,

$$\frac{c}{\sin 17°} = \frac{20}{\sin 101°} = \frac{20}{\sin 79°}$$

$$c = \frac{20 \sin 17°}{\sin 79°}$$

$$\log_{10} c = \log_{10} 20 + \log_{10} \sin 17° - \log_{10} \sin 79°$$
$$= 1.3010 + (.4659 - 1) - (.9919 - 1)$$
$$= .7750$$

Our chart for interpolation is

$$
10 - \left[\begin{array}{l} a - \left[\begin{array}{l} 5950 \\ c \end{array} \right. \\ \quad\; 5960 \end{array} \right.
\qquad\qquad
\begin{array}{l} 7745 \\ 7750 \end{array} \left] {-5} \right. \;{-7} \\ 7752
$$

Thus $a = $ ⁵⁰⁄₇ or approximately 7 and $c = 5.957$.

Figure F.3

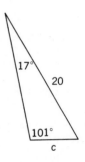

Example F.4   Use a law of tangents to find A and B in Fig. F.4.

$$\frac{\tan \frac{1}{2}(A - B)}{\tan \frac{1}{2}(A + B)} = \frac{a - b}{a + b}$$

Whence, $$\tan \frac{1}{2}(A - B) = \left(\frac{a - b}{a + b}\right) \tan \frac{1}{2}(A + B)$$

$$= \left(\frac{21 - 15}{21 + 15}\right) \tan \frac{1}{2}(180° - C)$$

$$= \left(\frac{6}{36}\right) \tan 76°$$

$$\log_{10} \tan \frac{1}{2}(A - B) = \log_{10} \tan 76° - \log_{10} 6$$
$$= .6032 - .7782$$
$$= .8250 - 1$$

Figure F.4

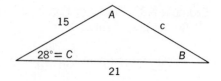

From Appendix C, $\qquad$ $\tfrac{1}{2}(A - B) = 33° \, 46'$

$$A - B = 67° \, 32'$$

Since $A + B = 152°$,

$$A = 109° \, 46'$$
$$B = 42° \, 14'$$

Example F.5 Use a law of tangents to find A in Fig. F.5.

$$\tan \frac{A}{2} = \frac{\rho}{s - a} = \frac{\left[\dfrac{(s - a)(s - b)(s - c)}{s}\right]^{1/2}}{s - a}$$

Now, $s = \tfrac{1}{2}(16 + 31 + 42) = 44.5$; therefore

$$s - a + 44.5 - 31 = 13.5$$
$$s - b = 44.5 - 16 = 28.5$$
$$s - c = 44.5 - 42 = 2.5$$

$$\log_{10} \tan \frac{A}{2} = \log_{10} \rho - \log_{10}(s - a)$$

$$= \tfrac{1}{2}(\log_{10}(s - a) + \log_{10}(s - b)$$
$$+ \log_{10}(s - c) - \log_{10} s) - \log_{10}(s - a)$$
$$= \tfrac{1}{2} \log_{10} 13.5 + \tfrac{1}{2} \log_{10} 28.5 + \tfrac{1}{2} \log_{10} 2.5 - \tfrac{1}{2} \log_{10} 44.5$$
$$- \log_{10} 13.5$$

$$= \tfrac{1}{2}(1.4548 + .3979 - 1.6484 - 1.1303)$$
$$= .5370 - 1$$

Thus, $\qquad\qquad\qquad\qquad A/2 = 19°$
$$A = 38°$$

Figure F.5

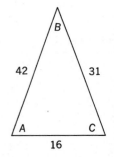

Appendix G
Answers
to Selected
Exercises

■ **EXERCISES 1.2**

1. (a) True. **(c)** False. **(e)** False. **(g)** True. **(i)** False. **(k)** False.
(m) True. **(o)** True. **(q)** True. **(s)** False. **(u)** False. **(w)** False.
(y) False. **(a′)** False. **(c′)** True. **(e′)** False. **(g′)** False. **(i′)** False.

2. $P \subset P$ is true.

3. (a) A. **(c)** A. **(e)** F. **(g)** F. **(i)** A. **(k)** $\{0, 1, 3, 5\}$. **(m)** F.
(o) $\{0, 1, 2, 3, 4\}$. **(q)** E. **(s)** D. **(u)** F. **(w)** E. **(y)** F.

5. (a) A. **(c)** B. **(e)** $\{1, 3\}$. **(g)** A. **(i)** $\varnothing$. **(k)** $\{1, 3\}$. **(m)** B.
(o) $\varnothing$. **(q)** C. **(s)** D. **(u)** D. **(w)** E. **(y)** E.

7. (a) $\varnothing$. **(c)** $\varnothing$. **(e)** $\{2, 4, 5\}$. **(g)** $\varnothing$. **(i)** B. **(k)** $\{5\}$. **(m)** $\varnothing$.
(o) C. **(q)** $\varnothing$. **(s)** $\varnothing$. **(u)** $\varnothing$. **(w)** $\varnothing$. **(y)** B.

9. (a) $\{0\}$. **(c)** $2, 4, 5$. **(e)** $\varnothing$.

10. (a) Area with horizontal shading. **(d)** Area with either horizontal or
vertical shading. **(g)** All shaded area. **(i)** Area with both horizontal and

diagonal shadings. **(k)** Area having all three types of shading. **(m)** Area having vertical but not horizontal shading. **(r)** Area having horizontal or vertical shading but not diagonal shading. **(u)** Area not shaded horizontally (to include the area not shaded at all). **(x)** The area having no shading or only diagonal shading. **(b′)** The area not having both horizontal and diagonal shading.

■ EXERCISES 1.3

1. (a) True. **(c)** False. **(e)** True. **(g)** True. **(i)** True. **(k)** True. **(m)** True. **(s)** True. **(q)** True. **(s)** True. **(u)** True.

2. (a) -2. **(c)** 0. **(e)** 21.

3. (a) 2. **(c)** 0. **(e)** 21.

4. (a) $7 + 3 = 10$.

5. (a) $x + 2 > 5$ implies $x > 3$ and conversely $x > 3$ implies $x + 2 > 3 + 2$ or $x > 5$. The solution set is $\{x : x > 3\}$.

6. (a) $\frac{1}{8}$. **(c)** $^{833}\!/_{500}$. **(e)** $^{2309}\!/_{999}$. **(g)** 1. **(i)** $\frac{1}{4}$.

7. Since $x^2 = (-x)^2$, both x and $-x$ are square roots of x^2. Choose the nonnegative one in each case.

9. (a) 81. **(c)** $-\frac{1}{8}$. **(e)** $\frac{1}{1296}$. **(g)** 3. **(i)** 2. **(k)** $\frac{1}{243}$. **(m)** $\frac{1}{3}$. **(o)** -5. **(q)** -79. **(s)** $x^{2(n+m)}$. **(u)** $2(3^n)$. **(w)** x/y if $y \neq 0$. **(y)** $(y + x)/xy$ if $x \neq 0, y \neq 0, x \neq y$.

10. (a) $1\frac{1}{15}$. **(c)** $^{13}\!/_{12}$. **(e)** $-\frac{4}{21}$. **(q)** $-\frac{9}{190}$. **(i)** 8. **(k)** $\frac{6}{5}$. **(m)** $-1\frac{5}{26}$. **(o)** $^{35}\!/_{68}$. **(q)** $^{2480}\!/_{301}$.

11. (a) $[a, c]$. **(c)** $\{a, b\}$. **(e)** $[c, b]$. **(g)** (a, b).

■ EXERCISES 1.4

1. (a) A is the set of all trigonometry students. **(c)** C is the set of all students studying algebra or trigonometry. **(e)** E is the set of all students studying trigonometry but not algebra. **(g)** G is the set of all students studying either trigonometry or algebra but not both.

3. (a) True. **(c)** True. **(e)** True. **(g)** False. **(i)** False. **(k)** False. **(m)** True.

4. (a) True. **(c)** True. **(e)** True. **(g)** False. **(i)** True. **(k)** True.

5. (a) $\{N, Z, Q, R\}$. **(d)** $\{Q, R\}$. **(i)** $\{I, R\}$. **(j)** $\{Q, R\}$.

6. (a) -3.   **(c)** 2.

7. (a) 3.   **(c)** 2.

8. Part of the proof follows from: $x < y$ implies $x + x < x + y$ and $(x + x)/2 < (x + y)/2$, which is to say that $x < (x + y)/2$.

9. (a) No; $x < 0$ implies $x < x/2 < 0$ by Exercise 8.

10. (a) $\frac{1}{50}$.   **(c)** $\frac{14}{9}$.

■ **EXERCISES 1.5**

1. (a) If $x \in A$, $x \in A$ or $x \in B$ so that $x \in A \cup B$.   **(c)** If $x \in A - B$, $x \notin B$ so that $x \in (A - B) \cap B$ is not possible.   **(e)** Follows from C, since $B - A \subset B$.   **(g)** If $x \in A - (B \cup C)$, then $x \in A$ but $x \notin B$ and $x \notin C$. Thus, $x \in A - B$ and $x \in A - C$; whence, $x \in (A - B) \cap (A - C)$, that is, $A - (B - C) \cap (A - B) \cap (A - C)$. On the other hand, if $x \in (A - B) \cap (A - C)$, $x \in A$ and not B and $x \in A$ and not C. Consequently, $x \in A$ but $x \notin C \cup B$. Necessarily, $x \in A - (B \cup C)$, indicating $(A - B) \cap (A - C) \subset A - (B \cup C)$. However, $A - (B \cup C) \subset (A - B) \cap (A - C)$ and $(A - B) \cap (A - C) \subset A - (B \cup C)$ implies that $A - (B \cup C) = (A - B) \cap (A - C)$.

2. (a) If $a < b$, there is a positive number d with $a + d = b$. Thus, $(a + c) + d = (b + c)$, meaning that $a + c < b + c$.   **(c)** If $a < b$, there is a positive number d with $a + d = b$. Then, $(a + d)c = bc$ and $(a + d)/c = b/c$, whence $ac + dc = bc$ and $a/c + d/c = b/c$. Since dc and d/c are positive, $ac < bc$ and $a/c < b/c$.

3. (c) Using a and b of 3, we see $-(|a| + |b|) \leqslant a + b \leqslant |a| + |b|$, whence $|a + b| \leqslant |a| + |b|(|a| + |b| = | |a| + |b| |)$.

4. (a) If d is positive, $a + d \neq a$, whence $a > a$ is not possible.

5. (a) Let a and b be, respectively, a lower and an upper bound for A.

6. (a) l.u.b. $= 1$, g.l.b. $= 0$.   **(c)** l.u.b. $= 1$, g.l.b. $= -1$.   **(e)** l.u.b. $= -1$, no g.l.b.

7. No.   No.

■ **EXERCISES 2.1**

1. (a) 5.   **(c)** 10.

2. (a) 5.   **(c)** $\sqrt{2}$.   **(e)** 6 2.   **(g)** $3\sqrt{2}$.

3. (a) Right triangle only.   **(c)** Right isosceles triangle.

4. Second coordinate zero; first coordinate zero.

5. Those lying below axis 1 have negative second coordinate, while those lying to the right of axis 2 have positive first coordinate.

6. (a) $\{(0, 0), (0, 1), (0, 2), (, 0), (1, 1), (1, 2), (2, 0), (2, 1), (2, 2)\}$.
(d) $\{(0, 1), (0, 3), (1, 1), (1, 3), (2, 1), (2, 3)\}$.

(g) $((0, 3, 1), ((0, 4, 1), ((0, 5), 1), ((1, 3), 1), ((1, 4), 1),$
$((1, 5), 1), ((0, 3), 3), ((0, 4), 3), ((0, 5), 3), ((1, 3), 3),$
$((1, 4), 3), ((1, 5), 3), ((2, 3), 1), ((2, 3), 3), ((2, 4), 1),$
$((2, 4), 3), ((2, 5), 1), ((2, 5), 3).$

(h) Like Exercise (g) except that the second element of the "internal" ordered pair is mated with the lone element to give elements of the form $(0, (3, 1))$ instead of $((0, 3), 1)$.

■ **EXERCISES 2.2**

2. (a) $y - 4 = (-\frac{1}{2})(x + 3)$. **(c)** $y = 1$. **(e)** $y - 2 = 4(x - 1)$.
(g) $y = 2x - 2$. **(i)** $y = 2x - 1$. **(l)** $y = 1 - x$.

4. $4y = 3x + 5$, perpendicular bisector.

5. Same slope, different lines. If two lines intersect, they have different slopes.

■ **EXERCISES 2.3**

1. (a) $(x - 1)^2 + (y - 1)^2 = 100$. **(c)** $(x - 1)^2 + (y + 1)^2 = 9$.
(e) $(x - 4)^2 + (y + 2)^2 = 4$. **(g)** $(x - 2)^2 + (y - 3)^2 = $
$[d((2, 3), (6, 6))]^2 = 25$.

2. Begin with $(x - 1)^2 + (y - 3)^2 = (x + 1)^2 + (y - 2)^2$ and remove the squared terms.

3. Begin in a way analogous to Exercise 2.

■ **EXERCISES 2.4**

1. (a) 5. **(c)** 2. **(e)** 10. **(g)** 6.

2. Yes for each of (a) through (d).

3. (a) $(2 - \sqrt{3})^{1/2}$. **(c)** Same as a. **(e)** $(2 - \sqrt{2})^{1/2}$. **(g)** $\sqrt{2}$. **(i)** $\sqrt{2}$.

4. Same.

7. Same position of the circumference.

8. (a) $x = 6$.   **(c)** $y - 3 = 2(x - 2)$.   **(e)** $y + 2 = (-\tfrac{2}{3})(x - 3)$.
(g) $y = \frac{1}{2}x + 2$.

9. (a) $x^2 + y^2 = 25$.   **(c)** $(x - 4)^2 + (y + 5)^2 = 9$.
(e) $(x + 4)^2 + (y + 5)^2 = 9$.

7. 10. (a) $\{(0, -2), (\tfrac{6}{5}, \tfrac{8}{5})\}$.

15. (a) $r_1 \cap r_2$, where $r_1 = \{(x, y) : x_1 = 0, 1 \leqslant y\}$ and
$r_2 = \{(x, y) : x = 0, y \leqslant 5\}$.   **(c)** $r_1 \cap r_2$, where $r_1 = \{(-5, y) : y \geqslant 4\}$
and $r_2 = \{(-5, y : y \leqslant 12\}$.   **(e)** $r_1 \cap r_2$, where r_1 emanates from $(2, 4)$
and passes through $(1, 6)$, while r_2 emanates from $(1, 6)$ and passes
through $(2, 4)$.

■ **EXERCISES 2.5**

1. (a) 55.   **(c)** 100.   **(e)** 8.

2. (a) m and m' are different.   **(c)** $3y + x = 6$.   **(e)** $2y + x = 0$.
(g) $y = 2$ and $y = -8$.

3. (a) $y^2 = 20x$.

4. h and k are given by $(h - 1)^2 + (k - 3)^2 = 25$.

5. $l_1 : y - x_2 = \dfrac{z_2 - x_2}{z_1 - x_1}(x - x_1)$. $l_2 : y - \dfrac{z_2 + x_2}{2} = m\left(x - \dfrac{x_1 + z_1}{2}\right)$.

Lines l_2, l_3, and l_4 will be parallel if and only if they are all horizontal.
No.

6. (a) The proof follows from the fact that $\sqrt{x} = 0$ if and only if $x = 0$.
(c) $\sqrt{26}$.   **(e)** $\sqrt{91}$.

7. (a) 3.   **(d)** 12.

■ **EXERCISES 3.1**

1. (a) Every element of X is mated to just one element from Y.
(c) For each element x of X there is exactly one choice for $f(x)$ in Y.
(e) Yes. Yes.

2. (a) No.   **(b)** g. h is **not** an even function, since $-1, -2$, and so forth
are not in A.

3. (a) *a.*   **(c)** *d.*   **(e)** *a.*   **(g)** *c.*

4. (a) *a.*   **(c)** *i.*   **(e)** *b.*   **(g)** *i.*   **(i)** *a.*   **(k)** *i.*

■ EXERCISES 3.2

1. f, h, t, and k have inverses.

2. Only $h:A \longrightarrow B$ of Exercise 3.1.1 has an inverse.

■ EXERCISES 3.3

1. (a)
0 $\longrightarrow$ ½
1 $\longrightarrow$ ⅓
2 $\longrightarrow$ ¼
3 $\longrightarrow$ ½
4 $\longrightarrow$ ⅕

(b) $f \circ g$ does not make sense.   (What is $f(½)$, for example?).

2. (a) *a.*   **(c)** *a.*   **(e)** ½.   **(g)** ¼.   **(i)** ½.   **(k)** ½.

3. $h \circ f, h \circ g, h \circ k, k \circ h$.

■ EXERCISES 3.4

1. $f:R \longrightarrow R$ is a 1:1 and an onto function.   **(a)** 1.   **(c)** 0.   **(e)** 3.
(g) 13.

2. f has an inverse $f(x) = 2x + 1$ implies $x = ½(f(x) - 1)$ so that $f^{-1}(x) = (x - 1)/2$.

3. $f[R] = \{x \in R : x \geqslant 0\}$.   **(a)** No.   **(b)** Yes.

4. (a) 0.   **(c)** 1.   **(e)** 4.

5. (a) $\{x \in R : x \geqslant 0\}$.   **(c)** R.

6. $\lambda^{-1}(x) = (x + 1)/3$.

7. (a) 2.   **(c)** 0.   **(e)** $\sqrt{7}$.   **(g)** 0.   **(i)** -1.   **(k)** -2.   **(m)** 0.   **(o)** 11.
(q) -4.

8. (a) ⅓.   **(c)** ⅔.   **(e)** -2.

9. (a) $\{x \in R : x \neq 1\}$.   **(c)** $\{x \in R : x \neq 0\}$.

10. **(a)** 1.   **(c)** ⅔.   **(e)** Not defined.   **(g)** 1.   **(i)** ⅙.   **(k)** ⅟30.

11. **(a)** $1/(x + 1)$.   **(c)** $\sqrt{x^2} = x$.

13. **(a)** n odd.   **(b)** n odd.   **(c)** n even.   **(d)** n odd.

14. **(a)** zero.   **(c)** 2.

15. **(a)** 6.   **(c)** 116.   **(e)** 6.   **(g)** 86.

16. **(a)** No zeros.   **(c)** No zeros.

17. Polynomials having only terms with odd powers of the variable are odd functions.

18. **(a)** Domain $= \{x \in R : x \neq -1\}$; $x = -1$ is a pole and $x = 0$ is the only zero.   **(c)** Domain $= \{x \in R : x \neq -2\}$; $x = -2$ is a pole and $x = -1$ is the only zero.   **(e)** The domain of μ is R, while $x = 0$ is a zero and μ has no poles.

19. **(a)** ½.   **(c)** Not defined.   **(e)** $-$½.   **(g)** Not defined.

21. **(a)** $f(x)f(y) = b^x b^y = b^{x+y} = f(x + y)$.
(c) $(f(x))^n = (b^x)^n = b^{nx} = f(nx)$.

22. **(a)** 1.   **(c)** 2.   **(e)** ⅟10.   **(g)** 4.   **(i)** 8.   **(k)** ⅛.   **(m)** $\sqrt{2}$.   **(o)** $2\sqrt{2}$.
(q) $1/\sqrt{10}$.

23. **(a)** 0.   **(c)** 1.   **(e)** ½.   **(g)** 2.   **(i)** 3.   **(k)** 0.   **(m)** 1.   **(o)** ½.   **(q)** 2.
(s) 2.

24. **(a)** $x = b^{f^{-1}(x)}$, while $y = b^{f^{-1}(y)}$, whence $xy = b^{f^{-1}(x)+f^{-1}(y)}$ and $f^{-1}(xy) = f^{-1}(x) + f^{-1}(y)$.   **(d)** Use the proof in (a) with the appropriate change in notation.

25. **(a)** 2.   **(c)** a.

■ **EXERCISES 3.5**

1. $A \times A$, $\varnothing$, $\{(0, 0)\}$, $\{(0, 1)\}$, $\{(1, 0)\}$, $\{(1, 1)\}$, $\{(0, 0), (0, 1)\}$;
$\{(0, 0), (1, 0)\}$, $\{(0, 0), (1, 1)\}$, $\{(0, 1), (1, 0)\}$, $\{(0, 1), (1, 1)\}$;
$\{(1, 0), (1, 1)\}$, $\{(0, 0), (1, 0), (0, 1)\}$, $\{(0, 0), (1, 0), (1, 1)\}$;
$\{(0, 0), (0, 1), (1, 1)\}$, $\{(0, 1), (1, 0), (1, 1)\}$.

3. **(a)** $A \times A$, $\{(0, 0), (1, 1)\}$, $\{(0, 0), (1, 0), (1, 1)\}$, $\{(0, 0), (0, 1), (1, 1)\}$.

4. **(a)** $A \times A$, $\varnothing$, $\{(0, 0)\}$, $\{(1, 1)\}$, $\{(0, 0), (1, 1)\}$, $\{(0, 1), (1, 1)\}$;
$\{(0, 0), (0, 1), (1, 0)\}$, $\{(1, 1), (0, 1), (1, 0)\}$.

5. (a) $A \times A$, $\varnothing$, $\{(0, 0)\}$, $\{(1, 0)\}$, $\{(0, 1)\}$, $\{(1, 1)\}$, $\{(0, 0), (1, 0)\}$; $\{(0, 0), (0, 1)\}$, $\{(0, 0), (1, 1)\}$, $\{(1, 1), (0, 1)\}$, $\{(1, 1), (1, 0)\}$; $\{(0, 0), (1, 1), (0, 1)\}$, $\{(0, 0), (1, 1), (1, 0)\}$.

6. (a) $A \times A$, $\{(0, 0), (1, 1)\}$.

7. (a) f is not a function. (Is $f(e) = 3$ or is $f(e) = 2$?) **(c)** h is a function but not $1:1$.

8. (a) $f = \{(0, a), (1, b), (2, c), (3, a), (4, b)\}$. **(c)** $f = \{(0, s), (1, u), (2, t), (3, r)\}$. **(e)** $f = \{(0, s), (1, u), (2, t), (3, r)\}$.

9. (c) $f^{-1} = \{(s, 0), (u, 1), (t, 2), (r, 3)\}$.

10. (a) $f = \{(x, 2x + 1) : x \in R\}$. **(c)** $\lambda^{-1} = \{(x, (x + 1)/3) : x \in R\}$. **(f)** $f^{-1} = \{(x, (x - 1)/2) : x \in R\}$.

12. No.

13. At most once.

■ EXERCISES 3.6

1. Yes.

3. In Exercise 1, $g|A = \{(0, a), (1, a), (2, a)\}$.

4. (a) Yes. **(b)** No.

5. (a) $f|A = \{(0, 0), (3, 4), (\frac{1}{2}, 7)\}$. **(b)** Yes. **(d)** $(f|A)^{-1} = \{(0, 0), (4, 3), (7, \frac{1}{2})\}$.

6. (b) $f|A$ is a constant function ($f(x) = 0$ except for $x = 0$ when $f(x) = -1$). **(c)** $f|(R - A)$ is a constant function.

7. (a) $g|Q$ is the identity function on Q.

8. Set of nonnegative real numbers.

9. (a) 6. **(c)** $\frac{1}{2}$. **(e)** 1. **(g)** 3. **(i)** 3. **(k)** -1.

10. (a) 1. **(c)** 2. **(e)** x. **(g)** $(x, 2x)$. **(h)** $(2x, 4x)$.

■ EXERCISES 3.7

1. (a) No. **(b)** Yes. **(c)** Yes. **(d)** No. **(h)** Yes, in each case. **(i)** $u[X] = \{a, b, c, e\}$, $u[A] = B$, $u^{-1}[Y] = X$, $u^{-1}[B] = \{0, 2, 3, 5, 6\}$, $u^{-1}[u[X]] = X$, $u^{-1}[u[A]] = \{0, 2, 3, 5, 6\}$, $u[u^{-1}[B]] = B$. **(k)** No. **(m)** Onto B. **(p)** $u|A = \{(3, b), (5, a), (6, r)\}$. **(q)** $(u|A)^{-1} = \{(6, 3), (a, 5), (e, 6)\}$. **(s)** c. **(u)** a. **(w)** c. **(y)** e.

2. (b) $x = 0$.   **(e)** -3.   **(g)** $\frac{1}{3}$.   **(i)** 1.   **(k)** 1.   **(m)** 0.   **(o)** -1.   **(q)** 1.

3. (a) $10x/3$.   **(c)** x^2.   **(e)** 0.   **(g)** $-\frac{10}{3}$.   **(i)** $\frac{8}{3}$.   **(k)** 0.   **(m)** 1.   **(o)** 9.
(q) 0.   **(s)** -18.

4. (a) $f \circ g(x) = x$.   **(c)** $(f + g)x = 2x$.   **(e)** $x^2 - 1 = (fg)(x)$.
(g) $f^{-1}(x) = g(x)$.   **(i)** R.   **(k)** R.   **(m)** R.   **(o)** R.   **(q)** R.   **(s)** R.
(u) R.   **(w)** $\{x \in R : x \geqslant -1\}$.   **(y)** R.

6. (a) $1/(x + 1)$.   **(c)** -1.   **(e)** $\frac{1}{2}$.   **(g)** Not defined.   **(i)** Domain for
$1/f$ is $\{x : x \neq -1\}$.

7. (g) 1.   **(h)** If $x^2 + y^2 = 1$, $-1 \leqslant x \leqslant 1$.   **(l)** $[-1, 1]$.   **(m)** $[-1, 1]$.

8. (a) Domain of f.   **(b)** Range of f.

10. (a) Polynomial function and rational function.   **(d)** Rational function.
(g) Neither polynomial nor rational.

12. (a) $\lambda(0) = 1$, $\lambda^{-1}(0)$ is not defined.

■ **EXERCISES 3.8**

1. If $mx_1 + b = mx_2 + b$, $mx_1 = mx_2$, whence $x_1 = x_2$ and f is $1:1$.
If $x \in R$, $(x - b)/m \in R$ with $f((x - b)/m) = m(x - b)/m + b = x - b + b = x$, implying that f is an onto function.

2. (c) Domain is R and the range is $\{x \in R : x \leqslant 20 + \sqrt{2}\}$.

8. (a) Yes.   **(b)** $(a, b) \in F$ implies that a and b have the same remainder,
say x, upon division by 7; x can be chosen to be among $\{0, 1, 2, 3, 4, 5, 6,$
and $7\}$.   Then $(a, b) \in F$ implies that (a, b) is in among $\boxed{0} \cup \boxed{1} \cup \boxed{2} \cup$
$\boxed{3} \cup \boxed{4} \cup \boxed{5} \cup \boxed{6}$.   **(c)** If $\boxed{a} = \boxed{b}$, $a - b$ has remainder 0 upon division
by 7 ($a = 7x + y$ and $b = 7z + y$ for some $x, y,$ and z.   Thus, $a - b = 7(x - y)$).   **(g)** $\boxed{6}$.   **(i)** $\boxed{4}$.

■ **EXERCISES 4.1**

1. (a) $\pi/2$.   **(c)** $3\pi/2$.   **(e)** $\pi/4$.   **(g)** $2\pi/3$.   **(i)** $5\pi/6$.   **(k)** $5\pi/4$.
(m) $5\pi/3$.   **(o)** $11\pi/6$.

2. (a) $-3\pi/2$.   **(c)** $-\pi/2$.   **(e)** $-7\pi/4$.   **(g)** $-4\pi/3$.   **(i)** $-7\pi/6$.
(k) $-3\pi/4$.   **(m)** $-\pi/3$.   **(o)** $-5\pi/6$.

3. (a) $5\pi/2$.   **(c)** $15\pi/2$.   **(e)** $5\pi/4$.   **(g)** $10\pi/3$.   **(i)** $25\pi/6$.   **(k)** $25\pi/4$.
(m) $25\pi/3$.   **(o)** $55\pi/6$.

These are five times those from Exercise 1.

6. (x, y) in quadrant I is given by $x > 0$ and $y > 0$.

■ **EXERCISES 4.2**

1. (a) $(1, 0)$.   **(c)** $(1, 0)$.   **(e)** $(1, 0)$.

2. (a) If $b \in [0, 2\pi)$, then $b = a$.

3. (a) Since $0 < 5\pi/24 < \pi/2$, $\alpha(5\pi/24)$ is in quadrant I.   **(c)** Since $\pi < 17\pi/16 < 3\pi/2$, $\alpha(17\pi/16)$ is in quadrant III.   **(e)** Since $3\pi/2 < 23\pi/12 < 2\pi$, $\alpha(23\pi/12)$ is in quadrant IV.   **(g)** Since $\pi/2 < 24\pi/53 < \pi$, $\alpha(24\pi/53)$ is in quadrant II.

4. (a) Since $\alpha(\pi/6) = (\sqrt{3}/2, \frac{1}{2})$, $\alpha(-\pi/6) = (\sqrt{3}/2, -\frac{1}{2})$.
(c) Since $\alpha(\pi) = (-1, 0)$, $\alpha(-\pi) = (-1, 0)$ also.   **(e)** Since $\alpha(3\pi/2) = (0, -1)$, $\alpha(-3\pi/2) = (0, 1)$.   **(g)** Since $\alpha(11\pi/6) = (\sqrt{3}/2, -1/2)$, $\alpha(-11\pi/6) = (\sqrt{3}/2, \frac{1}{2})$.

5. (a) IV.   **(c)** II.

6. (a) $(1, 0)$.   **(c)** $(\frac{1}{2}, \sqrt{3}/2)$.   **(e)** $(-\frac{1}{2}, \sqrt{3}/2)$.   **(g)** $(-\sqrt{3}/2, \frac{1}{2})$.
(i) $(-1/\sqrt{2}, -1/\sqrt{2})$.   **(k)** $(0, -1)$.   **(m)** $(1/\sqrt{2}, -1/\sqrt{2})$.
(o) $(\sqrt{3}/2, -\frac{1}{2})$.

7. (a) That arc in quadrant I together with $(1, 0)$ and $(0, 1)$.   **(c)** All except that arc in quadrant IV.   **(e)** That arc in quadrants II and III with $(0, 1)$ and $(0, -1)$.   **(g)** That arc in quadrant III with $(-1, 0)$ and $(0, -1)$.   **(i)** That arc in quadrant IV together with $(0, -1)$ and $(1, 0)$.

8. (a) $\alpha[[9\pi/4, 11\pi/4]]$ or $\alpha[[-7\pi/4, -5\pi/4]]$ for counterclockwise traversal and $\alpha[[3\pi/4, 9\pi/4]]$ for clockwise traversal.

9. (a) $\alpha(2\pi - a) = \alpha((2\pi - a) + 2\pi) = \alpha(-a)$.   **(c)** $\alpha(a + \pi) = \alpha((a + \pi) + 2n\pi) = \alpha(a + (2n + 1)\pi)$.

10. (a) $p_1 \circ \alpha(0) = p_1(1, 0) = 1$.   **(b)** $p_2 \circ \alpha(0) = p_2(1, 0) = 0$.
(c) $(1/p_1 \circ \alpha)(0) = 1/p_1(1, 0) = 1$.   **(d)** $(1/p_2 \circ \alpha)(0)$ is not defined.
(e) $(p_1 \circ \alpha/p_2 \circ \alpha)(0)$ is not defined.   **(f)** $(p_2 \circ \alpha/p_1 \circ \alpha)(0) = p_2 \circ \alpha(0)/p_1 \circ \alpha(0) = 0/1 = 0$.   **(g)** $[(p_1 \circ \alpha)^2 + (p_2 \circ \alpha)^2](0) = 1^2 + 0^2 = 1$.

12. (a) If $\alpha(a) = (\frac{3}{5}, \frac{4}{5})$: $p_1 \circ \alpha(a) = \frac{3}{5}$, $p_2 \circ \alpha(a) = \frac{4}{5}$, $(1/p_1 \circ \alpha)(a) = \frac{5}{3}$, $(1/p_2 \circ \alpha)(a) = \frac{5}{4}$, $(p_1 \circ \alpha/p_2 \circ \alpha)(a) = \frac{3}{4}$, $(p_2 \circ \alpha/p_1 \circ \alpha)(a) = \frac{4}{3}$, and $(p_1 \circ \alpha)^2 + (p_2 \circ \alpha)^2(a) = (\frac{3}{5})^2 + (\frac{4}{5})^2 = 1$.

14. A period for $p_1 \circ \alpha$ is 2π, while a period for $p_1 \circ \alpha/p_2 \circ \alpha$ is π (2π is also a period here).   See Exercise 15 of the exercises of this same section.

17. (a) $(1/\sqrt{2}, 1/\sqrt{2})$.   **(c)** $(\frac{5}{13}, \frac{12}{13})$, $(\frac{12}{13}, \frac{5}{13})$.   **(e)** $(\frac{7}{25}, \frac{24}{25})$, $(\frac{24}{25}, \frac{7}{25})$.

18. (a) $d(\alpha(7\pi/12), (0, 1)) = d(\alpha(\pi/12), (1, 0))$, whence $\alpha(7\pi/12) =$

$((1 - \sqrt{3})/2\sqrt{2}, (1 + \sqrt{3})/2\sqrt{2})$.   **(c)** $\alpha(143\pi/12) =$ $\alpha(12\pi - \pi/12) = \alpha(-\pi/12) = ((\sqrt{3} + 1)/2\sqrt{2}, (1 - \sqrt{3})/2\sqrt{2})$.

■ EXERCISES 4.3

2. If α (a) is in quadrant I or II, $\sin a > 0$ and conversely. Similarly, $\cos a > 0$ if and only if $\alpha(a)$ is in quadrant I or IV.

3. (a) $x = \pi/6$ or $5\pi/6$ are the special values. However, $x = \pi/6 + 2n\pi$ and $x = 5\pi/6 + 2n\pi$ for $n \in Z$ are valid choices. The total set can be given by $\{x \in R : x = (-1)^n \pi/6 + n\pi, n \in Z\}$.   **(c)** $\{x \in R : x = \pi/3 + 2n\pi$ or $x = 2\pi/3 + 2n\pi\} = \{x \in R : x = (-1)^n \pi/3 + n\pi\}$. **(e)** $\{x \in R : x = (-1)^n \pi/4 + n\pi\}$.   **(g)** $\{x \in R : \pi/2 + 2n\pi = x\}$.   **(i)** $\emptyset$.

4. If $\sin \alpha = x$ and $\cos a = y$, $\alpha(a) = (y, x)$ and $\alpha(-a) = (y, -x)$, whence $\sin(-a) = -x = -\sin a$ and $\cos(-a) = y = \cos a$. See Exercise 6.

5. $\alpha(a) = (y, x)$ and $\alpha(\pi/2 - a) = (x, y)$. Thus, $\sin(\pi/2 - a) = y = \cos a$ and $\cos(\pi/2 - a) = x = \sin a$. See Exercise 6.

6. $\alpha(a) = (x, y)$.

7. $\left(\dfrac{x}{(x^2 + y^2)^{1/2}}, \dfrac{y}{(x^2 + y^2)^{1/2}} \right)$ is on the unit circle and there is, then, an $a \in [0, 2\pi)$ with $\cos a = x/(x^2 + y^2)^{1/2}$ and $\sin a = y/(x^2 + y^2)^{1/2}$.

■ EXERCISES 4.4

1. (a) $-\frac{4}{5}$.   **(c)** $^{12}\!/_{13}$.   **(e)** $\frac{4}{5}$.   **(g)** $-^{12}\!/_{13}$.   **(i)** $-\frac{4}{5}$.   **(k)** $\frac{5}{13}$.   **(m)** $\frac{4}{5}$. **(o)** $-17/13\sqrt{2}$.

2. (a) $-^{36}\!/_{85}$.   **(b)** $^{77}\!/_{85}$.   **(c)** $-^{84}\!/_{85}$.   **(d)** $^{13}\!/_{85}$.

3. (a) $-^{24}\!/_{25}$.   **(b)** $-\frac{7}{25}$.   **(i)** $(\frac{3}{5}, -\frac{4}{5})$.   **(m)** $(^{77}\!/_{85}, -^{36}\!/_{85})$.

4. (a) $\cos \pi/5 = (1 - \sin^2 \pi/5)^{1/2}$.   **(c)** $\sin \pi/10 = (1 - \cos \pi/10)^{1/2}$.

5. (a) $\cos 3\pi/10 = \sin \pi/5 = .5878$.   **(c)** $\cos 3\pi/8 = \sin \pi/8 = .3827$. **(f)** $\sin 2\pi/5 = \cos \pi/10 = .9511$.

6. (a) $\alpha(b) = (\frac{4}{5}, \frac{3}{5})$.   **(c)** $\alpha(b) = (-1/\sqrt{2}, 1/\sqrt{2})$.   **(e)** $\alpha(b) = (^{15}\!/_{17}, \frac{8}{17})$.

7. (a) $\cos(23\pi/10) = \cos 3\pi/10$.   **(c)** $\cos 25\pi/8 = \cos(9\pi/8) = -\cos(\pi/8)$.   **(e)** $\cos 21\pi/5 = \cos \pi/5$.   **(g)** $\cos 83\pi/8 = \cos 3\pi/8 = \sin \pi/8$.   **(i)** $\cos 13\pi/40 = \cos(\pi/8 + \pi/5)$.

(k) $\cos \pi/40 = \cos (\pi/4 - 3\pi/10)$
$$= \cos \pi/4 \cos 3\pi/10 + \sin \pi/4 \sin 3\pi/10$$
$$= \cos \pi/4 \cos (\pi/5 - \pi/8) + \sin \pi/4 \sin (\pi/5 - \pi/8)$$

(m) $\cos 3\pi/5 = \cos (\pi/2 + \pi/10)$.

11. (b) $\sin 23\pi/6 = \sin (4\pi - \pi/6) = \sin (-\pi/6) = -\sin \pi/6$.
(e) $\sin 72\pi/5 = \sin (14\pi + 2\pi/5) = \sin 2\pi/5 = \cos \pi/10$.
(h) $\cos 3163\pi/16 = \cos 27\pi/16 = \cos (-5\pi/16) = \sin 3\pi/16$.

12. (a) $y = x$.   **(d)** $y = -x$.   **(f)** $\sqrt{3y} = x$.

14. Sine is odd; cosine is even.

■ **EXERCISES 4.5**

1. (a) $^{24}\!/_{25}$, $-\frac{7}{25}$.   **(c)** $^{240}\!/_{289}$, $-^{161}\!/_{289}$.   **(e)** $-^{120}\!/_{169}$, $-^{119}\!/_{169}$.

2. (a) $\sin a/2 = 1/\sqrt{5}$, $\cos a/2 = 2/\sqrt{5}$.   **(c)** $\sin a/2 = -3/\sqrt{34}$, $\cos a/2 = -5/\sqrt{34}$.   **(e)** $\sin a/2 = 2/\sqrt{13}$, $\cos a/2 = -3/\sqrt{13}$.
(g) $\sin a/2 = 7/5\sqrt{2}$, $\cos a/2 = 1/5\sqrt{2}$.

3. (a) $\frac{1}{2}(\sin 5a - \sin a)$.   **(c)** $\frac{1}{2}(\cos 7a + \cos 3a)$.
(e) $\frac{1}{2}(\cos 3a - \cos 7a)$.

4. (a) $\sqrt{2}/4$.   **(c)** $\frac{1}{2}(\frac{1}{2} - \sin \pi/8)$.

5. (a) $2 \sin 3a/2 \cos a/2$.   **(c)** $2 \cos 15a/12 \sin 5a/12$.   **(e)** $2 \sin 4a \sin a$.

6. (a) $-1/\sqrt{2}$.   **(c)** 1.11, approximately.

7. Slope $= \sin a/\cos a$, where $(x, y) = (r \cos a, r \sin a)$.

■ **EXERCISES 4.6**

In the exercises below, the indicated solution is but one method of approach.

1. $\sin (\pi/2 + a) = \sin \pi/2 \cos a + \cos \pi/2 \sin a = \cos a = \sin (\pi/2 - a)$. The substitution set is R and the restriction set is $\varnothing$.

4. $\sin a/\cos a + \cos a/\sin a = \sin^2 a/\cos a \sin a + \cos^2 a/\cos a \sin a = (\sin^2 a + \cos^2 a)/\cos a \sin a = 1/\cos a \sin a$.

9. $\dfrac{1}{\cos^2 a \sin^2 a} = \dfrac{\sin^2 a + \cos^2 a}{\cos^2 a \sin^2 a} = \dfrac{\sin^2 a}{\cos^2 a \sin^2 a} + \dfrac{\cos^2 a}{\cos^2 a \sin^2 a} =$
$1/\cos^2 a + 1/\sin^2 a$.   The restriction set A is $\{x \in R : x = n\pi/2\}$, while the substitution set is $R - A$.

15. Let $\alpha(b) = (\frac{3}{5}, \frac{4}{5})$.   Then $\cos b = \frac{3}{5}$ and $\sin b = \frac{4}{5}$, whence $3 \sin a + 4 \cos a = 5(\frac{3}{5} \sin a + \frac{4}{5} \cos a) = 5(\sin a \cos b + \cos a \sin b) = 5 \sin (a + b)$.

22. $\sin 2a / \cos a = 2 \sin a \cos a / \cos a = 2 \sin a$.   The restriction set is $A = \{x \in R : x = (2n + 1)\pi/2\}$ and the substitution is $R - A$.

30. $\dfrac{\sin a}{1 + \cos a} + \dfrac{1 + \cos a}{\sin a} = \dfrac{\sin^2 a + (1 + \cos a)^2}{\sin a(1 + \cos a)} =$

$\dfrac{1 + 2 \cos a + (\sin^2 a + \cos^2 a)}{\sin a(1 + \cos a)} = \dfrac{2(1 + \cos a)}{\sin a(1 + \cos a)} = 2/\sin a$.   The

restriction set is $A = \{x \in R : x = n\pi\}$ and the substitution set is $R - A$.

■ EXERCISES 4.7

2.

$\alpha(a)$	I	II	III	IV
$\tan a$	+	−	+	−
$\cot a$	+	−	+	−
$\sec a$	+	−	−	+
$\csc a$	+	+	−	−

4. $\tan 2a = 2 \tan a / 1 - \tan^2 a$, whence $\cot 2a = 1 - \tan^2 a/2 \tan a = \dfrac{(1 - \tan^2 a) \cot^2 a}{2 \tan a \cot^2 a} = \dfrac{\cot^2 a - 1}{2 \cot a}$

6. $\sec a = 1/\cos a = 1/\sin (\pi/2 - a) = \csc (\pi/2 - a)$; $\csc a = 1/\sin a = 1/\cos (\pi/2 - a) = \sec (\pi/2 - a)$.

7. (a) $\tan (-a) = \sin (-a)/\cos (-a) = -\sin a/\cos a = -\tan a$.

8. $\tan (a - b) = \tan (a + (-b)) = \dfrac{\tan a + \tan (-b)}{1 - \tan a \tan (-b)}$

which, by Exercise 7, gives

$$\tan (a - b) = \dfrac{\tan a - \tan b}{1 + \tan a \tan b}$$

9. (a) $\tan (a + n\pi) = \dfrac{\sin (a + n\pi)}{\cos (a + n\pi)} = \dfrac{(-1)^n \sin a}{(-1)^n \cos a} = \dfrac{\sin a}{\cos a} = \tan a$.

10. (a) $\sec (\pi/2 + a) = 1/\cos (\pi/2 + a) = 1/-\sin a = -\csc a$.
(c) $\tan a + \cot a = \sin a/\cos a + \cos a/\sin a = \sin^2 a/\cos a \sin a +$

$\cos^2 a/\cos a \sin a = 1/\cos a \sin a = \sec a \csc a.$ **(e)** $1 + \cot^2 a = 1 + \cos^2 a/\sin^2 a = (\sin^2 a + \cos^2 a)/\sin^2 a = \csc = a.$ **(h)** By (e) of this exercise, $\cot a/1 + \cot^2 a = \cot a/\csc^2 a = \sin^2 a \cot a = \sin^2 a(\cos a/\sin a) = \sin a \cos a.$ **(j)** $\sin x - \sin y =$

$$\frac{2 \cos \frac{1}{2}(x - y) \sin \frac{1}{2}(x - y)}{2 \cos \frac{1}{2}(x - y) \cos \frac{1}{2}(x - y)} = \tan \frac{1}{2}(x - y). \quad \textbf{(o)} \cos^2 a - \sin^2 a =$$

$$\cos^2 a \frac{(1 - \sin^2 a)}{\cos^2 a} = \frac{1 - \tan^2 a}{\sec^2 a} = \frac{1 - \tan^2 a}{1 + \tan^2 a}. \quad [\text{See (d) of this section.}]$$

11. (a) $\dfrac{\sin a}{1 + \cos a} = \dfrac{\sin a(1 - \cos a)}{1 - \cos^2 a} = \dfrac{\sin a(1 - \cos a)}{\sin^2 a} = 1/\sin a - \cos a/\sin a = \sec a - \cot a.$

■ EXERCISES 4.8

1. $\pi r/4,\ \pi r/3,\ \pi r/6.$

3. Consult the special value list.

4. (a) I. **(c)** IV. **(e)** II. **(g)** III.

5. (a) $(1, 0).$ **(d)** $(-\frac{1}{2}, \sqrt{3}/2).$ **(h)** $(0, 1).$

8. (a) $\sqrt{.2944}.$ **(c)** $1/.84.$ **(f)** $1.68 \sqrt{.2944}.$ **(j)** $\left(\dfrac{1 + \sqrt{.2944}}{2}\right)^{1/2}.$
(s) $\sqrt{.2944}.$ **(v)** $(\sqrt{.2944}, .84).$

9. $0 < a$ implies $0 < \cos a < 1$ and $0 < \sin a < 1$, whence $0 < \sin a/1 < \sin a/\cos a$ or $0 < a < \tan a.$

10. (g) $\dfrac{\sin a + \cos a}{\sec a + \csc a} = \dfrac{(\sin a + \cos a) \sin a \cos a}{(\sec a + \csc a) \sin a \cos a} =$

$\dfrac{\sin a + \cos a) \sin a \cos a}{\sin a + \cos a} = \sin a \cos a.$ **(i)** $2 \sin a + \sec a =$

$\sec a(2 \sin a \cos a + 1) = \sec a(2 \sin a \cos a + \sin^2 a + \cos^2 a) =$

$\sec a(\sin a + \cos a)^2.$ **(p)** $\dfrac{\sin a}{1 + \cos a} = \dfrac{2 \sin a/2 \cos a/2}{1 + 2 \cos^2 a/2 - 1} =$

$\dfrac{\sin a/2}{\cos a/2} = \tan a/2.$

11. (a) $2 \sin 8a \cos 2a.$ **(c)** $2 \cos 8a \cos 2a.$ **(h)** $2 \sin (x - h/2) \sin h/2.$

12. (c) $(\sqrt{2} \cos \pi/4, \sqrt{2} \sin \pi/4).$ **(e)** $(\sqrt{2}x \cos \pi/4, \sqrt{2}x \sin \pi/4).$
(f) $(\sqrt{2}|x| \cos 5\pi/4, \sqrt{2}|x| \sin 5\pi/4).$ **(k)** $(2x \cos \pi/6, 2x \sin \pi/6).$

■ EXERCISES 4.9

2. (Hint: Use the triangle inequality for distance.)

3. If $n = 1$, $np = p$ is a period so that the statement holds for $n = 1$. If the statement holds for $n = k$ (that is, $f(a + kp) = f(a)$), then $f(a + (k + 1)p) = f(a + kp + p) = f(a + kp) = f(a)$ and the theorem holds for $n = k + 1$. By mathematical induction, the theorem holds for all positive integers.

7. $(a, b) \sim (-\pi/2, \pi/2) \sim R$ implies, by transitivity, that $(a, b) \sim R$.

8. If $\sin a = \sin b$ for a and b in $[-\pi/2, \pi/2]$, then both a and b are in $[0, \pi/2]$ or both are in $[-\pi/2, 0]$, whence $\cos a = \cos b$ and $\alpha(a) = \alpha(b)$. Thus, $a = b$ and $\sin \|[-\pi/2, \pi/2]$ is 1:1. If $-1 \leqslant y \leqslant 1$, then $(\sqrt{1 - y^2}, y)$ is in quadrant I or II on the unit circle and is $\alpha(a)$ for some $a \in [-\pi/2, \pi/2]$. Thus, $\sin \|[-\pi/2, \pi/2]$ is onto $[-1, 1]$.

■ EXERCISES 5.1

1. (a) $\{x \in R : x = n\pi, \ n \in Z\}$.   **(c)** $\{x \in R : x = 7\pi/6 + 2n\pi \text{ or } x = 11\pi/6 + 2n\pi, \ n \in Z\} = \{x \in R : x = (-1)^{n+1}\pi/6 + n\pi\}$.
(e) $\{x \in R : x = 4\pi/3 + 2n\pi \text{ or } x = 5\pi/3 + 2n\pi\} = \{x \in R : x = (-1)^{n+1}\pi/3 + n\pi\}$.   **(g)** $\{x \in R : x = 3\pi/2 + 2n\pi\}$.   **(i)** $\{x \in R : x = 5\pi/4 + 2n\pi \text{ or } x = 7\pi/4 + 2n\pi\} = \{x \in R : x = (-1)^{n+1}\pi/4 + n\pi\}$.

2. (a) $\{x \in R : x = \pi/2 + n\pi\}$.   **(c)** $\{x \in R : x = \pm 4\pi/3 + 2n\pi\}$.
(e) $\{x \in R : x = \pm 5\pi/6 + 2n\pi\}$.   **(g)** $\{x \in R : x = (2n + 1)\pi\}$.
(i) $\{x \in R : x = \pm 3\pi/4 + 2n\pi\}$.

3. (a) $\{x \in R : x = n\pi\}$.   **(c)** $\{x \in R : x = 3\pi/4 + n\pi\}$.   **(e)** $\{x \in R : x = 2\pi/3 + n\pi\}$.   **(g)** $\{x \in R : x = 5\pi/6 + n\pi\}$.

4. (a) $\{x \in R : x = \pi/2 + n\pi\}$.   **(c)** $\{x \in R : x = 3\pi/4 + n\pi\}$.
(e) $\{x \in R : x = 5\pi/6 + n\pi\}$.   **(g)** $\{x \in R : x = 2\pi/3 + n\pi\}$.

5. (a) $\{x \in R : x = 2n\pi\}$.   **(c)** $\{x \in R : x = \pm\pi/4 + 2n\pi\}$.
(e) $\{x \in R : x = \pm\pi/6 + 2n\pi\}$.   **(g)** $\{x \in R : x = \pm\pi/3 + 2n\pi\}$.

6. (a) $\{x \in R : x = \pi/2 + 2n\pi\}$.   **(c)** $\{x \in R : x = \pi/4 + 2n\pi \text{ or } x = 3\pi/4 + 2n\pi\} = \{x \in R : x = (-1)^n \pi/4 + n\pi\}$.   **(e)** $\{x \in R : x = \pi/3 + 2n\pi \text{ or } x = 2\pi/3 + 2n\pi\} = \{x \in R : x = (-1)^n \pi/3 + n\pi\}$.
(g) $\{x \in R : x = \pi/6 + 2n\pi \text{ or } x = 5\pi/6 + 2n\pi\} = \{x \in R : x = (-1)^n \pi/3 + n\pi\}$.

7. (a) $\{x \in R : x = \pi/2 + 2n\pi\}$.   **(c)** $\emptyset$.   **(e)** $\{x \in R : x = 2\pi/3 + 2n\pi\}$.
(g) $\{x \in R : x = \pi/4 + n\pi/2\} = \{x \in R : x = (2n + 1)\pi/4\}$.
(i) $\{x \in R : x = n\pi/2\}$.

8. (a) $\{0\}$. **(c)** $\{\pi/6\}$. **(e)** $\{\pi/2\}$. **(g)** $\{-\pi/6\}$. **(i)** $\{-\pi/4\}$. **(k)** $\{\pi/4\}$. **(m)** $\{\pi/6\}$. **(o)** $\{\pi\}$. **(q)** $\{5\pi/6\}$. **(s)** $\{0\}$. **(u)** $\{\pi/4\}$. **(w)** $\{-\pi/3\}$. **(y)** $\{-\pi/6\}$.

9. (a) 3. **(c)** ¼. **(e)** $\pm^{15}\!/_{17}$. **(g)** $\pm\%$.

10. (a) $\{x \in R : x = \pi/6 + 2n\pi \text{ or } x = 5\pi/6 + 2n\pi\} = \{x \in R : x = (-1)^n \pi/6 + n\pi\}$. **(c)** $\{x \in R : x = \pi/3 + n\pi\}$.

11. (a) $\theta \in \arcsin \sqrt{2}/2 = \{x \in R : x = (-1)^n \pi/4 + n\pi\}$.
(c) $\theta \in \arccos \sqrt{3}/2 = \{x \in R : x = \pm\pi/6 + 2n\pi\}$.
(e) $a \in \arccos 1 \cup \arccos(-1) = \{x \in R : x = n\pi\}$.
(g) $a \in \arctan 1 \cup \arctan(-1) = \{x \in R : x = \pi/4 + n\pi/2\} = \{x \in R : x = (2n+1)\pi/4\}$. **(i)** $2\sin^2 a - \sin a - 1 = 0$,
$(2\sin a + 1)(\sin a - 1) = 0$, $\sin a = -½, 1$, $a \in \arcsin(-½) \cup \arcsin 1 = \{x \in R : x = (-1)^{n+1} \pi/6 + n\pi \text{ or } x = \pi/2 + 2n\pi\}$.
(k) $\sin^2 a - \cos^2 a = 0$, $2\sin^2 a - 1 = 0$, $\sin a = \pm 1/\sqrt{2}$,
$a \in \arcsin 1/\sqrt{2} \cup \arcsin(-1/\sqrt{2}) = \{x \in R : x = \pi/4 + n\pi/2\}$.
(l) $\cos a(\sin^2 a - 1) = 0$, $\cos a = 0$ or $\sin^2 a - 1 = 0$, whence $\sin a = \pm 1$.
Thus $a \in \arccos 0 \cup \arcsin 1 \cup \arcsin(-1) = \arccos 0 = \{x \in R : x = (2n+1)\pi/2\}$. **(n)** $\tan a = \cos a$, $\sin a = \cos^2 a = 1 - \sin^2 a$, $\sin^2 a + \sin a - 1 = 0$,

$$\sin a = \frac{-1 \pm \sqrt{1+4}}{2} = \frac{-1 \pm \sqrt{5}}{2}; a \in \arcsin \frac{-1+\sqrt{5}}{2} \cup \arcsin$$

$$\frac{-1-\sqrt{5}}{2}. \text{ However, } \arcsin \frac{-1-\sqrt{5}}{2} = \varnothing \text{ so that } a \in \arcsin \frac{-1+\sqrt{5}}{2}.$$

(r) $\sin a \tan a = 0$ implies $\sin a = 0$ or $\tan a = 0$. Thus, $a \in \arcsin 0 = \arctan 0 = \{x \in R : x = n\pi\}$. **(x)** $\sin 2\theta = \cos \theta$, $2\sin \theta \cos \theta = \cos \theta$, $(2\sin \theta - 1)\cos \theta = 0$, $\cos \theta = 0$ or $\sin \theta = ½$, $\theta \in \arccos 0 \cup \arcsin ½ = \{x \in R : x = (2n+1)\pi/2 \text{ or } x = (-1)^n\pi/6 + n\pi\}$. **(b')** $\theta \cos \theta - \theta - \cos \theta + 1 = 0$, $\theta(\cos \theta - 1) - (\cos \theta - 1) = 0$, $(\theta - 1)(\cos \theta - 1) = 0$, $\theta = 1$ or $\theta \in \arccos 1$. Thus $\theta = 1$ or $\theta \in \{x \in R : x = 2n\pi\}$.

12. (a) $(0, 1)$. **(c)** $-½, \sqrt{3}/2)$, $(½, -\sqrt{3}/2)$. **(e)** $(\sqrt{3}/2, ½)$, $(\sqrt{3}/2, -½)$. **(g)** $(-\frac{5}{13}, \frac{12}{13})$, $(-\frac{5}{13}, -\frac{12}{13})$. **(i)** $(^{24}\!/_{25}, -^7\!/_{25})$, $(-^{24}\!/_{25}, -^7\!/_{25})$.

■ **EXERCISES 5.2**

1. (a) 0. **(c)** $\pi/4$. **(e)** $\pi/6$. **(g)** 0. **(i)** $\pi/6$. **(k)** $-\pi/4$. **(m)** $-\pi/3$. **(o)** $5\pi/6$. **(q)** $3\pi/4$. **(s)** $-\pi/3$.

2. (a) ¼. **(c)** 6. **(e)** ⅗. **(g)** $-\pi/7$. **(i)** $\pi/5$. **(k)** $\pi/10$.

4. (a) $(\sqrt{3}/2, \frac{1}{2})$.   **(c)** $(1/\sqrt{2}, -1/\sqrt{2})$.   **(e)** $(-\frac{15}{17}, -\frac{8}{17})$.
(g) $(\frac{1}{2}, -\sqrt{3}/2)$.   **(i)** $(\sqrt{3}/2, \frac{1}{2})$.

■ **EXERCISES 5.3**

1. (a) $\frac{1}{2}$.   **(c)** 65.   **(e)** $\pi/12$.   **(g)** Not possible.   **(i)** Not possible.

2. (a) $\pi/4$.   **(c)** $\pi/3$.   **(e)** $\sqrt{3}/2$.   **(g)** $1/\sqrt{2}$.   **(i)** $2/\sqrt{3}$.   **(k)** $\frac{4}{5}$.   **(m)** $\frac{12}{13}$.
(o) $-\frac{7}{25}$.   **(q)** $\frac{12}{13}$.

3. (a) $(\sqrt{3} - 1)/2\sqrt{2}$.   **(c)** $\frac{36}{85}$.   **(e)** $-\frac{323}{36}$.   **(g)** $\frac{24}{25}$.
(i) $[(\sqrt{2} + 1)/2\sqrt{2}]^{1/2}$.   **(k)** $-\frac{336}{527}$.   **(m)** $\frac{1}{3}$.

4. (a) $0 < \tan^{-1} 6 < \pi/2$ and $0 < \tan^{-1} \frac{1}{6} < \pi/2$ so that $0 < \tan^{-1} 6 + \tan^{-1} \frac{1}{6} < \pi$, $\cos (\tan^{-1} 6 + \tan^{-1} \frac{1}{6}) = 6/\sqrt{37} \cdot 1/\sqrt{37} - 1/\sqrt{37} \cdot 6/\sqrt{37} = 0$, whence $\tan^{-1} 6 + \tan^{-1} \frac{1}{6} = \pi/2$.

The problem could be worked by noting that $\tan^{-1} \frac{1}{6} = \cot^{-1} 6$, whereby $\tan^{-1} 6$ and $\cot^{-1} 6$ (or $\tan^{-1} \frac{1}{6}$) are complementary numbers. Hence, add to $\pi/2$.   **(e)** $0 < \cos^{-1} \frac{12}{13} + \cos^{-1} \frac{24}{25} < \pi$ so that, if $x = \cos (\cos^{-1} \frac{12}{13} + \cos^{-1} \frac{24}{25})$, $\cos^{-1} x = \cos^{-1} \frac{12}{13} + \cos^{-1} \frac{24}{25}$. $\cos (\cos^{-1} \frac{12}{13} + \cos^{-1} \frac{24}{25}) = \frac{12}{13} \cdot \frac{24}{25} - \frac{5}{13} \cdot \frac{7}{25} = \frac{253}{325}$. The problem follows.   **(f)** Since $\sin^{-1} \frac{4}{5} = \cos^{-1} \frac{3}{5}$, $\sin^{-1} \frac{4}{5}$ and $\sin^{-1} \frac{3}{5}$ are complementary numbers (both lie in $(0, \pi/2)$) and necessarily $\sin^{-1} \frac{3}{5} + \sin^{-1} \frac{4}{5} = \pi/2$.

■ **EXERCISES 5.4**

1. (a) period $= 2\pi/5$, phase $= 0$, amplitude $= 3$.   **(c)** period $= \pi$, phase $= \pi/6$, amplitude $= \frac{1}{2}$.   **(e)** period $= 2\pi$, phase $= \pi/2$, amplitude $= 1$.   **(g)** period $= 6\pi/5$, phase $3\pi/5$, amplitude $= \frac{5}{3}$.
(i) period $= \pi$, phase $= (\frac{1}{2}) \sin^{-1} \frac{12}{13}$, amplitude $= 13$.

2. (a) $\frac{5}{2}\pi$.   **(c)** $1/\pi$.   **(e)** $\frac{1}{2}\pi$.   **(g)** $\frac{5}{6}\pi$.   **(i)** $1/\pi$.

3. See Figs. G.1, G.2, G.3, G.4, and G.5 for (a), (c), (e), (g), and (i), respectively.

4. (a) 24π.   **(c)** 12π.

■ **EXERCISES 5.5**

1. (a) arcsin 0.   **(c)** $\{x \in R : x = \pi/4 + 2n\pi\}$.   **(e)** $\varnothing$.   **(g)** $\varnothing$.
(i) $\{x \in R : x = -\pi/4 + 2n\pi\}$.   **(k)** arctan $1 = \{x \in R : x = \pi/4 + n\pi\}$.

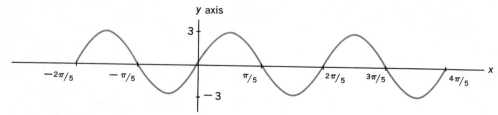

Figure G.1
$y = 3 \sin 5x$

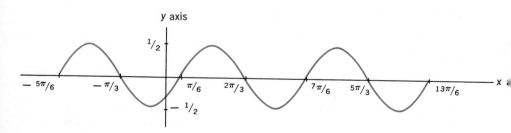

Figure G.2
$y = \frac{1}{2} \cos 2(x - \pi/6)$

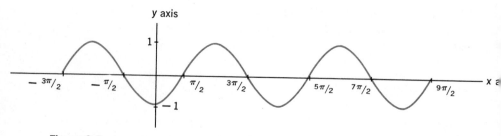

Figure G.3
$y = \sin (x - \pi/2)$

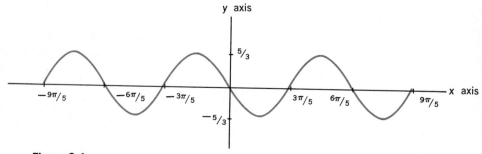

Figure G.4
$y = (\frac{5}{3}) \sin (5x/3 + \pi)$

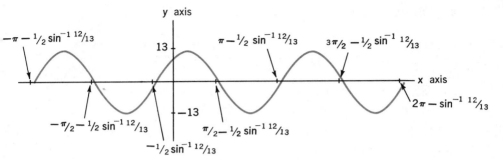

Figure G.5
$y = 5 \sin 2x + 12 \cos 2x$

(m) $\{x \in R : x = n\pi/2\}$.   **(o)** $\{x \in R : x = \pi/4 + n\pi \text{ or } x = 3\pi/4 + 2n\pi\}$.
(q) $\{x \in R : x = \pi/4 + 2n\pi\}$.   **(s)** $\{x \in R : x = -\pi/4 + 2n\pi\}$.

2. (a) π.   **(c)** $5\pi/6$.   **(e)** $\pi/2$.   **(g)** $7\pi/6$.   **(i)** $5\pi/4$.   **(k)** $-\pi/4$.
(m) $-\pi/6$.   **(o)** $-\pi$.   **(q)** $-5\pi/6$.   **(s)** π.   **(u)** $5\pi/4$.   **(w)** $2\pi/3$.   **(y)** $5\pi/6$.

4. (a) $x \in \{n\pi/2 : n \in Z\}$.   **(c)** $x \in \{\pi/4 + n\pi : n \in Z\}$.   **(e)** No solution.

5. (a) a if $-1 \leqslant a \leqslant 1$.   **(c)** a.   **(e)** $-\pi/6$.   **(g)** $\pi/4$.   **(i)** $\pi/7$.   **(k)** $^{15}/_{17}$.
(m) $^{5}/_{12}$.   **(o)** $^{15}/_{17}$.   **(q)** π.

7. (a) amplitude $= 3$, phase $= \frac{1}{2}$, period $= 2\pi/7$.   **(c)** amplitude $= \frac{3}{2}$,
phase $\frac{1}{3}$, period $= 2\pi/5$.   **(e)** amplitude $= \sqrt{10}$, phase $= \sin^{-1} 1/\sqrt{10}$,
period $= 2\pi$.   **(g)** amplitude $= 5$, phase $= (\sin^{-1} \frac{4}{5})/2$, period $= \pi$.

8. (a) $7/2\pi$.   **(c)** $5/2\pi$.   **(e)** $1/2\pi$.   **(g)** $1/\pi$.

■ **EXERCISES 5.6**

1. $[\pi/2, 3\pi/2]$ is one such interval.

4. $\tan \pi/2$ and $\tan (-\pi/2)$ are not defined.

5. $\cos |[-\pi/2, \pi/2]$ is not $1:1$.

6. $2\pi s/mp$, where s is the least common multiple of np and qm.

9. (a) $\sqrt{3}y = x$.   **(c)** $x = y$.   **(e)** $y = 0$.

13. G is the diagonal of $[-\pi/2, \pi/2] \times [-\pi/2, \pi/2]$, that is, $G = \{(x, x) : x \in [-\pi/2, \pi/2]\}$.

■ EXERCISES 6.1

1. (a) $(3, 0)^*$, $(-3, \pi)^*$. **(c)** $(1, \pi)^*$, $(-1, 0)^*$. **(e)** $(5, \sin^{-1} \frac{4}{5})^*$, $(-5, \sin^{-1} \frac{4}{5} + \pi)^*$. **(g)** $(17, \sin^{-1} \frac{15}{17})^*$, $(-17, \sin^{-1} \frac{15}{17} + \pi)^*$. **(i)** $(13, \pi - \sin^{-1} \frac{12}{13})^*$, $(-13, 2\pi - \sin^{-1} \frac{12}{13})^*$. **(k)** $(\sqrt{61}, \tan^{-1} \frac{5}{6})^*$, $(-\sqrt{61}, \pi + \tan^{-1} \frac{5}{6})^*$. **(m)** $(\sqrt{13}, \cos^{-1} (-\frac{2}{13}))^*$, $(-\sqrt{13}, \pi + \cos^{-1} (-\frac{2}{13}))^*$. **(o)** $(10, 5\pi/6)^*$, $(-10, 11\pi/6)^*$.

3. (a) The rectangular form of $(r, a)^*$ is $(r \cos a, r \sin a)$ and $p_1((r \cos a, r \sin a)) = r \cos a$.

4. y/x.

5. (a) $(3, 0)$. **(d)** $(-5, 0)$. **(g)** $(-\sqrt{2}, \sqrt{2})$. **(j)** $(0, -3)$. **(m)** $(-\frac{120}{13}, -\frac{50}{13})$.

6. If $y = mx$, $y/x = m$, (x, y) is given as $(r \cos a, r \sin a)$ for some $a \in R$ and $r = x^2 + y^2$. Thus $y/x = \tan a = m$, whence, if a is chosen as $\tan^{-1} m$, the equation is $a = \tan^{-1} m$.

■ EXERCISES 6.2

1. (a) $(4, 7)$. **(f)** $(5, -2)$. **(k)** $(0, 5)$. **(u)** $(\frac{1}{2}, \frac{1}{2})$.

2. (a) (a, b). **(b)** (a, b). **(c)** (a, b). **(d)** (a, b). **(e)** $(0, 0)$. **(i)** $(-1, 0)$. **(k)** $(0, -1)$. **(i)** $(1, 0)$.

5. $(0, 0)$.

6. $(1, 0)$.

9. (a) $(-bk, ak)$. **(c)** $(2a, 0)$. **(d)** $(0, 2b)$. **(e)** $(a^2 + b^2, 0)$.

10. If $(a, b) = \alpha(p)$, $\alpha(-p) = (a, -b)$.

11. (a) $(rs \cos (a + b), rs \sin (a + b))$. **(b)** $((r/s) \cos (a - b), (r/s) \sin (a - b))$. **(c)** $(r^2 \cos 2a, r^2 \sin 2a)$. **(d)** $(r^3 \cos 3a, r^3 \sin 3a)$.

■ EXERCISES 6.3

1. If $a \neq b$, $(a, 0) \neq (b, 0)$.

2. $i(a + b) = (a + b, 0) = (a, 0) + (b, 0) = i(a) + i(b)$, $i(ab) = (ab, 0) = a(b, 0) = ai(b) = (a, 0)(b, 0) = i(a)i(b)$, $i(0) = (0, 0)$ and, since i is $1:1$, $i(b) = (0, 0)$ implies $b = 0$.

6. $p_1(a, b)$ is the real part of (a, b) while $p_2(a, b)$ is the imaginary part.

7. $(a^2 + b^2)^{1/2}$.

■ **EXERCISES 6.4**

2. (a) $(0, -5832)$. **(c)** $(-1, 0)$. **(e)** $(16, 0)$. **(g)** $(0, 216)$.

5. (b) $re^{i\theta}se^{i\varphi} = r(\cos\theta, \sin\theta)s(\cos\varphi, \sin\varphi) = rs(\cos(\theta + \varphi),$
$\sin(\theta + \varphi)) = rse^{i(\theta+\varphi)}$.

6. (a) $3i$, $(0, 3)$. **(c)** $5\sqrt{3}/2 + 5i/2$, $(5\sqrt{3}/2, \frac{5}{2})$. **(e)** -2, $(-2, 0)$.

7. (a) $(1/\sqrt{2}, 1/\sqrt{2})$. **(c)** $(\sqrt{3}/2, \frac{1}{2})$.

■ **EXERCISES 6.5**

1. $(4, 0)$, $(-4, 0)$.

3. $(3\sqrt{3}/2, 3/2)$, $(-3\sqrt{3}/2, \frac{3}{2})$, $(0, -3)$.

5. $10(\cos\pi/8, \sin\pi/8)$, $10(\cos 9\pi/8, \sin 9\pi/8)$.

7. $(\sqrt{2}/2)(\sqrt{3}, 1)$, $(\sqrt{2}/2)(-\sqrt{3}, -1) = (-\sqrt{3}/2)(3, 1)$.

9. $3e^{3\pi i/8}$, $3e^{7\pi i/8}$, $3e^{11\pi i/8}$, $3e^{15\pi i/8}$.

■ **EXERCISES 6.6**

1. (a) $(3/\sqrt{2}, -3/\sqrt{2})$. **(c)** $(-13, 0)$. **(e)** $(^{64}\!/_5, ^{48}\!/_5)$. **(g)** $(-^{96}\!/_{17}, ^{180}\!/_{17})$.
(i) $(r\cos\theta, r\sin\theta)$.

2. (a) $(65, \sin^{-1} 12/13)^*$. **(c)** $(85, \sin^{-1} 12/13)^*$. **(e)** $(30, 0)^*$.
(g) $(102, 2\pi/3)^*$.

3. (a) Line given by $y = x$ in rectangular coordinates. **(c)** Circle $(x-32)^2 + y^2 = 32^2$. **(e)** Circle $x^2 + y^2 = 36$. **(g)** Spiral. (See Fig. G.6.)

4. (a) $3\sqrt{2}$, $\pi/4$, 3, 3. **(c)** 4, $\pi/3$, 2, $2\sqrt{3}$. **(e)** 16, π, -16, 0. **(g)** $\sqrt{5}$.
$\tan^{-1}(-\frac{1}{2})$, 2, -1. **(i)** $3\sqrt{2}$, $-\pi/4$, 3, -3.

5. (a) $8 - i$. **(c)** $3 - 2i$. **(e)** 2. **(f)** $-1 + i$. **(i)** 1. **(k)** $-^{8}\!/_{13} + 27i/13$.

6. (a) $18i$. **(c)** $-8 + 8\sqrt{3}i$. **(e)** 256. **(g)** $3 - 4i$. **(i)** $-18i$.

7. (c) 64. **(e)** -4096.

8. (c) $\sqrt{3} + i$, $-\sqrt{3} - i$. **(e)** $4i$, $-4i$.

9. (e) $\sqrt{2} + \sqrt{2}i$, $-\sqrt{2} + \sqrt{2}i$, $-\sqrt{2} - \sqrt{2}i$, $\sqrt{2} - \sqrt{2}i$.

■ EXERCISES 6.7

1. Write $(r, \theta)^*$, $(s, \Phi)^*$ in rectangular coordinates and calculate the distance.

2. See Figs. G.7, G.8, and G.9 for graphs of (a), (c), and (e), respectively.

3. If $k = 0$, the problem is trivial. If $k \neq 0$, the two intersect only at points (s, φ), where $2\varphi \in$ **arccos** ¼. Thus, since $r = k$ is a circle of radius $k (\neq 0)$ about the origin, the points of intersection must be $(k/2, (½) \cos^{-1} ¼)^*$. It is also possible to solve for the points of intersection by rewriting the equations in "rectangular" form.

5. (a) $z = (8\sqrt{2} + 8\sqrt{2}i)^{1/2} - 1$ for both square roots of $8\sqrt{2} + 8\sqrt{2}i$.
(c) $z = 3 + 4i$.

6. (a) Circle of center a and radius 2.   **(c)** Interior of the circle from (a).
(e) Line given by $x = 2$.   **(g)** "Left half plane."   **(i)** Line $y = x$ except for $(0, 0)$.   **(k)** The ray from $(0, 0)$ through $(½, \sqrt{3}/2)$ except for the point $(0, 0)$.

■ EXERCISES 7.1

1. See Fig. G.10 for (a) to (d).

2. (a) $\{⅛ + n : n \in Z\}$.   In fact, if $\theta \in R$, $\theta + n$ describes all measures of angles coterminal with θ.

3. (a) $\pi/4$, $\pi/4 + 2n\pi$.   **(d)** $5\pi/4$, $5\pi/4 + 2n\pi$.   **(i)** $\pi/6$, $\pi/6 + 2n\pi$.
(j) $\pi/3$, $\pi/3 + 2n\pi$.   **(k)** $2\pi/3$, $2\pi/3 + 2n\pi$.

Figure G.6

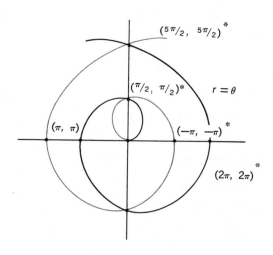

Figure G.7

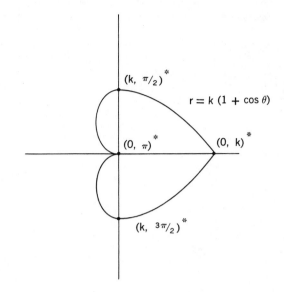

Figure G.8

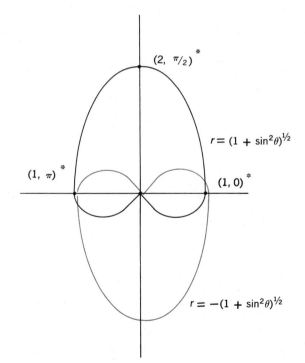

Figure G.9

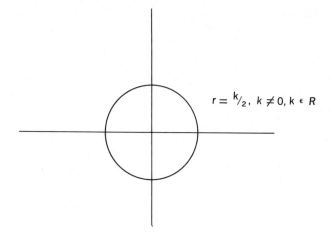

$r = {}^k/_2,\ k \neq 0,\ k \in R$

Figure G.10

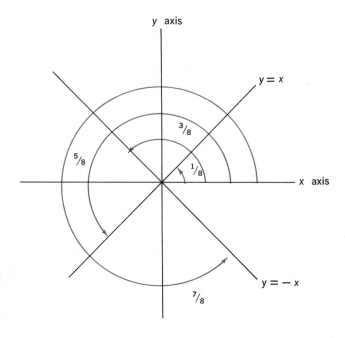

y axis

$y = x$

$^3/_8$

$^5/_8$

$^1/_8$

x axis

$y = -x$

$^7/_8$

4. (a) $45°,\ 45° + n360°$ **(i)** $30°,\ 30° + n360°$.   **(j)** $60°,\ 60° + n360°$.

5. (a) I.   **(c)** II.   **(e)** III.   **(g)** I.   **(i)** III.

6. (a) $1/2\pi,\ (180/\pi)°$.   **(c)** $3/2\pi,\ (540/\pi)°$.

8. (a) I.   **(c)** III.   **(e)** IV.   **(g)** II.   **(i)** III.

9. (a) $^7/_{36},\ 7\pi/18$.   **(c)** $^{103}/_{180},\ 103\pi/90$.   **(e)** $-^7/_{90},\ -7\pi/45$.

11. **(a)** 2. **(c)** π. **(e)** $\pi/2$. **(g)** $2\pi/3$. **(i)** $5\pi/3$. **(k)** $10\pi/3$.

15. **(a)** $y = 0$. **(c)** $x = y$. **(e)** $x = 0$. **(g)** $y = -x$. **(l)** $y = 0$.

16. **(a)** $k = r$, $k = \pi r/180$, or $k = 2\pi r$ depending on whether Δw is given in terms of radians, degrees, or rotations, respectively. **(b)** Same as k in (a).

17. **(a)** 8 ft/sec. **(b)** 15 ft/sec.

18. **(a)** 6 rad/sec. **(b)** 475 rad/hr.

19. $\theta r^2/2$.

■ **EXERCISES 7.2**

1. The tangent function is undefined for angles of measure $(2n + 1)\pi/2$ radians.

2. The tangent function is not defined for angles of measure $(2n + 1)90°$ or $(2n + 1)/4$ rotations.

4. $x/r = \cos\theta$, $y/r = \sin\theta$, whence $x = r\cos\theta$ and $y = r\sin\theta$.

8. **(a)** $\sin\theta \csc\theta = (y/r)(r/y) = 1$, where (x, y) is on the terminal side of θ, $y \neq 0$, and $r = (x^2 + y^2)^{1/2}$. **(d)** Using x, y, and r as in (a), $\sin^2\theta + \cos^2\theta = y^2/r^2 + x^2/r^2 = (y^2 + x^2)/r^2 = r^2/r^2 = 1$.

9. **(a)** $\sin\theta = \frac{4}{5}$,
$\cos\theta = \frac{3}{5}$,
$\tan\theta = \frac{4}{3}$,
$\csc\theta = \frac{5}{4}$,
$\sec\theta = \frac{5}{3}$,
$\cot\theta = \frac{3}{4}$.

(h) $\sin\theta = 0$,
$\cos\theta = \pm 1$,
$\tan\theta = 0$,
$\csc\theta$ is not defined,
$\sec\theta = \pm 1$,
$\cot\theta$ is not defined.

(j) $\sin\theta = 2/\sqrt{5}$,
$\cos\theta = 1/\sqrt{5}$,
$\tan\theta = 2$,
$\csc\theta = \sqrt{5}/2$,
$\sec\theta = \sqrt{5}$,
$\cot\theta = \frac{1}{2}$.

(n) $\sin\theta = 5/\sqrt{26}$,
$\cos\theta = 1/\sqrt{26}$,
$\tan\theta = 5$,
$\csc\theta = \sqrt{26}/5$,
$\sec\theta = \sqrt{26}$,
$\cot\theta = \frac{1}{5}$.

10. **(a)** $0°$. **(c)** $45°$. **(f)** $120°$. **(h)** $150°$. **(m)** $270°$. **(q)** $360°$.

12. **(a)** $\sin 1860° = \sqrt{3}/2$,
$\cos 1860° = \frac{1}{2}$,
$\tan 1860° = \sqrt{3}$,
$\csc 1860° = 2/\sqrt{3}$,
$\sec 1860° = 2$,
$\cot 1860° = 1/\sqrt{3}$.

(f) $\sin(-585°) = 1/\sqrt{2}$,
$\cos(-585°) = -1/\sqrt{2}$,
$\tan(-585°) = -1$,
$\csc(-585°) = \sqrt{2}$,
$\sec(-585°) = -\sqrt{2}$,
$\cot(-585°) = -1$.

13. (a) $(16 \cos 45°, 16 \sin 45°)$. **(c)** $(64 \cos \pi/6, 64 \sin \pi/6)$.
(e) $(20 \cos 0°, 20 \sin 0°)$. **(n)** $(2 \cos 330°, 2 \sin 330°)$.

14. (a) $(16 \cos (-315°), 16 \sin (-315°))$.
(c) $64 \cos (-11\pi/6, 64 \sin (-11\pi/6))$.
(e) $(20 \cos (-360°), 20 \sin (-360°))$. **(n)** $(2 \cos (-30°), 2 \sin (-30°))$.

15. Tan $45° = 1$ implies $y/x = 1$, whence $y = x = 1$. Since $r = (x^2 + y^2)^{1/2}$, $r = \sqrt{2}$. $\beta = 180° - 90° - 45° = 45°$.

16. $\tan 30° = 1/\sqrt{3} = y/x$, whence $x = \sqrt{3}y = \sqrt{3}$.
$r = (x^2 + y^2)^{1/2} = 2$. $\beta = 90° - 30° = 60°$.

17. $\tan 60° = \sqrt{3} = y/x$, whence $\sqrt{3}x = y = 1$ and $x = 1/\sqrt{3}$.
$r = (x^2 + y^2)^{1/2} = 2/\sqrt{3}$. $\beta = 90° - 60° = 30°$.

18. $\sin \eta =$ opposite/hypotenuse, $\cos \eta =$ adjacent/hypotenuse, $\tan \eta =$ opposite/adjacent, $\csc \eta =$ hypotenuse/opposite, $\sec \eta =$ hypotenuse/ adjacent, $\cot \eta =$ adjacent/opposite.

19. (a) $c = \sqrt{a^2 + b^2} = \sqrt{5}$; $A = \sin^{-1} 1/\sqrt{5} = 26° \, 30'$, approximately; $B = 90° - A = 63° \, 30'$, approximately. **(h)** $a = 45, b = 45\sqrt{3}, A = 30°$.
(m) $b = c \sin B = 18$, approximately, $a = 6.1$, approximately, $A = 18° \, 40'$.

21. 18.6 ft, approximately.

23. 47 and 17.1, approximately.

25. Approximately $15° \, 20'$, $56° \, 10'$, and $108° \, 30'$.

27. 4905 ft.

29. 61.75 ft.

32. (a) $(7\sqrt{3}/2, \frac{1}{2})$. **(c)** $-8/\sqrt{2} - 8i/\sqrt{2}$. **(e)** $15\sqrt{3}/2 - 15i/2$.
(g) $(.7257, 2.9109)$. **(i)** $(-19.972, -1.046)$.

33. $360°/n$ or $2\pi/n$ radians.

34. (a) $45°$, $\pi/4$ radians. **(c)** $3\pi/4$ radians. **(e)** $-\pi/6$ radians.

35. θ radians.

■ EXERCISES 7.3

1. $B = 55° \, 23'$, $C = 88° \, 37'$; $c = 25.5$ and $B = 124° \, 27'$; $C = 19° \, 33'$, $c = 8.539$.

3. $A = 30°$, $a = 10.95$, $b = 46.32$.

5. No triangle.

7. 207 ft, approximately.

■ **EXERCISES 7.4**

1. $c = 14.8$, $A = 20°$, approximately, $B = 147°$.

3. $b = 17$, $A = 71°\ 8'$, $C = 58°\ 52'$.

5. $a = 69.9$, $B = 2°$, $C = 81°$.

7. $A = 58°\ 36'$, $B = 41°\ 6'$, $C = 80°\ 18'$.

9. $A = 19°\ 44'$, $B = 28°\ 10'$, $C = 132°\ 6'$.

■ **EXERCISES 7.5**

1. $A = 19°\ 6'$, $B = 28°\ 8'$, $C = 132°\ 46'$.

4. $A = 20°\ 17'$, $B = 35°\ 17'$, $C = 124°\ 26'$, $c = 35.69$.

7. Right triangle $30° - 60° - 90°$ with $c = \sqrt{3}$.

10. $B = 6°$ or $174°$, $C = 171°$ or $3°$, $c = 2.989$ or 1.

12. No triangle.

15. 32.56 ft.

17. $52°$, $128°$, 56.9.

■ **EXERCISES 7.6**

1. (a) IV. **(d)** IV. **(g)** I. **(j)** I. **(m)** I.

2. (a) $(-540/7)°$, $-3/14$. **(d)** $300°$, $\%$.

3. (g) $-281\pi/180$, $-28\frac{1}{360}$. **(j)** $5\pi/36$, $\frac{5}{12}$.

4. (m) $13\pi/3$, $780°$.

6. (a) $\sin(-3\pi/7) = -.9739$,
  $\cos(-3\pi/7) = .2224$,
  $\tan(-3\pi/7) = -4.384$,
  $\csc(-3\pi/7) = -1.026$,
  $\sec(-3\pi/7) = 4.498$,
  $\cot(-3\pi/7) = -.2285$.

  (g) $\sin(-281°) = .9816$,
  $\cos(-281) = .1908$,
  $\tan(-281°) = 5.145$,
  $\csc(-281°) = 1.019$,
  $\sec(-281°) = 5.241$,
  $\cot(-281°) = .1944$.

7. (a) IV. **(d)** III. **(g)** I.

8. (a) 0. **(d)** 1. **(g)** 1.557. **(j)** .2867. **(m)** $-\sqrt{3}$. **(p)** -1.

9. (a) $9\pi/2$. **(d)** $\frac{3}{2}$. **(g)** 6π. **(j)** $12\pi/5$.

10. (a) $\{x : x = 213° + n360°, n \in Z\}$.   **(d)** $\{x : x = 2\pi/3 + 2n\pi, n \in Z\}$.
(g) $\{x : x = \pi/4 + 2n\pi, n \in Z\}$.

11. (a) $(\sqrt{5}\cos\theta, \sqrt{5}\sin\theta)$, where $\theta = \cos^{-1} 1/\sqrt{5}$.
(d) $(5\cos\theta, 5\sin\theta)$, $\theta = \cos^{-1}(-4/5)$.   **(g)** $(\sqrt{10}\cos\theta, \sqrt{10}\sin\theta)$,
$\theta = \cos^{-1}(-3/\sqrt{10})$.

13. ⅑ rad.

15. $585\pi/6$ ft/min.

18. They experience the same increase.

19. 2410.

21. (a) and (b).

23. 254 miles.

■ EXERCISES 7.7

6. (f) $[(5 - \sqrt{5})/8]^{1/2}$.   **(g)** $(\sqrt{5} + 1)/4$.   **(h)** Quotient of (f) and (g) above.

■ EXERCISES 8.1

4. (a) $\mathbf{v}_x = (4, 0)$, $\mathbf{v}_y = (0, 1)$.   **(c)** $\mathbf{v}_x = (\frac{1}{2}, 0)$, $\mathbf{v}_y = (0, -\frac{2}{3})$.   **(e)** $\mathbf{v}_x = (0, 0)$, $\mathbf{v}_y = (0, 1)$.   **(g)** $\mathbf{v}_x = \mathbf{v}_y = (0, 0)$.

5. (a) $(1, 10)$.   **(c)** $(12, 1)$.   **(e)** $(11, -22)$.

7. $(1, 1) = \mathbf{i} + \mathbf{j}$, $(a, b) = a\mathbf{i} + b\mathbf{j}$.   If $\mathbf{v} = (a, b)$, $\mathbf{v}_x = a\mathbf{i}$ and $\mathbf{v}_y = b\mathbf{j}$.

8. (a) $\mathbf{i} + 2\mathbf{j}$.   **(c)** $2\mathbf{i} - 3\mathbf{j}$.   **(e)** $\sqrt{2}\mathbf{i} + \mathbf{j}$.   **(g)** $0\mathbf{i} + 0\mathbf{j}$.

9. $(a, b) + (c, d) = (a\mathbf{i} + b\mathbf{j}) + (c\mathbf{i} + d\mathbf{j}) = (a + c)\mathbf{i} + (b + d)\mathbf{j}$.

10. $(a, b) = (a/\sqrt{2} + b/\sqrt{2})(1/\sqrt{2}, 1/\sqrt{2}) + (-a/\sqrt{2} + b/\sqrt{2})(-1/\sqrt{2}, 1/\sqrt{2})$.

11. $k_1(a, b) + k_2(c, d)$ lies on the same line as (a, b) and (c, d). Thus, if (e, f) is not on this line, (e, f) cannot be so written.

12. This follows from Exercise 11, since (a, b) and $(0, 0)$ lie on the same line through the origin.

■ EXERCISES 8.2

1. (a) 2. **(c)** $-\frac{3}{4}$. **(e)** 0. **(g)** $\sqrt{5}$. **(i)** 1. **(k)** $|k|\sqrt{13}$.

2. (a) 90°. **(c)** 30°. **(e)** $\cos^{-1} 68/5\sqrt{185}$. **(g)** $\cos^{-1}(-20/\sqrt{13 \cdot 37})$.
(i) $\cos^{-1} 2/\sqrt{5}$. **(j)** 0.

3. (a) $(0, 1), (0, -1)$. **(c)** $(-1/\sqrt{2}, 1/\sqrt{2}), (1/\sqrt{2}, -1/\sqrt{2})$.
(e) $(-\sqrt{3}/2, \frac{1}{2}), (\sqrt{3}/2, -\frac{1}{2})$.

4. (a) $6/\sqrt{2}\,\mathbf{i} + 6/\sqrt{2}\,\mathbf{j}$. **(c)** $\sqrt{3}\,\mathbf{i} - \mathbf{j}$. **(e)** $\sqrt{3}/2\,\mathbf{i} + \frac{1}{2}\,\mathbf{j}$.
(g) $\frac{3}{5}\,\mathbf{i} + \frac{6}{5}\,\mathbf{j}$. **(i)** $-\frac{24}{13}\,\mathbf{i} + \frac{10}{13}\,\mathbf{j}$.

5. (a) 36. **(c)** 4. **(e)** 1. **(g)** 4. **(i)** 4.

6. (a) $\pi/4$. **(c)** $-\pi/6$. **(e)** $\pi/6$. **(g)** $\sin^{-1}\frac{3}{5}$. **(i)** $\pi - \tan^{-1}\frac{5}{12}$.

7. $\mathbf{v}/|\mathbf{v}| \cdot \mathbf{v}/|\mathbf{v}| = \mathbf{v} \cdot \mathbf{v}/|\mathbf{v}|^2 = |\mathbf{v}|^2/|\mathbf{v}|^2 = 1$.

■ EXERCISES 8.3

1. (a) $(1, 0)$. **(c)** $(-1, 0)$. **(e)** $(1/\sqrt{10}, -3/\sqrt{10})$.
(g) $(a/(a^2 + b^2)^{1/2}, b/(a^2 + b^2)^{1/2})$ if $(a, b) \neq (0, 0)$.

2. (a) The projection of $(1, 2)$ onto $(6, 0)$ is $(1, 0)$. The projection of $(1, 2)$ onto $(0, 4)$ is $(0, 2)$. The projection of $(1, 2)$ onto $(-5, 0)$ is $(1, 0)$. The projection of $(1, 2)$ onto $(-3, -4)$ is $(\frac{33}{5}, \frac{44}{5})$. The projection of $(1, 2)$ onto $(1, -3)$ is $(-5/\sqrt{10}, 15/\sqrt{10})$. The projection of $(1, 2)$ onto $(7, 24)$ is $(\frac{77}{5}, \frac{264}{5})$. The projection of $(1, 2)$ onto (a, b) is $(a(a + 2b)/(a^2 + b^2)^{1/2}, b(a + 2b)/(a^2 + b^2)^{1/2})$.

4. The vector itself.

5. $|\mathbf{v}| \cos \alpha$, where α is the angle between $\mathbf{v}$ and the unit vector in the direction θ.

■ EXERCISES 8.4

1. See Fig. G.11.

2. See Fig. G.12 for $3\mathbf{v}_1$ and $-\mathbf{v}_1$.

3. See Fig. G.13 for $\mathbf{v}_1 + \mathbf{v}_2$.

4. See Fig. G.14 for (a).

5. See Fig. G.15 for (a).

6. See Fig. G.16 for the vector given in relation to $\mathbf{v}_1$.

Figure G.11

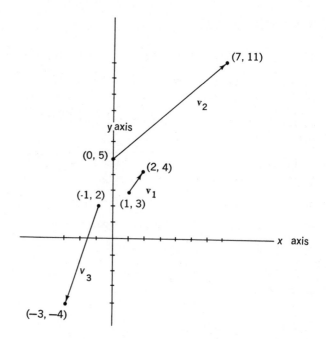

Figure G.12

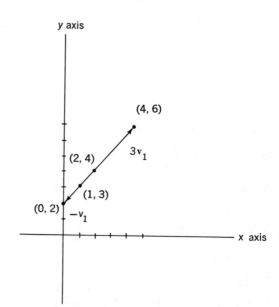

Figure G.13

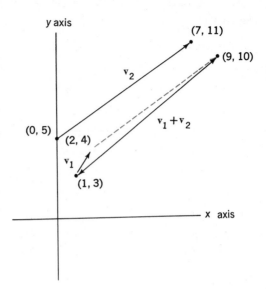

Figure G.14

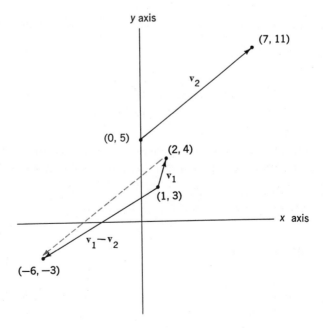

Figure G.15

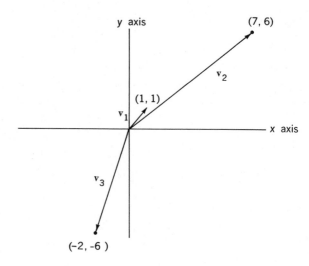

Figure G.16

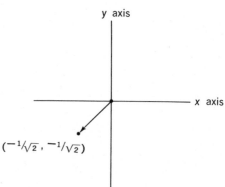

7. **(a)** All vectors lying on the line given by $y - 2 = x - 1$ and emanating from $(1, 2)$. Each vector is from $(1, 2)$ to $(1 + a, 2 + a)$ for some real number a.

8. The horizontal component of v_1 is a vector from $(1, 3)$ to $(2, 3)$, while the vertical component can be thought of as a vector from $(1, 3)$ to $(1, 4)$.

9. **(a)** $(2 \cos 81°, 2 \sin 81°)$. **(c)** $(-2, -2)$.
(e) $(-1 + 2\sqrt{3} \cos 56°, -2 + 2\sqrt{3} \sin 56°)$.

■ **EXERCISES 8.5**

2. 489.9 mph, 2° 52′.

4. 37 mph, 14⁷⁄₁₂ miles.

6. $\sqrt{34}$, 50 ft lb, 68 ft lb.

8. 95.85 ft/sec.

10. If he aims at a point $\sqrt{3}$ miles upstream from the mouth, he will end up directly across from where he started. He should not head more than 2.83 miles upstream.

■ **EXERCISES 8.6**

1. (a) $v_x = (0, 0)$, $v_y = (0, 0)$.   **(d)** $v_x = (3, 0)$ $v_y = (0, -4)$.
(g) $v_x = (16, 0)$, $v_y = (0, 24)$.

2. The projections are for the vector $(3, -4)$: **(a)** $(3, 0)$.
(c) $(\frac{1}{2}) -\frac{1}{2})$.   **(e)** $(0, -4)$.

3. If $z = (x, y)$; $R(z) = z_x$, $I(z) = z_y$.

5. (a) 0.   **(c)** b.   **(e)** $a^2 + b^2$.

6. (a) 0.   **(c)** 1.   **(e)** $5\sqrt{2}$.   **(g)** $8\sqrt{13}$.

7. (c) $(1, 0)$ and $(-1, 0)$.   **(e)** $(1/\sqrt{2}, 1/\sqrt{2})$ and $(-1/\sqrt{2}, -1/\sqrt{2})$.
(g) $(3/\sqrt{13}, 2/\sqrt{13})$ and $(-3/\sqrt{13}, -2/\sqrt{13})$.

8. (a) $\tan^{-1} \frac{4}{3}$.   **(d)** $15°$.

9. (a) $(-8, -6)$, $(8, 6)$.   **(c)** $(0, 1)$, $(0, -1)$.   **(e)** $(-1, \frac{1}{2})$, $(1, -\frac{1}{2})$.

11. The resultant vector is $(5 - 4\sqrt{2}, 5\sqrt{3} - 4\sqrt{2})$.

13. He should head upstream with a heading of $60°$.

Index